Conjugate Problems
in Convective
Heat Transfer

HEAT TRANSFER
A Series of Reference Books and Textbooks

Editor

Afshin J. Ghajar

Regents Professor
School of Mechanical and Aerospace Engineering
Oklahoma State University

1. Engineering Heat Transfer: Third Edition, *William S. Janna*
2. Conjugate Problems in Convective Heat Transfer, *Abram S. Dorfman*

Upcoming titles include:

Introduction to Thermal and Fluid Engineering, *Allan D. Kraus, James R. Welty, and Abdul Aziz*

Conjugate Problems
in Convective
Heat Transfer

Abram S. Dorfman

CRC Press
Taylor & Francis Group
Boca Raton London New York

CRC Press is an imprint of the
Taylor & Francis Group, an **informa** business

CRC Press
Taylor & Francis Group
6000 Broken Sound Parkway NW, Suite 300
Boca Raton, FL 33487-2742

First issued in paperback 2018

© 2010 by Taylor and Francis Group, LLC
CRC Press is an imprint of Taylor & Francis Group, an Informa business

No claim to original U.S. Government works

ISBN-13: 978-1-4200-8237-1 (hbk)
ISBN-13: 978-1-138-37271-9 (pbk)

Library of Congress Cataloging-in-Publication Data

Dorfman, A. Sh. (Abram Shlemovich)
    Conjugate problems in convective heat transfer / Abram S. Dorfman.
        p. cm. -- (Heat transfer)
    Includes bibliographical references and index.
    ISBN 978-1-4200-8237-1 (hardcover : alk. paper)
    1. Heat--Transmission. 2. Heat--Convection. 3. Fluid dynamics. I. Title.

TJ260.D64 2009
621.402'25--dc22   2009018850

Visit the Taylor & Francis Web site at
http://www.taylorandfrancis.com

and the CRC Press Web site at
http://www.crcpress.com

In Memory

*In memory of our dearest, lovely,
loving daughter, Ella Fridman.*

Contents

Part II: Theory and Methods

List of Examples

Chapter 9

Chapter 10

Preface

The conjugate nature is an inherent feature of any heat transfer problem, because the intention of any heat transfer process is an interaction of at least two mediums or subjects. This inherent property specifies the fact that heat transfer processes began to be studied as conjugate problems soon after such a possibility was realized with the advent of computers. Apparently, the same consideration gives an understanding why the number of publications in this area continues to grow like an avalanche.

The development of the conjugate heat transfer approach is important not only in that conjugate formulation is one of fundamental nature access, but also that it has a wide range of applications. The advanced conjugate modeling of convective heat transfer phenomena is now used extensively in different applications. Beginning with simple examples in 1965 through the 1970s, this approach is now used to create models of various device operational and technological processes from relatively simple procedures to complex, multistage, nonlinear processes.

I have studied the conjugate convective heat transfer systematically for more than 40 years, beginning in about 1965, and initially this book was intended as a sum of my achievements. However, in the process of working, it became clear that a real value of any result can be understood only from the point of view of the contemporary knowledge in the area of interest, and the book gradually has been transformed into a review of the body of publications. Thus, finally, the book has become one of the first consistent presentations of the current situation in conjugate convective heat transfer. I hope that I have been sufficiently successful in meeting and resolving such a challenge.

This book begins with a short introduction, "What is conjugate heat transfer like?" and closes by discussing the question, "Should any heat transfer problem be considered as a conjugate?" Three parts and ten chapters incorporate the theory, the methods of study, and the applications of conjugate convective heat transfer. More than a hundred examples of solved typical conjugate problems in different areas from early publications to many contemporary results are considered. Although it is obvious that the choice of such examples is random and depends on preferences, background, and insight, I hope that the presented pattern of examples offers the reader a basic understanding of the current situation in the area of conjugate convective heat transfer.

The first part of the book consists of the approximate analytical solutions of conjugate heat transfer problems (Chapter 2) and methods of estimation of heat transfer from nonisothermal surfaces (Chapter 1), which were basically used in the early works to get approximate conjugate solutions. Two types of approximate solutions are considered. The first includes solutions based on the approximate energy equation when the approximate velocity distribution across the thermal boundary layer is used. Another group contains solutions with results that could not be refined or that are difficult to upgrade. Most examples of this type of solution are taken from early works, yet some results obtained recently are included as well, as these approximate methods are still in use due to their simplicity and comprehensibility.

This first part of the book, presents the initial period of conjugate heat transfer development, and the remainder includes the theory, the methods of solutions, the applications, and the examples that fit the modern level of knowledge in this area.

The second part contains five chapters (Chapters 3 through 7). These chapters represent theory (Chapters 3 through 5); modern analytical (Chapter 6); and numerical (Chapter 7) methods of conjugate convective heat transfer investigation and corresponding examples of solutions. Chapters 3 and 4 incorporate the exact and accurate approximate solutions of the thermal boundary layer equations for arbitrary nonisothermal surfaces in the laminar (Chapter 3) and turbulent (Chapter 4) flows. Solutions for both the laminar and turbulent incompressible flows with zero and nonzero pressure gradients are given in two forms: in the differential form in the series of consecutive derivatives of the temperature; and in the integral form with the influence function of the unheated zone. In the case of laminar flow, the other exact solutions for arbitrary nonisothermal surfaces are obtained: for the cases of heat flux distribution and unsteady temperature distribution; for high-speed compressible flow; for power law non-Newtonian fluids; for fluids with significant energy dissipation; and for axisymmetric and rotation bodies.

The general theory of nonisothermal and conjugate heat transfer developed on the basis of the solutions presented in Chapters 3 and 4 is outlined in Chapter 5. The general properties of nonisothermal and conjugate convective heat transfer are formulated, and the basic parameters determining the intensity of heat transfer are indicated. The effect of different factors is studied, including the temperature head distribution, the temperature head and pressure gradients, the flow regime, the body shape, and the Biot number. Reynolds analogy and gradient analogy (between pressure and temperature head gradients) are investigated, and conditions are formulated when such analogies are valid. A phenomenon of heat flux inversion that is similar to separation of flow is studied, and the like and unlike properties of both phenomena are indicated. The deformation of the temperature head profile in the thermal boundary layer as a physical reason for the existence of heat inversion is analyzed. On the basis of this analysis, it is shown that there are temperature head distributions that correspond to surfaces with zero heat transfer. Such a surface can be theoretically generated, if the temperature head profile has a zero angle tangent at each point of the surface. In the last section of this chapter, the technique of optimization of heat transfer in flows over bodies is provided.

In Chapter 6, the general method for solving conjugate heat transfer problems is developed. Both the differential and integral forms for surface heat flux obtained in Chapter 3 are used. It is shown that the differential expression for heat flux is, in fact, the general condition of the third kind, which transforms in the usual condition of the third kind for the case of an isothermal surface when all of the derivatives become zero. By taking into account some derivatives, a more accurate solution can be obtained, and it becomes possible to estimate an error, which can occur when the condition of the third kind is used, and then to decide whether there is a need for a conjugate problem solution. The method of reducing the steady or unsteady conjugate heat transfer problems to an equivalent problem for a conduction equation with a general boundary condition of the third kind is offered, which is similar to a well-known approach with the usual condition of the third kind. Using both the differential and integral expressions for surface heat flux, the method of successive differential–integral approximations is suggested in which integral expression is used to refine the result obtained by the differential expression, and vice versa. The simple method for calculating the conjugate heat transfer from thermally thin bodies using universal tabulated functions is proposed for both the steady and unsteady cases. It is shown that the singularity in temperature distribution at the leading edge of the plate is defined by the Reynolds number exponent in the Nussel number equation for an isothermal surface, so that the series describing the temperature distribution at the leading edge is in integer powers of one-half and one-fifth for laminar and turbulent flows, respectively. The other cases are considered, and a general rule for determining the character of singularity in the temperature distribution at the leading edge is derived. It is also proved that the Biot number is a conjugate criterion defining the level of thermal coupling between the fluid and the solid, and that the various relations suggested by different authors are in fact the other forms of the Biot number.

Sections 6.7 through 6.10 in Chapter 6 discuss the following analytical methods currently used for solving conjugate heat transfer problems: integral transforms and similar methods, asymptotic series in eigenfunctions, superposition, Green's function, and perturbation methods. Examples are presented for each method.

Chapter 7 contains an analysis of contemporary numerical methods for solving conjugate heat transfer problems. The relation and mutual importance of numerical and analytical approaches are discussed. Different finite-difference and finite-element methods are described and compared using the weighted residual approach. The difficulties in computing the convection–diffusion terms, velocities, and pressure, and methods of resolving these problems are discussed. The central-difference and upwind scheme as well as some others are compared to show their advantages and shortcomings. The SIMPLE and SIMPLER software are described. The existing numerical approaches for subdomain solutions conjugation, such as simultaneous solving of one large set of equations for fluids and solids, iterative schemes, and using the superposition principle for conjugation are analyzed. Some modern numerical approaches to improve the convergence of the iterative conjugation methods, such as the multigrid or meshless techniques, are discussed as well. A number of examples of the numerical solutions of the conjugate heat transfer of the flows in tubes, channels, and around and inside the bodies are presented to illustrate the use of the described methods.

The third part of the book is composed of three chapters describing applications in the thermal treatment of material (Chapter 8), in the technological processes (Chapter 9), and in the manufacturing equipment operation (Chapter 10). Numerous conjugate heat transfer problems of simulation of various industrial processes are presented including:

- Protection of the reentry rocket by a thermal shield (Example 10.10).
- Heat transfer from a moving plate (Examples 8.1 and 8.2).
- The thermal process in the microchannel of challenging a heat exchanger (Example 10.7).
- The Czochralski crystal growth process (Example 9.4).
- The optical fiber coating process (Example 8.7).
- The rewetting process in nuclear reactors in case of an emergency accident (Examples 10.11 and 10.12).
- The cooled turbine blades (Examples 10.8 and 10.9).
- The multidimensional freeze-drying process of food (Example 9.9).

Different approaches are used. Mostly, the finite-difference and finite-element methods are employed because the problems are usually complicated. However, some analytical and other numerical methods are used as well. In particular, these include:

- The special method of separation variables is used in the problem of rewetting the surfaces by falling film (Example 10.11).
- The integral method is employed in the problem of rewetting semi-infinite plates (Example 10.12).
- The method of reduction of the conjugate heat transfer problem to an equivalent heat conduction problem (Chapter 3) is applied in the problem of drying a moving strip of material (Example 8.2).

Abram Dorfman
Ann Arbor, Michigan

ACKNOWLEDGMENTS

During my life, many people have supported, encouraged, and helped me, and I wish to express my appreciation and thanks to them: to my graduate students and coauthors, O. Grechannyy, O. Didenko, O. Lipovetskaiy, B. Davidenko, V. Novikov, V. Gorobets, and V. Vishnevskyy; to my friends and colleagues from the Ukrainian Academy of Science, Professors E. Diban, O. Gerashenko, M. Stradomskyy, and G. Selyvin; and to Professor A. Ginevskyy from the Zhukovsky Central Aerohydrodynamics Institute and to Professor Z. Shulman from the Belorussia Institute of Heat and Mass Transfer.

I am also grateful to my senior colleagues Professor V. Tolubiskyy from the Ukrainian Academy of Science and Professor B. Petuchov from the High Temperature Institute for their interest in my work, as well as to Professor A. Leontev from the Novosebirian Institute of Heat and Mass Transfer for his friendliness and support.

My sincere thanks to my American friends Professor H. Merte Ju, Professor S. Meerkov, and Professor M. Kaviany from the University of Michigan for encouraging discussions and for helping me to integrate into American society, and especially to my young friend Professor A. McGaughey from Carnegie Mellon University, Pittsburgh, Pennsylvania, for his useful suggestions and help in my English and computer work.

I also wish to express my appreciation to Professor R. Viskanta from Purdue University, West Lafayette, Indiana, and Professor J. Sucec from the University of Maine for their patience and kindness. Although we have never met, we have been aware of one anothers' studies for many years, and I inundated them with questions when I came to America.

I received significant help from the staff of the Art and Engineering and Science Shapiro libraries of the University of Michigan. My thanks go especially to Verena Ward and Alena Verdiyan from the Art and Engineering library and to reference assistants from both libraries.

There are several people whose humanity and experience made it possible for me to write this book: first, doctors from the Medical Center of the University of Michigan, who granted me additional years after my infarct in 2000, Professor R. Prager from the Division of Adult Cardiac Surgery, Professor M. Grossman from the Cardiovascular Center, and nurses C. Martinez, L. McCrumb, and C. Harris from his team, and my family physician, Dr. S. Gradwohl.

Other Americans who have made significant contributions to this book are my first English tutors, because I came with zero English. They are still my friends — Jim Brayson, Mary Catherwood, Gail Dumas, and Mary Bohend.

Friends of our family, Dr. Brian Schapiro and his wife Margo; Dr. Stephen Saxe and his wife Kim; Dr. Barbara Appelman; and Mrs. Sallie Abelson helped us in many ways during the time that I worked on this book, and I take this opportunity to express my love to them. A special thanks to my young friends: high school student David Schapiro and University of Michigan student Zach Renner who help me with my computer work.

I have greatly benefited from my family during my work and I am thankful to my daughter Ella, son-in-law Isaac, grandson Alex, and granddaughter-in-law Shirley for their advise and for helping me find solutions to my computer and English problems. I am especially thankful to my dear wife Sofia for your tolerance and understanding; and the difficulties of being my wife for almost 60 years.

Finally, I wish express my appreciation to the staff of Taylor & Francis/ CRC Press. Personally, to Senior Editor — Mechanical, Aerospace, Nuclear and Energy Engineering, Jonathan W. Plant, for his interest in publishing this book, to his assistants Amber Donley and Arlene Kopeloff for their advise and help in many different ways. I am especially grateful to Linda Leggio whose careful editing improved my English and general quality of the manuscript text.

The Author

Abram S. Dorfman, Ph.D., was born in 1923 in Kiev, Ukraine, in the former Soviet Union. He graduated from the Moscow Institute of Aviation in 1946, as an Engineer of Aviation Technology. From 1946 to 1947, he worked in the Central Institute of Aviation Motors (ZIAM) in Moscow. From 1947 to 1990, Dr. Dorfman studied fluid mechanics and heat transfer at the Institute of Thermophysics of the Ukrainian Academy of Science in Kiev, first as a junior scientist from 1947 to 1959, then as a senior scientist from 1959 to 1978, and finally as a leading scientist from 1978 to 1990. He earned a Ph.D. with a thesis entitled, "Theoretical and Experimental Investigation of Supersonic Flows in Nozzles," in 1952. In 1978, he received a Doctor of Science degree, which was the highest scientific degree in the Soviet Union, with a thesis and a book, *Heat Transfer in Flows around the Nonisothermal Bodies*. From 1978 to 1990, he was associate editor of *Promyshlennaya Teploteknika*, which was published in English as *Applied Thermal Science* (Wiley). Dr. Dorfman was also an advisor to graduate students for many years.

In 1990, he emigrated to the United States and continues his research as a visiting professor at the University of Michigan in Ann Arbor (since 1996). During this period, he has published several papers in leading American journals. He is listed in *Who's Who in America 2007*.

Dr. Dorfman has published more than 130 papers and two books in fluid mechanics and heat transfer in Russian (mostly) and in English. More than 50 of his papers published in Russian have been translated and are also available in English. Since 1965, he has been systematically studying conjugate heat transfer.

Introduction

What Is Conjugate Heat Transfer Like?

Exact formulation of convective heat transfer problems requires using the boundary condition of the fourth kind. Such a boundary condition consists of the conjugation of the body and fluid temperature fields at their interface. Because a conjugation procedure is complicated, instead of it, the boundary condition of the third kind has been used since the time of Newton:

$$q_w = h(T_w - T_\infty) \tag{I.1}$$

The heat transfer coefficient, h, is usually determined experimentally, and no well-found theoretical approach was available until the last few decades. Because the value of the heat transfer coefficient depends on the wall temperature distribution, data for the isothermal wall ($T_w = const$) heat transfer coefficient have been used for practical calculations. This simplified approach in which the effect of actual wall temperature distribution is neglected was acceptable before the advent of the computer, when the required calculation accuracy was not as high.

Intense interest in the conjugate convective heat transfer arose at the end of 1960s, and beginning at that time, many convective heat transfer problems have been considered using a conjugate, coupled, or adjoint formulation. These three equivalent terms correspond to a situation when the solution domain consists of two or more subdomains in which studied phenomena are described by different differential equations. After solving the problem in each of the subdomains, these solutions should be conjugated. The same procedure is needed if the problem is governed by one differential equation, but the subdomains have different materials or other properties.

For example, heat transfer between a body and a fluid flowing past it is a conjugate problem, because the heat transfer inside the body is governed by the elliptic Laplace equation or by the parabolic differential equation, while the heat transfer inside the flowing fluid is governed by the elliptic Navier-Stokes equation or by the parabolic boundary layer equation. The solution of such a problem gives the temperature and heat flux distributions along the body–fluid interface, and there is no need for a heat transfer coefficient. Moreover, the heat transfer coefficient can be calculated using these results. Another example of a conjugate problem is the transient heat transfer between a hot plate and a cooling thin fluid film flowing along it. In such a case, the plate at each moment is divided into a wet part covered with moving film and a dry one until uncovered. Although heat transfer in both portions of the plate is governed by the same conduction equation, the solutions obtained for each of the parts should be conjugated, because the thermal properties of wet and dry parts are different.

There are many conjugate problems in other areas of science. For instance, studying subsonic–supersonic flows required conjugation, because the subsonic flow is governed by elliptic or parabolic differential equations, while the supersonic flow is described by the hyperbolic differential equation [1]. Combustion theory and biological processes are two other examples. Every combustion process has two areas containing fresh and burnt gases with different properties [2]. In biology, diffusion processes usually proceed simultaneously in qualitatively different areas (e.g., in membranes) and, therefore, require a conjugation procedure [3].

Analogous mathematical problems with a similar formulation, usually called mixed problems, began to be considered much earlier than these physical systems. The most famous mixed parabolic/hyperbolic equation was investigated by Tricomi in 1923. This equation is now widely used for subsonic–supersonic flows [1].

References

1. Manwell, A. R., 1979. Ticomi equation, with applications to the theory of plane transonic flow. *Research Notes in Mathematics*, No. 35. Pitman, London.
2. Dorfman, A. S., 2005. Combustion stabilization by forced oscillations in a duct, *SIAM J. Appl. Math.* 65: 1175–1199.
3. Kleinstreuer, C., 2006. *Biofluid Dynamics*. Taylor & Francis, Boca Raton, FL.

Nomenclature

$\mathrm{Bi} = \dfrac{h\Delta}{\lambda_w}$

Biot number

$\mathrm{Br} = \dfrac{\lambda\Delta}{\lambda_w L}\,\mathrm{Pr}^m\,\mathrm{Re}^n$

Brun number

C_1, C_2

Exponents of influence functions in integral forms determining surface heat flux and temperature

$C_f = \dfrac{\tau_w}{\rho U_\infty^2}$

Friction coefficient

$2\mathrm{St}/\mathrm{Cf}$

Reynolds analogy coefficient

c, c_p

Specific heat and specific heat at constant pressure J/kg K

$\hat{c} = \rho c \Delta$

Thermal capacity J/m²K

D, D_h

Diameter and hydraulic diameter m

D_m

Diffusion coefficient, m²/s

$\mathrm{Ec} = \dfrac{U^2}{c_p \theta_w}$

Eckert number

$f(\xi/x)$

Influence function of unheated zone at temperature jump

$f_q(\xi/x)$

Influence function of unheated zone at heat flux jump

$\mathrm{Fo} = \dfrac{\alpha t}{L^2}$

Fourier number

$G = \dfrac{\mu_0 V^2}{\lambda_w \Delta T}$

Griffith number, Example 8.6

$Gr = \dfrac{\beta \theta_w g L^2}{v^2}$ Grashof number

g_k, h_k Coefficients of series in differential forms determining surface temperature and heat flux

g Gravitational acceleration, m/s^2

h, h_m Heat and mass transfer coefficients, W/m^{2}K

k Specific heat ratio or turbulence energy in $k - \omega$ model

K_τ, K_q Constants in power rheology laws for non-Newtonian fluids

$Kn = \dfrac{l}{D_h}$ Knudsen number, Example 10.7

l Body length or mixing length or free path, m

L Characteristic length, m

$Le = \dfrac{D_m}{\alpha}$ Lewis number, Example 8.8

$Ls = \dfrac{\lambda_w h}{\rho_w^2 c_w^2 U_w^2 \Delta}$ Leidenfrost number, Example 10.12

$Lu = \dfrac{\rho c}{\rho_w c_w}$ Luikov number, Section 6.6.1

$M = \dfrac{U}{U_{sd}}$ Mach number

M Moisture content, kg/kg

n, s Exponents in power rheology law for non-Newtonian fluids

$Nu = \dfrac{hL}{\lambda}, Nu = \dfrac{hL^{s+1}}{K_q U^s}$ Nusselt number for Newtonian and non-Newtonian fluids

p Pressure, Pa

$\text{Pe} = \dfrac{UL}{\alpha}$ — Peclet number for Newtonian and non-Newtonian fluids

$\text{Pr} = \dfrac{v}{\alpha}, \text{Pr} = \dfrac{\rho c_p U^{1-s} L^{1+s}}{K_q}$ — Prandtl number for Newtonian and non-Newtonian fluids

q, q_v — Heat flux, W/m², and volumetric heat source, W/m³

r/s — Exponent in heat transfer coefficient expression for isothermal surface

$\text{Ra} = \dfrac{\beta \theta_w g L^3}{v \alpha}$ — Rayleigh number

$\text{Re} = \dfrac{UL}{v}, \text{Re} = \dfrac{\rho U^{2-n} L^n}{K_\tau}$ — Reynolds number for Newtonian and non-Newtonian fluids

$\text{Sc} = \dfrac{v}{D_m}$ — Schmidt number, Example 9.1

$\text{Sh} = \dfrac{h_m}{\rho c_p D_m}$ — Sherwood number, Example 9.1

$\text{St} = \dfrac{h}{\rho c_p U}$ — Stanton number

$\text{Sk} = \dfrac{4\sigma T_\infty^4 L}{\lambda_\infty}$ — Starks number, Example 7.15

$\text{Ste} = \dfrac{c_p \Delta T}{\Lambda}$ — Stephan number, Example 9.2

t — Time, s

T — Temperature, K

u, v, w — Velocity components; u, v parts in integrating by parts

U — Velocity on outer edge of boundary layer

$$u_\tau = \sqrt{\tau_w/\rho}$$ Friction velocity

$$u^+ = u/u_\tau, \, y^+ = yu_\tau/v$$ Variables in wall law

$$x, y, z$$ Coordinates

Greek Symbols

α Thermal diffusivity m²/s

β Dimensionless pressure gradient in self-similar solutions and turbulent equilibrium boundary layer or volumetric thermal expansion coefficient, 1/K

$$\chi_t = \frac{h}{h_*}, \, \chi_p = \frac{h_m}{h_{m*}}$$ Nonisothermicity and nonisobaricity coefficients, Example 8.8

$$\chi_f = \frac{C_f}{C_{f*}}$$ Nonisotachicity coefficient, Section 5.2

$\delta, \delta^*, \delta^{**}$ Boundary layer thicknesses, m

δ, δ_{ij} Delta function and Kronecker delta

Δ Body or wall thickness, m

κ Constant determining mixing length

λ Thermal conductivity, W/mK

Λ Latent heat, J/kg or λ_s/λ

μ Viscosity, kg/s m

v Kinematic viscosity, m²/s

ξ Unheated zone length, m

$\theta = T - T_\infty$ Temperature excess

$\theta_w = T_w - T_\infty$	Temperature head
ρ	Density, kg/m³
σ	Stefan-Boltzmann constant, W/m²K⁴
τ	Shear stress, N/m²
Φ, φ	Prandtl-Mises-Görtler variables
ψ	Stream function, m²/s
ω	Specific dissipation rate in $k - \omega$ model

Some of these symbols are also used in different ways, as indicated in each case.

Subscripts

av	Average
ad	Adiabatic
as	Asymptotic
bl	Bulk
e	End or effective or entrance
i	Initial; inside
L	At $x = L$
m	Mass average, or mean value, or moisture
o	Outside
s	Solid
sd	Sound

q	Constant heat flux
t	Thermal
tb	Turbulent
w	Fluid–solid interface
ξ	After jump
∞	Far from solid
$*$	Isothermal

Superscipts

$+, -$	From both sides of interface
$+, ++$	Wall law and other variables for turbulent boundary layer

---

Overscores

$\bar{o}, \tilde{o}$	Dimensionless; or transformed or auxiliary

Part I

Approximate Solutions

1

Analytical Methods for the Estimation of Heat Transfer from Nonisothermal Walls

In this chapter, the approximate analytical methods for calculating heat transfer from arbitrary nonisothermal walls are analyzed. Most of the early solutions of conjugate heat transfer problems are based on those approximate methods. Moreover, approximate analytical methods are frequently used at present due to their simplicity and easy physical interpretation of the results.

1.1 Basic Equations

The majority of practically important problems of flow and heat transfer are characterized by high Reynolds (Re) and Peclet (Pe) numbers. In such a case, viscosity and conductivity are significant only in thin boundary layers, and the system of Navier-Stokes and energy equations is simplified to boundary layer equations. For laminar steady-state flow of an incompressible fluid with constant thermophysical properties, these equations are [1]

$$\frac{\partial u}{\partial x} + \frac{\partial v}{\partial y} = 0, \quad u\frac{\partial u}{\partial x} + v\frac{\partial u}{\partial y} - U\frac{dU}{dx} - v\frac{\partial^2 u}{\partial y^2} = 0,$$

$$u\frac{\partial T}{\partial x} + v\frac{\partial T}{\partial y} - \alpha\frac{\partial^2 T}{\partial y^2} - \frac{v}{c_p}\left(\frac{\partial u}{\partial y}\right)^2 = 0 \tag{1.1}$$

The solutions of system (1.1) must satisfy the boundary conditions at the body surface and far away from the body:

$$y = 0 \quad u = v = 0, \quad T = T_w(x), \quad y \to \infty \quad u \to U, \quad T \to T_\infty \tag{1.2}$$

Prandtl and Misis independently transformed the boundary layer equations into the form that is close to that of the heat-conduction equation [1]. This form is obtained using new variables x, stream function ψ, and $Z = U^2 - u^2$:

$$u = \frac{\partial \psi}{\partial y}, v = -\frac{\partial \psi}{\partial x}, \quad \frac{\partial Z}{\partial x} - vu\frac{\partial^2 Z}{\partial \psi^2} = 0, \quad \frac{\partial T}{\partial x} - \alpha\frac{\partial}{\partial \psi}\left(u\frac{\partial T}{\partial \psi}\right) - \frac{v}{c_p}u\left(\frac{\partial u}{\partial \psi}\right)^2 = 0$$

$$\psi = 0 \quad Z = U^2(x), \ T = T(x), \quad \psi \to \infty \quad Z \to 0, \ T \to T_\infty$$

$$\tag{1.3}$$

3

Equation (1.3) has a singularity on the surface such that the derivatives with respect to ψ become infinite. Because the velocity near the wall is an analytic function, one gets

$$u \sim c_1 y + c_2 y^2 + .,, \quad \psi = \int_0^y u \, dy \sim \frac{c_1}{2} y^2 + \frac{c_2^2}{3} y^3 + \cdots \quad (1.4)$$

Hence, $y \sim \psi^{1/2}$, the derivative $\partial y / \partial \psi \sim \psi^{-1/2}$, and $\partial y / \partial \psi \to \infty$ when $\psi \to 0$.

1.2 Self-Similar Solutions of the Boundary Layer Equations

If the body has a shape of a wedge with opening angle $\pi\beta$, then the velocity of the potential of an ideal fluid flow obeys a power law distribution. If the temperature head along the wedge also obeys a power law distribution,

$$U(x) = Cx^m, \quad m = \beta/(2-\beta), \quad T_w(x) - T_\infty = C_1 x^{m_1} \quad (1.5)$$

and then the partial differential equation (Equation 1.1) reduces to ordinary differential equations. In this case, it is said that the boundary layer equations have self-similar solutions. Physically, this means that the velocity and the temperature head distributions in different sections of the boundary layer are identical, so that single velocity and single temperature head profiles form if they are plotted in the self-similar variables:

$$\frac{u}{U} = \varphi'(\eta), \quad \frac{T-T_w}{T_\infty - T_w} = \theta(\eta) \quad \eta = y\sqrt{\frac{(m+1)Cx^{m-1}}{2v}} \quad \varphi = \psi\sqrt{\frac{m+1}{2vCx^{m+1}}} \quad (1.6)$$

Equation (1.1) in the variables φ, θ, and η transform into ordinary differential equations:

$$\varphi''' + \varphi\varphi'' + \beta(1-\varphi'^2) = 0 \quad \theta'' + \mathrm{Pr}\,\varphi\theta' + \frac{2m_1}{m+1}\mathrm{Pr}\,\varphi'(1-\theta) = 0 \quad (1.7)$$

Equation (1.7) is numerically integrated and tabulated [1]. Using these results, the friction coefficient, the Nusselt number, and the heat flux are determined:

$$C_f = \frac{\sqrt{2(m+1)}\varphi''(0)}{\mathrm{Re}_x^{1/2}}, \quad \mathrm{Nu}_x = \sqrt{\frac{m+1}{2}}\theta'(0)\sqrt{\mathrm{Re}_x}, \quad q_w = C_1\lambda\sqrt{\frac{(m+1)C}{2v}}\theta'(0)x^{m_1+\frac{m-1}{2}}$$

$$(1.8)$$

TABLE 1.1

Self-Similar Solutions

Thermal Boundary Condition		β	m	$\sqrt{2(m+1)}\varphi''(0)$	m_1	$\sqrt{\dfrac{m+1}{2}}\theta'(0)$		
						$\text{Pr}\to 0$	$\text{Pr}=1$	$\text{Pr}\to\infty$
Plate	$T_w = const.$	0	0	0.664	0	$0.564\text{Pr}^{0.5}$	$0.332\text{Pr}^{1/3}$	$0.339\text{Pr}^{1/3}$
Plate	$q_w = const.$	0	0	0.664	1/2	$0.885\text{Pr}^{0.5}$	$0.453\text{Pr}^{1/3}$	$0.458\text{Pr}^{1/3}$
Stagnation	$T_w = const.$	$\pi/2$	1	2.466	0			
point	$q_w = const.$	$\pi/2$	1	2.466	0	$0.791\text{Pr}^{0.5}$	$0.570\text{Pr}^{1/3}$	$0.661\text{Pr}^{1/3}$

It is seen that in self-similar flows, the heat flux changes along the wall also according to the power law with exponent $m_2 = m_1 + (1-m)/2$. Thus, heat flux is constant on the wall if $m_1 = (1-m)/2$. Hence, on the plate ($m = 0$) with constant heat flux, the temperature varies as $\sim x^{1/2}$, while the stagnation point surface ($m = 1$) in such a case is isothermal.

Example 1.1: The Effect of Boundary Conditions

Table 1.1 shows that boundary conditions significantly affect the value of the Nusselt number. Thus, on the stagnation point surface, the heat transfer coefficients for the surfaces with $T_w = const.$ and $q_w = const.$ are equal, while for the plate these coefficients differ remarkably. For the medium and high Pr, the difference between two coefficients is 36%, whereas for the low Prandtl numbers it reaches 57%.

1.3 Solutions of the Boundary Layer Equations in the Power Series

Self-similar solutions are applicable to only a narrow class of problems. In general, when the velocity $U(x)$ and temperature head $T_w - T_\infty = \theta_w(x)$ are arbitrary functions, the boundary layer equations may be solved by approximate methods. In 1908, Blasius was the first to suggest such a method [1]. The method is based on the assumption that the velocity and temperature head distributions around the symmetrical body can be presented in the form of a power series:

$$U(x) = u_1 x + u_3 x^3 + u_5 x^5 + \cdots, \qquad \theta_w(x) = \theta_{w0} + \theta_{w2}x^2 + \theta_{w4}x^4 + \cdots \qquad (1.9)$$

The solutions presented in the same form of power law series,

$$\psi(x,y) = \psi_1(y)x + \psi_3(y)x^3 + \psi_5(y)x^5 + \cdots$$

$$T - T_\infty = \theta(x,y) = \theta_0(y) + \theta_2(y)x^2 + \theta_4(y)x^4 + \cdots \qquad (1.10)$$

being substituted into boundary layer equations give the system of ordinary differential equations defining the coefficients of these series. Howarth [1] shows that the variable $\eta = y\sqrt{u_1/x}$ transforms equations for series coefficients into a universal form that is independent on a specific problem. These coefficients and corresponding derivatives are tabulated [2,3] and can be used to calculate the friction and heat transfer for any series (Equation 1.9). For air (Pr = 0.7), the Nusselt number is given as (D is a characteristic length) [3]:

$$
\mathrm{Nu} = \frac{D}{\theta_w}\sqrt{\frac{u_1}{\nu}}\left\{\theta_{w0}\left[0.4959 + 0.4764\frac{u_3}{u_1}x^2 + \frac{u_3}{u_1}\left(1.054 + 0.6678\frac{u_3^2}{u_1 u_3}\right)x^4\right]\right.
$$

$$
\left. + \theta_{w2}\left[0.852x^2 + 06678\frac{u_3}{u_1}x^4 + \cdots\right] + \theta_{w4}[1.054x^4 + \cdots] + \cdots\right\}
$$

(1.11)

The more general case of an asymmetrical body when a power law series (Equation 1.9) contains all values of exponents is considered in the literature [3].

Computational experience shows that the methods based on the power law extension yield satisfactory results basically at a relatively small distance from the stagnation point. In order to obtain satisfactory accuracy at a large distance from the stagnation point, a large number of universal functions is required. This is because the number of universal functions needed for expressing coefficients in final series increases rapidly with the number of terms. This problem is especially severe for the asymmetrical bodies. At the same time, in the simpler case of symmetrical flow over a body, universal functions for the shear stress have already been calculated up to terms containing x^{11}, and even the velocity profiles on the circular cylinder including a point of separation are calculated [1].

Görtler [1] suggested power law series solutions in special variables:

$$
\Phi = \frac{1}{\nu}\int_0^x U(\zeta)d\zeta,, \qquad \eta = \frac{yU}{\nu\sqrt{2\Phi}}, \qquad \varphi = \frac{\psi}{\nu\sqrt{2\Phi}}, \qquad \theta = \frac{T - T_w}{T_\infty - T_w} \qquad (1.12)
$$

Converting Equation (1.1) to these variables, one obtains

$$
\frac{\partial^3\varphi}{\partial\eta^3} + \varphi\frac{\partial^2\varphi}{\partial\eta^2} + \beta(\Phi)\left[1 - \left(\frac{\partial\varphi}{\partial\eta}\right)^2\right] = 2\Phi\left[\frac{\partial^2\varphi}{\partial\Phi\partial\eta}\frac{\partial\varphi}{\partial\eta} - \frac{\partial\varphi}{\partial\Phi}\frac{\partial^2\varphi}{\partial\eta^2}\right] \qquad (1.13)
$$

$$
\frac{1}{\mathrm{Pr}}\frac{\partial^2\theta}{\partial\eta^2} + \varphi\frac{\partial\theta}{\partial\eta} + 2\Phi\beta_t(\Phi)\frac{\partial\varphi}{\partial\eta}(1-\theta) = 2\Phi\left[\frac{\partial\varphi}{\partial\eta}\frac{\partial\theta}{\partial\Phi} - \frac{\partial\varphi}{\partial\Phi}\frac{\partial\theta}{\partial\eta}\right]
$$

(1.14)

$$
\beta(\Phi) = \frac{2}{U^2}\frac{dU}{dx}\int_0^x U(\zeta)d\zeta \qquad \beta_t(\Phi) = \frac{2\nu}{U\theta_w}\frac{d\theta_w}{dx}
$$

If the velocity $U(x)$ and the temperature head $\theta_w(x) = T_w(x) - T_\infty$ are power law functions (Equation 1.5), then β and β_t are constant. In this case, φ and θ are functions only of variable Φ, and the Görtler Equations (1.13) and (1.14) become identical to the self-similar Equation (1.7). In general, when $U(x)$ and $\theta_w(x)$ are arbitrary functions, β and β_t are presented as power series in variable Φ. In this case, the solution of Equations (1.13) and (1.14) can be introduced in the form of a similar power law series in Φ. Then, the ordinary differential equations for coefficients of the Görtler series are obtained, which are analogous to these for the Blasius series. The Görtler series are used for solving a number of problems. In particular, these series are used to study forced [3] and natural [4] convective heat transfer from nonisothermal surfaces. It is found that the Görtler series converge more rapidly than the Blasius power series, but their convergence is also not always satisfactory.

1.4 Integral Methods

Integral methods are widely used because of their relative simplicity, which follows from replacing the differential equations of the boundary layer by the integral relationships. In this case, the conservation laws are not satisfied in each point of the flow, but only integrally, on the average for each cross-section through the boundary layer.

To obtain the integral relationships, Equation (1.1) is rewritten in the form

$$\frac{\partial(Uu)}{\partial x} + \frac{\partial(Uv)}{\partial y} - u\frac{dU}{dx} = 0 \qquad \frac{\partial(u^2)}{\partial x} + \frac{\partial(uv)}{\partial y} - U\frac{dU}{dx} - v\frac{\partial^2 u}{\partial y^2} = 0 \qquad (1.15)$$

The first part of Equation (1.15) is obtained by multiplication of the continuity Equation (1.1) by U. The second part is a sum of continuity Equation (1.1), multiplied by u, and momentum Equation (1.1). Integrating across the boundary layer the difference between the first and second parts of Equation (1.15) and taking into account that the derivative $(\partial u/\partial y)$ at $y \to \infty$ and the product $v(U - u)$ at $y = 0$ and $y \to \infty$ turn to zero, one obtains the integral boundary layer momentum equation. The integral boundary layer energy equation is obtained in a similar manner by integrating across the boundary layer the difference between the energy Equation (1.1) and continuity Equation (1.1) multiplied by $(T - T_\infty)$:

$$\frac{d}{dx}(U\delta^{**}) + U\frac{dU}{dx}\delta^* = \frac{\tau_w}{\rho} \qquad \frac{d}{dx}(U\theta_w\delta_t^{**}) = \frac{q_w}{\rho c_p} \qquad (1.16)$$

$$\delta^* = \int_0^\infty \left(1 - \frac{u}{U}\right)dy \qquad \delta^{**} = \int_0^\infty \frac{u}{U}\left(1 - \frac{u}{U}\right)dy \qquad \delta_t^{**} = \int_0^\infty \frac{u}{U}\frac{T - T_\infty}{T_w - T_\infty} \qquad (1.17)$$

Here $\delta^*, \delta^{**}, \delta_t^{**}$ are displacement, momentum, and energy thicknesses. Because the integral boundary layer equations satisfied the conservation laws on the average for each cross-section through boundary layer, they do not yield any information on the velocity and temperature distributions across the boundary layer. For this reason, the velocity and temperature profiles are found in integral methods by selecting suitable sets of functions satisfying the boundary conditions.

**Example 1.2: Friction and Heat Transfer from
an Arbitrary Nonisothermal Plate [5]**

The velocity and temperature distributions are specified as the four power polynomials. Ten coefficients of these polynomials are selected in such a manner as to satisfy the following ten conditions on the plate, as well as on the outer edge of the velocity $(y = \delta)$ and thermal $(y = \delta_t)$ layers.

$$y = 0, \qquad u = 0, \qquad T = T_w(x); \qquad \frac{\partial^2 u}{\partial y^2} = \frac{\partial^2 T}{\partial y^2} = 0,$$

$$y = \delta, \qquad u = U, \qquad \frac{\partial u}{\partial y} = \frac{\partial^2 u}{\partial y^2} = 0; \qquad y = \delta_t, \qquad T = T_\infty, \qquad \frac{\partial T}{\partial y} = \frac{\partial^2 T}{\partial y^2} = 0$$

$$(1.18)$$

The third conditions at $y = 0$ follow from the boundary layer equation for gradientless flow, and the other conditions are obvious. Simple calculations give the profiles

$$\frac{u}{U} = 2\eta - 2\eta^3 + \eta^4 = F(\eta) \quad \eta = \frac{y}{\delta} \qquad \frac{T - T_w}{T_\infty - T_w} = 2\eta_t - 2\eta_t^3 + \eta_t^4 = F_t(\eta_t), \quad \eta_t = \frac{y}{\delta_t}$$

$$(1.19)$$

Then, Equation (1.16) yields the differential equations, defining the thicknesses of the velocity and thermal boundary layers. The dynamic thickness is obtained simply:

$$\delta \frac{d\delta}{dx} = \frac{630}{37} \frac{v}{U}, \qquad \delta_t \frac{d}{dx}(\theta_w \delta_t^{**}) = \frac{2\alpha\theta_w}{U} \qquad \delta = 5.83\sqrt{vx/U} \qquad (1.20)$$

Two cases must be considered to get the thermal boundary layer thicknesses: when the thermal boundary layer is thinner $(\mathrm{Pr} \geq 1)$ and thicker $\mathrm{Pr} \to 0$ than the velocity boundary layer. Taking into account that in the second case outside of the velocity boundary layer, the function $F_t(\eta_t) = 1$, relations for the first and second cases are obtained $(\varepsilon = \delta_t/\delta)$:

$$\frac{\delta_t^{**}}{\delta_t} = \frac{2\varepsilon}{15} + \frac{3\varepsilon^3}{140} + \frac{\varepsilon^4}{180}, (\varepsilon \leq 1), \quad \frac{\delta_t^{**}}{\delta_t} = \frac{3}{10} - \frac{3}{10\varepsilon} + \frac{2}{15\varepsilon^2} - \frac{3}{140\varepsilon^4} - \frac{1}{180\varepsilon^5} (\varepsilon > 1)$$

$$(1.21)$$

Retaining here only the two first terms and putting the restricted expressions into the second Equation (1.20) gives two ordinary differential equations. The solution of these equations yields two expressions defining the thickness of the thermal boundary layers for $(\Pr \geq 1)$ and $\Pr \to 0$:

$$\frac{\delta_t}{\delta} = \frac{0.871}{\Pr^{1/3}[\theta_w(x)]^{1/2} x^{1/4}} \left[\int_0^x \frac{[\theta_w(x)]^{3/2}}{x^{1/4}} dx \right]^{1/3} \tag{1.22}$$

$$\frac{\delta_t}{\delta} = \frac{0.626}{\Pr^{1/2} \theta_w(x) x^{1/2}} \left[\int_0^x [\theta_w(x)]^2 dx \right]^{1/2} \tag{1.23}$$

Comparing these results with self-similar solutions gives the estimation of accuracy.

Nusselt numbers corresponding to Equation (1.8) for a power law temperature head are

$$\mathrm{Nu}_x = 0.358(2m_1 + 1)^{1/3} \Pr^{1/3} \mathrm{Re}_x^{1/2}, \qquad \mathrm{Nu}_x = 0.548(2m_1 + 1)^{1/2} \Pr^{1/2} \mathrm{Re}_x^{1/2}$$

$$\tag{1.24}$$

Then, for surfaces with $\theta_w =$ const. $(m_1 = 0)$ and $q_w =$ const. $(m_1 = 1/2)$, one obtains: 0.358,0.548 and 0.452,0.775. for $(\Pr \geq 1)$ and $\Pr \to 0$, respectively. According to data from Table 1.1, the largest difference, about 12%, is in the case of $q_w =$ const. and $\Pr \to 0$.

The case when both functions $\theta_w(x)$ and $U(x)$ are arbitrary is examined in [6].

Integral methods employing polynomials generally yield satisfactory results in cases when the profiles of velocity and temperature in the boundary layer do not exhibit singularities and can be accurately described by polynomials. This condition generally holds at relatively small velocity gradients in the mainstream and small temperature heads. In the case of large, especially of large negative, gradients (divergent flows and a temperature head falling along the flow), approximation of velocity and temperature profiles in a boundary layer by polynomials becomes unsatisfactory.

In these cases, significantly better results are obtained when the velocity and temperature distributions across the layer are described by a set of profiles corresponding to the exact solution of some particular problem of the boundary layer, for example, by a set of profiles obtained via self-similar solutions. Such a method for the velocity boundary layer was developed by Kochin and Loytsyanskiy [7]. Subsequently, this method for heat transfer coefficients was developed by Skopets [8], Zysina-Molozhen et al. [9], and Drake [10].

The Kochin-Loytsyanskiy method is based on an assumption that at arbitrary $U(x)$ and $\theta_w(x)$, the velocity and temperature distributions across the layer are described by function (1.6) that depends on parameters, just as in the case of self-similar flows:

$$u_x/U = \varphi'(\eta, \beta), \qquad (T - T_w)/(T_\infty - T_w) = \theta(\eta, \beta, m_1, \mathrm{Pr}), \qquad \eta = y\sqrt{U'/\beta v}$$

(1.25)

The parameters β and m_1 are selected in such a manner that the integral relation (1.16) is satisfied in each section across the boundary layer. For this purpose, function (1.25) is substituted into Equation (1.16). This yields two differential equations:

$$\frac{df}{dx} = \frac{U''}{U'}f + \frac{U'}{U}F(f), \qquad \frac{df_t}{dx} = \left(\frac{U''}{U'} - 2\frac{\theta'_w}{\theta_w}\right)f_t + \frac{U'}{U}F_t(f, f_t)$$

(1.26)

which define, at specified $U(x)$ and $\theta_w(x)$, functions $f(x) = \delta^{**2}U'/v$ and $f_t(x) = \delta_t^{**2}U'/v$, termed the velocity and thermal shape parameters, respectively. Functions $F(f)$ and $F_t(f, f_t)$ are evaluated from tables of self-similar solutions. For known $F(f)$ and $F_t(f, f_t)$, Equation (1.26) can be solved numerically. As a result, the thicknesses δ^{**} and δ_t^{**} are determined, and then the friction coefficient and Nusselt number are obtained:

$$C_f = (v/U\delta^{**})\zeta(f) \qquad \mathrm{Nu}_x = (x/\delta_t^{**})\zeta_t(f, f_t)$$

(1.27)

where $\zeta(f)$ and $\zeta_t(f, f_t)$ are tabulated functions [7].

Methods for calculating heat transfer coefficients from nonisothermal surfaces were also developed by Ambrok [11], Shuh [12], Spalding [13], and Seban [14]. A greater number of studies, which are surveyed by Spalding and Pun [15], are concerned with heat transfer from isothermal surfaces.

Integral methods are extensively used for calculating coefficients of friction and heat transfer in turbulent flows. In this case, however, experimental data must be used for defining the functions describing the velocity and temperature distributions in the layer and for defining the relationships contained in the integral equations. Many methods are suggested. Most of these pertain to calculating friction coefficients. Much less is done on heat transfer from isothermal surfaces, and relatively few studies are concerned with heat transfer from nonisothermal surfaces. Surveys of such methods are presented, among others, in [1,16–19].

One of the simplest methods for calculating friction and heat transfer in turbulent flow is based on the power functions for the velocity and temperature distributions:

$$\frac{u}{U} = \left(\frac{y}{\delta}\right)^{1/7}, \qquad \frac{T - T_\infty}{T_w - T_\infty} = \left(\frac{y}{\delta_t}\right)^{1/7}$$

(1.28)

These functions describe rather satisfactorily the distribution of these parameters in the turbulent layer, but they do not represent their distribution in the viscous sublayer. Because of this, the skin friction and the heat transfer cannot be calculated by differentiating function (1.28), as in the case of a laminar layer. In this case, the friction and heat transfer coefficients are determined using correlations of numbers of experimental data.

Using the well-known law of the wall $u/u_\tau = 8.74(yu_\tau/v)$, the constants $C = 0.045$ and $n = -1/4$ in the general relation $C_f = C\operatorname{Re}_\delta^n$ may be defined as follows [19]. Transforming the first part of Equation (1.28) to variables of law of the wall leads to the following expression:

$$u/u_\tau = (C/2)^{-4/7} \operatorname{Re}_\delta^{-(4n+1)/7} (yu_\tau/v)^{1/7}, C_f = 0.045 \operatorname{Re}_\delta^{-1/4} \tag{1.29}$$

which should be identical with this law. This occurs only if $4n+1=0$ when the Reynolds number exponent becomes zero and $(C/2)^{-4/7} = 8.74$, giving Equation (1.29).

The heat transfer coefficient can be estimated using the following ratio for the turbulent layer out of the viscous sublayer: $\tau/q = \lambda_{tb}(\partial T/\partial y)/\mu_{tb}(\partial u/\partial y)$. Calculating derivatives from Equation (1.28) and assuming that turbulent Prandtl number $\operatorname{Pr}_{tb} = 1$ and that $\tau = \tau_w$ and $q = q_w$ in the vicinity of the wall, one gets

$$St = 0.0225 \operatorname{Pr}^{-1/4} \operatorname{Re}_\delta^{-3/4} (\delta/\delta_t)^{1/7} \tag{1.30}$$

where $\operatorname{Pr}^{-3/4}$ is added to take into account the Prandtl number effect. Using this expression and relation (1.29) for the friction coefficient, one obtains from the integral relationship (1.16) two differential equations:

$$\frac{d\delta^{5/4}}{dx} - \frac{115}{28} \frac{1}{U} \frac{dU}{dx} \delta^{5/4} = 0.289 \left(\frac{v}{U}\right)^{1/4} \tag{1.31}$$

$$\frac{d\varepsilon^{9/7}}{dx} + \frac{8}{9} \frac{1}{U\theta_w \delta} \frac{d(U\theta_w \delta)}{dx} \varepsilon^{9/7} = \frac{0.260}{\operatorname{Pr}^{3/4} \delta^{5/4}} \left(\frac{v}{U}\right)^{1/4} \tag{1.32}$$

Integrating these equations [16] yields δ and $\varepsilon = \delta_t/\delta$. Then, Equation (1.29) and Equation (1.30) can be used for calculating the friction and heat transfer coefficients.

Integral methods based on power law approximation (1.28) as well as methods employing polynomials are suitable in those cases when the velocity and temperature profiles across the layer do not exhibit singularities. The law of the wall is valid for a whole boundary layer at small Reynolds numbers. At high Reynolds numbers, the velocity distribution across the layer is described by a composite function that consists of the law of the wall for the inner and the velocity defect law,

$$\frac{U-u}{u_\tau} = f\left(\frac{y}{\delta}\right) \tag{1.33}$$

for the outer parts of the turbulent boundary layer. (See details in Section 4.1.) Integral methods employing such composite profiles were developed by Patankar and Spalding [17].

In calculating the coefficients of heat transfer from nonisothermal walls, extensive use is made of a method based on the conservative nature of the laws of heat transfer, written in the form of relationships between the local parameters of the boundary layer [20–23]. Theoretical analysis and experiments [22,23] show that one can neglect the effect of the pressure gradient and thermal boundary condition on the law written as

$$St = C \, Re_{\delta_t^{**}}^n \, Pr^m \tag{1.34}$$

This makes it possible to use, in the general case, this equation derived for the simplest case of gradientless flow and constant wall temperature. Substitution of Equation (1.34) into the second integral relationship (1.16) yields an equation that, when integrated, gives

$$Re_{\delta_t^{**}} = \frac{1}{\theta_w} \left[C(1-n) Re_L \, Pr^m \int_0^{x/L} \theta_w^{1-n} d\left(\frac{x}{L}\right) \right]^{\frac{1}{1-n}} \tag{1.35}$$

The constants $C = 0.0128$, $n = -1/4$, $m = -0.6$ are obtained from workup of a large volume of experimental data. Once $Re_{\delta_t^{**}}$ is defined, St is found from Equation (1.34).

This approach allows determination of the solution of a large number of problems in a rather simple manner and with accuracy sufficient for practical needs [20,21,24]. However, it is shown [21–23] that the relative error in St, resulting from neglecting the effect of boundary conditions on the governing relationship, can be estimated by the inequality

$$\frac{\Delta St}{St} \leq 0.1 \frac{Re_{\delta_t^{**}}}{St \, Re_L \, \theta_w} \frac{d\theta_w}{d(x/L)} \tag{1.36}$$

1.5 Method of Superposition

The equation of the thermal boundary layer is linear, so one can apply to it the principle of superposition. According to this principle, the solution for the general case of an arbitrary nonisothermal wall can be obtained as a sum of particular solutions, found for some simple law of the wall temperature variations. One such simple law is stepwise change in temperature downstream of a nonheated zone. In this case, the wall temperature remains equal to the free stream temperature to some point $x = \xi$ and at this point changes suddenly to some constant value T_w. If the solution of this problem is known, then for arbitrary wall temperature distribution, the solution is given by the following integral [25,26]:

$$\frac{T-T_{\infty}}{T_w-T_{\infty}} = \theta(x,y.\xi), \quad T-T_{\infty} = \int_0^x \theta(x,y,\xi)\frac{dT_w}{d\xi}d\xi + \sum_{k=1}^i \theta(x,y,\xi_k)\Delta T_{wk} \quad (1.37)$$

Here the integral pertains to the continuous part, and the summation pertains to breaks in the temperature distribution, and ΔT_{wk} is a temperature jump in a point with coordinate ξ_k.

Differentiating Equation (1.37) with respect to y and using the influence function $f(x, \xi)$ yields the expression for the heat flux at the arbitrary nonisothermal wall:

$$q_w = h_* \left[\int_0^x f(x,\xi)\frac{dT}{d\xi}d\xi + \sum_{k=1}^i f(x,\xi_k)\Delta T_{wk} \right] \qquad h_\xi/h_* = f(x,\xi) \quad (1.38)$$

Here the influence function describes the effect of the nonheated zone on the coefficient of heat transfer h_ξ downstream of the sudden temperature change. Expression (1.38) is valid for any flow mode, and because successful methods for determining h_* are available, the main difficulty consists in calculating the influence function.

Example 1.3: Calculating the Influence Function [25,26]

For the elementary case of laminar gradientless flow and $Pr \geq 1$, the influence function is calculated by the integral method. The velocity and temperature profiles across the layer were approximated by third-degree polynomials. In this case, the thickness δ and the ratio of $\varepsilon = \delta_t/\delta$ are calculated by the same differential equation (Equation 1.16) as when using polynomials (Equation 1.18) but with different coefficients:

$$\delta\frac{d\delta}{dx} = \frac{140}{13}\frac{v}{U}, \qquad \delta\varepsilon\frac{d(\delta\varepsilon^2)}{dx} = 10\frac{\alpha}{U} \quad (1.39)$$

Substituting δ obtained from the first equation into the second equation yields a differential equation with a solution (with $13/14 \approx 1$) that gives the influence function in the following form:

$$\varepsilon^3 + 4\varepsilon^2 x\frac{d\varepsilon}{dx} = \frac{13}{14\,Pr} \qquad f(\xi/x) = [1-(\xi/x)^{3/4}]^{-1/3} \quad (1.40)$$

A more general result, pertaining to arbitrary gradient flows but limited to high Pr, is due to work by Lighthill [27]. He used the Prandtl-Mises form of thermal boundary Equation (1.3) and assumed that the velocity in the thermal boundary layer is proportional to y or, according to Equation (1.4), to $\psi^{1/2}$: $u = \tau_w y/\mu = (2\tau_w\psi/\mu)^{1/2}$. The correctness of this assumption improves at higher Pr, as the thermal layer becomes thinner compared with the velocity layer. Substituting the relation for velocity into Prandl-Mises Equation (1.3) gives

$$\frac{\partial T}{\partial x} = \frac{1}{Pr} \left(\frac{2\tau_w \nu}{\rho} \right)^{1/2} \frac{\partial}{\partial \psi} \left(\psi^{1/2} \frac{\partial T}{\partial \psi} \right) \tag{1.41}$$

Solution of this equation for the heat flux in the case of stepwise temperature ΔT_w and a varying temperature $T_w(x)$ is given by Lighthill [27]:

$$q_w = 0.5384\lambda \left(\frac{Pr}{\rho \nu^2} \right)^{1/3} \sqrt{\tau_w(x)} \left(\int_{\xi}^{x} \sqrt{\tau_w(z)} dz \right)^{-1/3} \Delta\theta_w \tag{1.42}$$

Example 1.4: Accuracy of Linear Velocity Distribution [27]

To estimate the exactness of linear velocity distribution across the thermal boundary layer, Lighthill calculated the Nusselt number for self-similar solution and uniform surface temperature. In this case, Equation (1.8) becomes

$$\frac{Nu_L}{\sqrt{Re_L}} = 0.807 \, Pr^{1/3} [\varphi'(0)]^{2/3} \left[\frac{3}{2}(m+1) \right]^{-2/3} \tag{1.43}$$

Table 1.2 compares the results obtained by this equation with self-similar solutions for $Pr = 0.7$. The agreement is best for $m = 0$, because in this case the derivatives of velocity at the wall vanish, and the velocity profile is close to linear. For $m > 0$, the error is greater because $\partial^2 u/\partial y^2 < 0$. For $m < 0$, the profile has a point of inflextion, $\partial^2 u/\partial y^2$ is positive for small y and negative for large y, giving too small a heat transfer rate.

Using Equation (1.42) and expression for shear stress, one can calculate the influence function. In particular, for self-similar flows it is found [16,26]:

$$f(\xi/x) = [1 - (\xi/x)^{(3/4)(m+1)}]^{-1/3} \tag{1.44}$$

TABLE 1.2

Accuracy of the Lighthill Formula (Equation 1.42)

m	−0.0904	−0.0654	0	1/9	1/3	1	4	→ ∞
Equation (1.43)		0.497	0.601	0.648	0.653	0.587	0.398	−0.921
Self-similar data		0.541	0.585	0.596	0.576	0.495	0.325	−0.741
Error (%)		−8	+3	−9	−13	−18.5	−22.5	+24

Equation (1.43) becomes exact as $\Pr \to \infty$. In the second limiting case, when $\Pr \to 0$, the thickness of the velocity boundary layer tends to zero, so that the longitudinal velocity over the entire thermal layer can be taken equal to the free-stream velocity: $u = U(x)$. The equation of the thermal boundary layer then becomes simpler, and for self-similar flows, the influence function can be presented in a form similar to that of Equation (1.44):

$$U(x)\frac{\partial T}{\partial x} = \alpha \frac{\partial^2 T}{\partial y^2}, \qquad f(\xi/x) = [1 - (\xi/x)^{m+1}]^{-1/2} \qquad (1.45)$$

The influence functions for turbulent gradientless flow at $\Pr \approx 1$ are similar:

$$f(\xi/x) = [1 - (\xi/x)^{9/10}]^{-1/9} \qquad f(\xi/x) = [1 - (\xi/x)]^{-0.114} \qquad (1.46)$$

The first expression can be derived theoretically by the integral method, using power law distributions (Equation 1.28) for velocities and temperatures [26]; the second one is obtained in Reference [21] by using the heat transfer law (Equation 1.34) and solving an integral equation. Both influence functions in Equation (1.46) yield virtually identical results that are in satisfactory agreement with data obtained by several experiments [28]. A stronger effect of the unheated zone is described by the two other functions [29]:

$$f(\xi/x) = [1 - (\xi/x)^{39/40}]^{-7/39} \qquad f(\xi/x) = [1 - (\xi/x)]^{-0.2} \qquad (1.47)$$

The method of superposition can also be used for solution of the inverse problem, when the temperature head distribution $\theta_w(x)$, which corresponds to known heat flux distribution, $q_w(x)$ should be estimated. The first part of Equation (1.38) shows that such an inverse problem for continuity temperature head is reduced to the Volterra integral equation, with the influence functions for gradientless self-similar flow in the following form:

$$\int_0^x f(x,\xi)\frac{dT_w}{d\xi}d\xi = \frac{q_w(x)}{h_*(x)} \qquad f(\xi/x) = [1 - (\xi/x)^{C_1}]^{-C_2} \qquad (1.48)$$

Introducing the new variables, $z = x^{C_1}, \zeta = \xi^{C_1}$ and $F(z) = q_w/h_*z^{C_2}$, one reduces Equation (1.48) to the generalized Abel equation with the solution

$$\int_0^z \frac{T_w'(\zeta)d\zeta}{(z-\zeta)^{C_2}} = F(z) \qquad \frac{dT_w}{dz} = \frac{1}{(C_2-1)(-C_2)}\frac{d}{dz}\int_0^z \frac{F(\zeta)}{(z-\zeta)^{1-C_2}}d\zeta \qquad (1.49)$$

Returning to variables x and ξ leads to the equation defining the temperature head [26]:

$$T_w - T_\infty = \frac{C_1}{\Gamma(C_2)\Gamma(1-C_2)} \int_0^x \left[1 - \left(\frac{\xi}{x}\right)^{C_1}\right]^{C_2-1} \left(\frac{\xi}{x}\right)^{C_1(1-C_2)} \frac{q_w(\xi)}{h_\bullet(\xi)\xi} d\xi \qquad (1.50)$$

1.6　Solutions of the Boundary Layer Equations in the Series of Shape Parameters

The solutions of boundary layer equations can be represented in the form of series in powers of perturbations of homogenous conditions at the outer edge of the layer and on the surface of the body. In this case, the first term of the series corresponds to the solution of the given problem for homogeneous conditions — that is, for constant free-stream velocity $U(x)$ and constant wall temperature $T_w(x)$. The subsequent terms of the series represent the effect of the perturbation of the homogeneous conditions and contain the sequence of derivatives of functions $U(x)$, and $T_w(x)$, (or $\theta_w(x)$) with respect to longitudinal coordinates. The series are constructed using dimensionless combinations of variables, which, depending on the number of the derivative they contain, are called the first, second, and so forth, shapes parameters.

One uses either sets of shape parameters that depend only on derivatives, or those that also correct for the effect of previous history by incorporating the thicknesses of the layer. The solution of equations of the velocity boundary layer in the form of series of shape parameters of the first type

$$f_k = \frac{x^k}{U}\frac{d^k U}{dx^k}, \qquad f_{tk} = \frac{x^k}{\theta_w}\frac{d^k \theta_w}{dx^k} \qquad (1.51)$$

is derived in [30]. An analogous solution for the equations of the thermal boundary layer is derived in [31]. These solutions are represented by multiple series containing nonlinear combinations of the shape parameters:

$$\frac{\psi}{\sqrt{vxU}} = \psi_0(\eta) + \psi_1(\eta)f_1 + \psi_2(\eta)f_2 + \psi_{11}(\eta)f_1^2 + \psi_3(\eta)f_3 + \psi_{12}(\eta)f_1 f_2 + \cdots \qquad (1.52)$$

$$\theta = \frac{T - T_\infty}{T_w - T_\infty} = T_0(\eta) + T_1^{(d)}(\eta)f_1 + T_1^{(t)}(\eta)f_{t1} + T_2^{(d)}(\eta)f_2 + T_2^{(t)}(\eta)f_{t2} + \cdots \qquad (1.53)$$

Here $\eta = y\sqrt{U/vx}$, bottom indices of functions $\psi(\eta)$ and $T(\eta)$ give the number, whereas the upper ones indicate the type of shape parameters (velocity or thermal).

Substitution of these series into boundary layer equations yields a set of ordinary differential equations defining the coefficients $\psi(\eta)$ and $T(\eta)$. The friction and heat transfer coefficients $(C_f/2)\,\mathrm{Re}_x^{1/2}$ and $\mathrm{Nu}_x\mathrm{Re}_x^{-1/2}$ are presented by the same series in which functions $\psi(\eta)$ and $T(\eta)$ have been replaced by the corresponding values of derivatives at $\eta = 0$: $\psi''(0)$ and $[-T'(0)]$. Equations defining functions $\psi(\eta)$ and $T(\eta)$ do not incorporate the distributions of $U(x)$ and $T_w(x)$, which vary from one problem to another. As a result, they are universal and can be tabulated.

Analogous results are obtained by using shape parameters of the second type, which additionally contain the characteristic thicknesses of the layer:

$$f_k = \left(\frac{\delta^{**2}}{\nu}\right)^k U^{k-1}\frac{d^k U}{dx^k}, \qquad f_{tk} = \left(\frac{\delta_t^{**2}}{\nu}\right)^k \frac{U^k}{\theta_w}\frac{d^k \theta_w}{dx^k} \qquad (1.54)$$

In this case, the solution of the boundary layer equations is represented by the previous series (Equations 1.52 and 1.53). Only the scales for variables are different: $U\delta^{**}/B$ and δ^{**}/B, where B is a normalizing constant, for ψ and η, respectively. As in the case when shape parameters of the first type are used, the friction and heat transfer coefficients $(C_f/2)(\mathrm{Re}_{\delta^{**}}/B)$ and $\mathrm{Nu}_x\delta^{**}/xB$, are given by series (1.52) and (1.53) in which the functions $\psi(\eta)$ and $T(\eta)$ are replaced by the corresponding values of the derivatives at $\eta = 0$: $\psi''(0)$ and $[-T'(0)]$. The results of integration of the corresponding equations defining coefficients $\psi(\eta)$ and $T(\eta)$ and the values of derivatives needed for calculations are given in the book by Loytsyanskiy [32] and in the paper by Oka [33], for the velocity and thermal layers, respectively.

To be able to calculate shape parameters (Equation 1.54), one must first find the thicknesses δ^{**} and δ_t^{**}. As the first approximation, this can be done by means of the standard integral relationships. Then, δ^{**} and δ_t^{**} are refined using the integral equations based on the share stress and heat transfer flux calculated by series (Equations 1.52 and 1.53) [32,33]. The advantage of shape parameters of the second kind (Equation 1.54) is that in this case the initial thickness of the layer can be arbitrary, whereas when the shape parameters of the first type (Equation 1.51) are used, the initial thickness of the layer is always zero. This pattern makes it possible to joint different regions of the boundary layer using δ^{**} and δ_t^{**}.

The effectiveness of the shape parameter methods is frequently limited by the rates of convergence of the series. These difficulties can be circumvented by reducing the initial equations of the boundary layer to universal form [34]. In this case, one can evaluate the universal equations numerically and compile tablets that can then be used in specific calculations, without having to resort to series. To be able to reduce the boundary layer equations to a universal form, one must use the shape parameters as independent variables.

This approach eliminates difficulties due to series convergence, but other problems arise. These are related to numerical integration of differential equations with a large number of independent variables, and also to the use of tables with many entries. So far, one cannot expect to obtain solutions of universal equations with not more than three variables [1]. This means that for the velocity equation, one can construct either a single-parameter solution that is a function of the first shape parameter, or a two-parameter solution that is a function of the first two shape parameters. The first of these solutions is equivalent to the exact solution of boundary layer equations with a linear velocity distribution $U(x)$ in the free stream, whereas the second is equivalent to the exact solution for a parabolic distribution of $U(x)$. Some results of calculations are presented by Loytsyanskiy [34].

References

1. Schlichting, H., 1979. *Boundary Layer Theory*. McGraw-Hill, New York.
2. Goldstein, S., 1938. *Modern Developments in Fluid Dynamics*. Clarendon Press, Oxford.
3. Frösling, N., 1958. Calculation by series expansion of the heat transfer in laminar boundary layer at nonisothermal surfaces, *Arch. Physik*. 14: 143–151.
4. Chellwppa, A. K., 1988. Buoyancy-induced boundary layer flow over a nonisothermal horizontal cylinder, *Amer. Soc. Mechan. Eng., Heat Transfer Division. Publication HTD* 105: 137–143.
5. Love, G., 1957. An approximate solution of the laminar heat transfer along a plate with arbitrary distribution of the surface temperature, *J. Aeron. Sci.* 24: 920–921.
6. Dienemann, W., 1953. Berechnung des Wärmeübergang an laminar unströmten Körpern mit ortsveränder Wandtemperatur, *Z. Angew. Math.* 33: 89–109.
7. Loytsyanskiy, L. G., 1962. *The Laminar Boundary Layer* (in Russian). Fizmatgiz Press, Moscow (Engl. trans., 1966).
8. Skopets, M. B., 1953. An Approximate method for calculating the laminar boundary layer in a compressible gas in the presence of heat transfer, *Zh. tekhn. fiz.* 23: 76–92.
9. Zysina-Molozhen, L. M., Zisin, L. V., and Polyak, M. P., 1974. *Heat Transfer in Turbomachines* (in Russian). Mashinostroyeniye Press, Leningrad.
10. Drake, R. M., 1953. Calculation method for three dimensional rotationally symmetrical laminar boundary layer with arbitrary free stream velocity and arbitrary wall temperature variation, *J. Aeron. Sci.* 20: 309–316.
11. Ambrok, G. S., 1957. Effect of variability of wall temperature on heat transfer in the case of a laminar boundary layer, *Zh. tekhn. fiz.* 27: 812–821.
12. Shuh, H., 1954. Ein neues Verfahrenzum Berechnen des Wärmeüberganges in ebenen und rotatationssy-mmetrischen laminaren Grenzchichten bei Konstanten und veründerlicher Wandtemperatur, *Forschg. Ing. Wes.* 20: 37–47.
13. Spalding, D. B., 1958. Heat transfer from surfaces of nonuniform temperature, *J. Fluid Mech.* 4: 22–32.
14. Seban, R. A., 1950. Calculation method for two dimensional laminar boundary layer with arbitrary wall temperature variation, *Univ. Calif., Inst. Eng. Res. Rept. No. 1*, pp. 2–12.

15. Spalding, D. B., and Pun, W. M., 1962. A review of methods for predicting heat-transfer coefficients for laminar uniform-property boundary layer flows, *Int. J. Heat and Mass Transfer* 5: 239–250.

16. Romanenko, P. N., 1976. *Heat and Mass Transfer and Friction in Gradient Flow of Liquid* (in Russian). Energiya Press, Moscow.

17. Patankar, S. V., and Spalding, D. B., 1970. *Heat and Mass Transfer in Boundary Layers.* Intertext, London.

18. Kestin, J., and Richardson, P. D., 1963. Heat transfer across turbulent incompressible boundary layers, *Int. J. Heat and Mass Transfer* 6: 147–189.

19. Ginzburg, I. P., 1970. *Theory of Drag and Heat Transfer* (in Russian). Leningrad University Press, Leningrad.

20. Kutateldze, S. S., 1970. *Fundamentals of Heat Transfer Theory* (in Russian). Mashgiz Press, Moscow.

21. Kutateladze, S. S., and Leontev, A. I., 1972. *Heat and Mass Transfer and Friction in Turbulent Boundary Layer* (in Russian). Energiya Press, Moscow.

22. Iyevlev, V. M., 1952. Certain aspects of the hydrodynamic theory of heat transfer in incompressible flows, *Dokl. Akad. Nauk SSSR* 86: 1077–1080.

23. Iyevlev, V. M., 1952. Certain aspects of the hydrodynamic theory of heat transfer in gas flows, *Dokl. Akad. Nauk SSSR* 87: 23–24.

24. Kutateladze, S. S., 1973. *Wall Turbulence* (in Russian). Nauka Press, Novosibirsk.

25. Eckert, E. R. G., and Drake, R. M., 1959. *Heat and Mass Transfer.* McGraw-Hill, New York.

26. Kays, W. M., 1980. *Convective Heat and Mass Transfer.* McGraw-Hill, New York.

27. Lighthill, M. L., 1950. Contributions to the theory of heat transfer through a laminar boundary layer, *Proc. Roy. Soc. London A* 202: 359–377.

28. Reynolds, W. C., Kays, W. M., and Kline, S. T., 1960. A summary of experiments on turbulent heat transfer from a nonisothermal flat plate, *Trans. ASME, J. Heat Transfer Ser. C* 4: 341–348.

29. Mironov, B. P., Vasechkin, V. N., and Yarugina, N. I., 1977. Effect of an upstream adiabatic zone on heat transfer in a subsonic and supersonic downstream boundary layer at different flow histories. In *Teplomassoobmen–V (Heat and Mass Transfer–V [Proceedings of the Fifth All-Union Conference on Heat and Mass Transfer]),* Pt. 1: 67–97. Minsk [Engl. trans. 1977. *Heat Transfer — Soviet Research* 9: 57–65].

30. Shkadov, V. Ya., 1962. Concerning a solution of the problem of the boundary layer, *Izv. Akad. Nauk SSSR Otd. techn. Nauk. Mekhanika i Mashinostroenie* 3: 173–175.

31. Bubnov, V. A., and Grishmanovskaya, K. N., 1964. Concerning exact solutions of problems of a nonisothermal boundary layer in an incompressible liquid (in Russian). *Trudy Leningrad. Politekhn. Inst.* 230: 77–83.

32. Loytsyanskiy, L. G., 1973. *Fluid Mechanics* (in Russian). Nauka Press, Moscow.

33. Oka, S.,1968. Calculation of the thermal laminar boundary layer of the incompressible fluid on a flat plate with specified variable surface temperature. In *Teplo-i massoperenos (Heat and Mass Transfer [Proceedings of the 3rd All-Union Conference on Heat and Mass Transfer]),* 9: 74–91. Minsk.

34. Loytsyanskiy, L. G., 1965. Universal equations and parametric approximations in the theory of the laminar boundary layer, *Prikl. Mat. i Mekh.* 29: 80–87.

2

Approximate Solutions of Conjugate Problems in Convective Heat Transfer

Beginning in the 1960s, when computers came into use and it became clear that the conjugate heat transfer formulation was a natural, comprehensive, and powerful approach, numerous solutions of conjugate heat transfer problems were published. Analytical and numerical methods of different levels of accuracy were used, however, as not only early works, but frequently contemporary studies as well are based on the approximate analytical methods outlined in Chapter 1.

In this chapter, typical examples of such approximate analytical solutions published during the last 50 to 55 years are presented and discussed. Two types of approximate solutions are considered. The first are solutions based on the approximate energy equation, when, for example, linear velocity distribution across the thermal boundary layer or slug flow model is used. Another group of problems considered in this chapter includes the solution for which approximate results could not be refined or were difficult to upgrade. Such solutions are obtained, in particular, by integral methods or by using series that slowly converge.

Exact and more accurate solutions of conjugate convective heat transfer problems are considered in following chapters.

2.1 Formulation of a Conjugate Problem of Convective Heat Transfer

In general, the conjugate problem of convective heat transfer in the boundary layer approximation is governed by a system of four equations: continuity, momentum, and energy equations for a fluid and a conduction equation for a body. In the case of a two-dimensional heat transient problem for incompressible fluid, this system is

$$\frac{\partial u}{\partial x} + \frac{\partial v}{\partial y} = 0 \tag{2.1}$$

$$\frac{\partial u}{\partial t} + u\frac{\partial u}{\partial x} + v\frac{\partial u}{\partial y} = -\frac{1}{\rho}\frac{dp}{dx} + g\beta(T - T_\infty) + v\frac{\partial^2 u}{\partial y^2} \tag{2.2}$$

$$\frac{\partial T}{\partial t} + u\frac{\partial T}{\partial x} + v\frac{\partial T}{\partial y} = \alpha\frac{\partial^2 T}{\partial y^2} + \frac{v}{c_p}\left(\frac{\partial u}{\partial y}\right)^2 \tag{2.3}$$

$$\rho_s c_s \frac{\partial T_s}{\partial t} = \lambda_s \left(\frac{\partial^2 T_s}{\partial x^2} + \frac{\partial^2 T_s}{\partial y^2} \right) + q_v \qquad (2.4)$$

The boundary and conjugate condition are

$$y = 0 \quad u = v = 0, \qquad y \to \infty \quad u \to U, \quad T \to T_\infty \qquad (2.5)$$

$$y = 0 \quad T_s = T_w, \quad \lambda_s \left(\frac{\partial T_s}{\partial y} \right)_{y=0} = \lambda \left(\frac{\partial T}{\partial y} \right)_{y=0} \qquad (2.6)$$

The initial conditions are usually specified for each particular problem.

In the case of forced heat transfer, the second term in the right-hand part of Equation (2.2) should be omitted, and the first should be substituted by $U(dU/dx) = - (1/\rho)(dp/dx)$. This relation follows from Equation (2.2), written for the free stream of the boundary layer. In the case of natural convective heat transfer, the first term in the right-hand part of Equation (2.2) should be omitted.

There are two cases when the problem for the body can be simplified. If the body is thin ($\Delta \ll L$) and its thermal resistance is comparable with that of the heat transfer agent, ($\mathrm{Bi} = h\Delta/\lambda_s \approx 1$), then the longitudinal conduction can be neglected. In such a case, the temperature distribution across a body thickness is described approximately by the one-dimensional conduction equation. That form follows from Equation (2.4) when the derivative with respect to the longitudinal coordinate is taken to be zero. If, for example, there are no heat sources inside a body, in the case of a steady-state problem, the temperature distribution across its thickness is close to linear, and, hence, there is no need to solve the conduction equation for the body.

Another case of simplification is when the body can be considered as a thermally thin object. This means that the body thermal resistance is small, so at any instant, the body temperature is practically constant over its thickness. In such a case, it is reasonable to integrate the conduction equation (2.4) across body thickness to get an averaged form:

$$\frac{1}{\alpha_s} \frac{\partial T_{av}}{\partial t} - \frac{\partial^2 T_{av}}{\partial x^2} + \frac{q_{w1} + q_{w2}}{\lambda_s \Delta} - \frac{(q_v)_{av}}{\lambda_s} = 0 \qquad (2.7)$$

2.2 The Case of Linear Velocity Distribution across the Thermal Boundary Layer

There are at least three situations when this assumption is reasonable: when a fluid is characterized by a large Prandtl number, in the case of existence of an unheated starting length upstream of a heated section, and for the

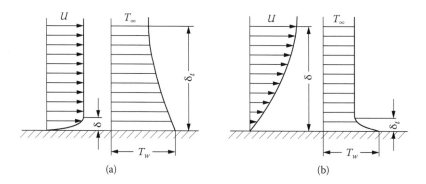

FIGURE 2.1
The effect of a Prandtl number on the thicknesses of velocity and thermal boundary layers: (a) $Pr \to 0$, b; (b) $Pr \to \infty$.

entrance region of a tube or a channel with a fully developed velocity profile. In the first case, the thickness of the thermal boundary layer is small in comparison with that of the velocity boundary layer because a large Prandtl number indicates that fluid viscosity is large and, hence, the influence of the wall goes far into the fluid (Figure 2.1). Consequently, the velocity boundary layer becomes thick, and the distribution of velocity across a thin thermal boundary layer can be approximately substituted by a tangent.

In the other case, the thermal boundary layer starts to develop inside the velocity boundary layer (Figure 2.2), which grows when it flows along the unheated part of a plate. As a result, the thickness of the thermal boundary layer is much smaller than the thickness of the velocity boundary layer, and the velocity distribution across the thermal boundary layer is close to linear.

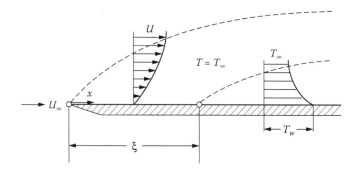

FIGURE 2.2
The effect of an unheated zone on the development of the velocity and thermal boundary layers.

A similar situation takes place in the entrance region of a tube or a channel when the thermal boundary layer starts to develop inside the fully hydrody-namically developed flow.

Example 2.1: An Incompressible Flow with a Large Prandtl Number Past Plate with a Heat Source [1,2]

In the case of a steady-state problem for a plate, the system (2.1) to (2.4) has the self-similar solution with $\beta = m = 0$ (Section 1.1). In this case, according to Equation (1.6) and Table 1.1, the linear velocity profile can be presented as $\varphi(\eta) = c\eta = 0.664\eta$, where $\varphi = \psi/\sqrt{vUx}$ and $\eta = (y/2)\sqrt{U/vx}$.

The energy equation for fluid is used in the Prandtl-Mises form Equation (1.3), and the conjugate condition (2.6) is presented as follows:

$$\frac{\partial T}{\partial x} - \alpha \frac{\partial}{\partial \psi}\left(u\frac{\partial T}{\partial \psi}\right) = 0 \qquad u\frac{\partial T}{\partial \psi}\bigg|_{\psi=0} = -\Lambda S(x) \tag{2.8}$$

where $\Lambda = (\lambda_s/\lambda)$ and $S(x) = -(\partial T/\partial y)_{y=0}$. Using these two relations and a new variable $z = (2/3)\psi^{3/4}$, one obtains a differential equation with boundary condition

$$\frac{4}{b}x^{1/4}\frac{\partial T}{\partial x} = \frac{\partial^2 T}{\partial z^2} + \frac{1}{3z}\frac{\partial T}{\partial z} + \frac{3^{1/3}b}{2^{4/3}\alpha}\frac{z^{1/3}}{x^{1/4}}\left(\frac{\partial T}{\partial z}\right)_{z=0}$$

$$= -\Lambda S(x) \qquad b = \frac{\alpha\sqrt{c/2}U^{3/4}}{v^{1/4}} \tag{2.9}$$

where the fluid temperature far away from the plate is taken to be zero, and c is constant in the velocity distribution. Equation (2.9) can be trans-formed to an integral equation [1,2]:

$$T(x) = \frac{3^{1/3}\alpha\Lambda}{2\Gamma(2/3)b^{2/3}}\int_0^x \frac{S(y)}{(x^{3/4}-y^{3/4})^{2/3}}dy \tag{2.10}$$

The solution of conduction equation (2.4) for the plate is obtained by a standard method. For example, in the case of the heat source (Equation 2.11), one obtains

$$q(x,y) = q_0\varepsilon(l-x) \qquad \varepsilon(x) = \begin{cases} 1 & x > 0 \\ 0 & x < 0 \end{cases} \tag{2.11}$$

$$T(x) = \frac{q_0}{\lambda_s}\varepsilon(l-x)\left[x\left(1-\frac{x}{2}\right)-\frac{l^2}{2}\right] - \frac{1}{\Delta}\int_0^\infty G(x,y)S(y)dy$$

$$-\frac{1}{\pi}\int_0^\infty \ln\left[\frac{1-\exp\left(-\pi\frac{x+y}{\Delta}\right)}{1-\exp\left(\pi\frac{|x-y|}{\Delta}\right)}\right]S(y)dy, \qquad G(x,y) = \begin{cases} y & y < x \\ x & y > x \end{cases} \tag{2.12}$$

The system of two integral equations (Equations 2.10 and 2.12) determines two unknown functions $T(x)$ and $S(x)$. The asymptotic solution of this system for $x/\Delta \gg 1$ is obtained by Perelman [1,2]. The first terms of the series defining the surface temperature and the Nusselt number are

$$T(x) = \frac{q_0 \Delta}{\lambda_s} \frac{2^{1/3}}{3^{2/3} c^{1/3}} gx \left[1 - \frac{1}{2^{5/3} 3^{2/3} c^{1/3}} g \frac{\Delta}{x} - \frac{7}{5 \cdot 2^{4/3} 3^{4/3} c^{1/4}} g^2 \left(\frac{\Delta}{x} \right)^2 + \cdots \right] \quad (2.13)$$

$$\frac{Nu_x}{Re_x^{1/2}} = \frac{3^{2/3} c^{1/3} Pr^{1/3} g}{2^{1/3} g_1} - \frac{3^{1/3}}{5 \cdot 2^{5/3} c^{1/3}} g g_1 \left(\frac{\Delta}{x} \right)^2 + \cdots. \quad (2.14)$$

$$g = \left[\frac{\Gamma(1/3)}{\Gamma(2/3)} \right]^2 \frac{\alpha}{Re_x^{1/2} Pr^{1/3}} \qquad g_1 = \frac{\alpha}{Re_x^{1/2} Pr^{1/3}}$$

These results are applicable for $(\alpha / Re_x^{1/2} Pr^{1/3}) \ll x \le l, Pr > 1, Re_x < 3 \cdot 10^5$.

The same method of solution is used in two other early studies. Kumar and Bartman [3] considered the same problem for compressible flow over a radiating plate. Kumar [4] solved the problem for a semi-infinite porous block with injection streamlined by an incompressible fluid.

Example 2.2: Transient Heat Transfer in the Thermal Entrance Region of a Parallel Plate Channel with the Fully Developed Flow [5]

First, a simple problem for a channel of height $2R$ is considered using mass average velocity u_m (slug flow). A flow and channel are initially at constant temperature T_0 when suddenly the temperature of the channel walls becomes an arbitrary function of position and time, and the fluid inlet temperature begins varying arbitrarily with time. The governing equation and boundary conditions are

$$\frac{\partial \theta}{\partial t} + u_m \frac{\partial \theta}{\partial x} = \alpha \frac{\partial^2 \theta}{\partial y^2} \quad (2.15)$$

$t = 0, \ x > 0, \ 0 \le y \le R, \ \theta = 0, \qquad x = 0, \ t > 0, \ 0 \le y \le R, \ \theta = \theta_e(t)$

$t > 0, \ x > 0, \ y = R, \ \partial\theta/\partial y = 0, \qquad y = 0, \ \theta = \theta_w(x,t)$

Here, $\theta = T - T_0$ and θ_e are the inlet temperature excess.

The next simplification is the use of the quasi-steady approach when it is assumed that the steady-state relations are valid at each instant of time (Section 6.6). Physically, this means that the heat transfer process in the fluid is slow in comparison with that in the walls. The previous system of equations becomes

$$(\tau = t - x/u_m)$$

$$u_m \frac{\partial \theta}{\partial x} = \alpha \frac{\partial^2 \theta}{\partial y^2} \quad (2.16)$$

$$x = 0, \ 0 \leq y \leq R, \ \theta = \theta_e(\tau), \qquad x > 0, \ y = R, \ \partial\theta/\partial y = 0,$$

$$y = 0, \ \theta = \theta_w(x, \tau + x/u_m)$$

This problem is solved by the Laplace transform to obtain the expression for surface heat flux using a new variable $\sigma = t - x/u_m + \xi/u_m$:

$$q_w(\tau, t) = \frac{\lambda}{R} \int_0^t \frac{\sum_{n=-\infty}^{\infty} (-1)^n \exp\{-[R^2 n^2/\alpha(t-\sigma)]\}}{\sqrt{\pi\alpha(t-\sigma)/R^2}} \frac{\partial\theta_w}{\partial\sigma}(\tau, \sigma) d\sigma \quad \text{for} \ \ t \leq x/u_m \qquad (2.17)$$

$$q_w(\tau, t) = \frac{\lambda}{R} \int_0^x \frac{\sum_{n=-\infty}^{\infty} (-1)^n \exp\{-[R^2 u_m n^2/\alpha(x-\xi)]\}}{\sqrt{\pi\alpha(x-\xi)/u_m R^2}} \frac{\partial\theta_w}{\partial\xi}(\xi, \tau) d\xi \quad \text{for} \ \ t \geq x/u_m \qquad (2.18)$$

It is asserted that these equations will hold approximately in nonslug flows if the kernels in the integrands come from the appropriate nonslug flow solutions. Thus, if a solution for the surface heat flux in steady, non-slug flow over an isothermal surface is

$$q_w(x) = G(x)\theta_{w0}, \qquad (2.19)$$

where θ_{w0} is a constant wall temperature excess, then Equations (2.17) and (2.18) can be written for $t < x/u_m$ and $t > x/u_m$, respectively, as

$$q_w = \int_0^t G[u_m(t-\sigma)] \frac{\partial\theta_w}{\partial\sigma}(\tau, \sigma) d\sigma \qquad q_w = \int_0^t G(x-\xi) \frac{\partial\theta_w}{\partial\xi}(\xi, \tau) d\xi \qquad (2.20)$$

Expression (2.20) take into approximate account both thermal history and finite thermal capacity of the fluid, while other authors who used the quasi-steady approach focused only on the effect of thermal capacity [6] or on the effect of thermal history [7,8].

Sucec [5] used expression (2.20) to solve two transient heat transfer problems. He considered a fully developed flow in a parallel plate duct with outside insulated walls. In the first case, a transient is initiated by step change in the fluid inlet temperature; in the second case, the unsteadiness is caused by a fluid inlet temperature that varies sinusoidally with time. For a thermally thin plate, the energy balance is used in the form that follows from Equation (2.7):

$$\frac{\partial\theta_w}{\partial t} + \frac{q_w}{\rho_w c_{pw}\Delta} = 0 \qquad (2.21)$$

The linear velocity profile is chosen to approximate the actual velocity distribution in the entrance region of a duct with fully developed flow. Using the mass average velocity, u_m, and Lighthill's formula (1.42) for the steady-state heat flux, one obtains function $G(x)$ in Equation (2.19), where $\tilde{L} = \alpha x / u_m R^2$ is the nondimensional distance along a duct:

$$u_x(y) = 3u_m(y/R) \qquad G(x) = 0.5384(3)^{1/3}(\lambda/R)\tilde{L}^{-1/3} \qquad (2.22)$$

If T_i is an initial temperature, $\theta_w = T_w - T_i$, and hence, $\theta_w = 0$ at $t = 0$, then it is seen by inserting the first part of Equation (2.20) into Equation (2.21) that $\theta = 0$ for the domain $t < x/u_m$. Inserting Equation (2.22) into the second part of Equation (2.20), and then Equation (2.21) gives the equation for surface temperature distribution, after noting that $\partial\theta_w/\partial t = \partial\theta_w/\partial\tau$:

$$\frac{\partial\theta_w}{\partial\tau} + \frac{0,538(3)^{1/3}\lambda}{\rho_w c_{pw} R\Delta} \int_0^{\tilde{L}} \frac{1}{(\tilde{L}-\xi)^{1/3}} \frac{\partial\theta_w}{\partial\xi}(\xi,\tau)d\xi = 0 \quad \text{for} \quad \tau \geq 0 \qquad (2.23)$$

Applying the Laplace transform to solve this equation subjected to side conditions

$$\tau = 0, \ \tilde{L} > 0, \ \theta_w = 0, \qquad \tilde{L} = 0, \ \tau > 0, \ \theta_w = \theta_e \qquad (2.24)$$

gives the wall temperature and heat flux for $\tau \geq x/u_m$:

$$\frac{\theta_w}{\theta_e} = 1 - \frac{3}{2}\int_f^\infty z(z-f)^2 Ai(z)dz, \quad \frac{q_w R\tilde{L}^{1/3}}{3(0.538)\Gamma(2/3)\lambda\theta_e} = -\int_f^\infty z(z-f)^2 Ai(z)dz$$

$$Ai(z) = \frac{1}{\pi}\sqrt{\frac{z}{3}} K_{1/3}\left(\frac{2}{3}z^{3/2}\right) \qquad f = \frac{0.538^{1/3}\Gamma(2/3)\lambda\tau}{\rho_w c_{pw}\Delta R\tilde{L}^{1/3}} \qquad (2.25)$$

where $Ai(z)$ and $K_{1/3}$ are, respectively, the Airy function and the modified Bessel function of second kind. For domain $t < x/u_m$, the surface heat flux and both temperatures are zero.

The other problem is solved by the same procedure. Equation (2.23) is integrated subjected to the same side conditions with $\theta_e = \Delta T_0 \sin\omega t$, where ΔT_0 is the amplitude of the periodical inlet temperature. The results are given in [5].

To ascertain the accuracy and limitation of the suggested method, the obtained results are compared to the numerical solutions. It is found that an improved quasi-steady approach, which attempts to take into account both thermal history and thermal capacity effects, predicts the finite-difference results reasonable well and does significantly better than the standard quasi-steady method. Yet the improved quasi-steady approach is more difficult to apply than the standard quasi-steady solution.

A linear velocity distribution across the thermal boundary layer is also used for studying heat transfer between two fluid flows separated by a plate without thermal resistance [9].

2.3 The Case of Uniform Velocity Distribution across the Thermal Boundary Layer (Slug Flow)

The other limiting case corresponds to a fluid with small Prandtl number. In this case, the thermal boundary layer is thick in comparison with the velocity boundary layer because a small Prandtl number indicates that fluid viscosity is small. Hence, the influence of the wall is confined to a region of fluid close to the wall, and the velocity across the flow is approximately equal to the free stream velocity (Figure 2.1). For some fluids, like mercury and other liquid metals, such an assumption is close to reality.

Another reason to consider the velocity components to be constant across the thermal boundary layer is to simplify the energy equation. Because of the relative simplicity it affords, this assumption is used by a number of investigators. Perelman [10] and Rizk et al. [11] studied conjugate heat transfer to laminar flow from heated blocks, Luikov [12] investigated the case of a plate with a heat source, and Viskanta and Abrams [13] used the slug flow assumption to study heat transfer between two concurrent and countercurrent streams separated by a thin plate.

Several researchers used the slug flow approximation to study transient coupled heat transfer when the unsteadiness is caused by different sources: by flow with time-varying temperature, by step temperature change of free stream at the leading edge of a plate, and by unsteady sources in the wall. Sparrow and DeFarias [14], Siegel and Perlmutter [15], and Namatame [16] considered coupled heat transfer in the channel flows, and Sucec [17–19] and Chambre [20] solved conjugate problems of heat transfer to flows over plates. Travelho and Santos [21] studied the effect of wall heat capacity in a circular duct for the case of periodically varying inlet temperature of the laminar flow.

Example 2.3: Transient Heat Transfer from a Plate of Appreciable Thermal Capacity with Heat Generation of q_0 to Turbulent Flow [6]

The governing equation is the energy equation (2.3) with additional eddy diffusivity ε_H that is estimated using the expression of frictional stress:

$$\frac{\partial T}{\partial t} + u\frac{\partial T}{\partial x} + v\frac{\partial T}{\partial y} = \frac{\partial}{\partial y}\left[(\alpha + \varepsilon_H)\frac{\partial T}{\partial y}\right] \qquad \frac{\tau_w}{\rho U^2} = 0.0228\,\mathrm{Re}_\delta^{-1/4} \qquad (2.26)$$

and 1/7 power law (Equation 1.28) of velocity distribution. Assuming that turbulent Prandtl number equals unity and using an approach similar to that described in Section 1.4, total momentum and heat diffusivity are obtained as

$$\alpha + \varepsilon_H = 0.179 U^{27/35} v^{8/35} x^{-3/35} \tag{2.27}$$

Because the plate is assumed to have appreciable thermal capacity but negligible thermal resistance, the initial, boundary, and conjugate conditions are as follows:

$$t = 0, \; x = 0, \; y \to \infty, \; T = 0, \; \lim_{y \to 0}\left[-\rho c(\alpha + \varepsilon_H)\frac{\partial T}{\partial y}\right]$$

$$= q_0 - \frac{1}{2}\rho_w c_w \Delta\left(\frac{\partial T}{dt}\right)_{y=0} \quad t > 0, \; x > 0 \tag{2.28}$$

For early times when the thickness of the thermal layer is thinner than that of the laminar sublayer, the eddy diffusivity can be neglected in comparison with the molecular diffusivity. In this case, the transient energy equation (2.26) reduces to the laminar slug flow equation that has a solution for the case of a step heat source that has two forms: for small and at the far downstream position where $\vartheta \leq X$, and for large and at the leading (upstream) position where $\vartheta \geq .X$, respectively:

$$\frac{T_w(\vartheta, X)}{T_{RS}} = 2\sqrt{\frac{\vartheta}{\pi}} + \exp(\vartheta)\operatorname{erfc}(\sqrt{\vartheta}) - 1, \tag{2.29}$$

$$\frac{T_w(\vartheta, X)}{T_{RS}} = 2\sqrt{\frac{X}{\pi}} + \exp(\vartheta)\operatorname{erfc}\left(\frac{\vartheta + X}{2\sqrt{X}}\right) - \operatorname{erfc}\left(\frac{\vartheta - X}{2\sqrt{X}}\right) \tag{2.30}$$

Here, $\vartheta = t/B^2$, $X = x/B^2U$, $T_{RS} = q_0\sqrt{\alpha B}/\lambda$ is reference temperature; $B = \hat{c}\sqrt{\alpha}/\lambda$ is the thermal capacity parameter; and $\hat{c} = (1/2)\rho_w c_w \Delta$ is plate thermal capacity. These results are valid for times less than 1 ms. For times greater than 1 ms, the thermal boundary layer penetrated into the region represented by the 1/7 power law (1.28). The solution of transient turbulent energy equation (2.26) together with Equations (2.27) and (2.28) is found by two steps considering several simpler problems. The first is the case of zero thermal capacity. Two equations are solved for this case:

$$\frac{\partial T}{\partial t} = \frac{\partial}{\partial y}\left[(\alpha + \varepsilon_H)\frac{\partial T}{\partial y}\right] \qquad u\frac{\partial T}{\partial x} + v\frac{\partial T}{\partial y} = \frac{\partial}{\partial y}\left[(\alpha + \varepsilon_H)\frac{\partial T}{\partial y}\right] \tag{2.31}$$

The first equation corresponds to small but greater than 1 ms times, for which convective terms have little effect. This equation is solved with initial condition $t = 0, T = 0$. The second equation (2.31) corresponds to large values of time when transient effects should die away, and the heat transfer characteristics should approach those of a steady-state condition. Consequently, the initial condition for this equation should be $x = 0, T = 0$. Introducing a new variable x/U, one can see that in the case of the slug flow, the second equation (2.31) is the same as the first. Thus, both equations have similar solutions:

$$\frac{\rho c U T_w(t, x)}{q_0} = 30.3 \text{Re}_x^{0.2} \left(\frac{tU}{x} \right)^{0.125}, \quad \tau_x = \frac{tU}{x} \le 1$$

$$\frac{\rho c U T_w(t, x)}{q_0} = 30.3 Re_x^{0.2}, \quad \tau_x \ge 1$$

(2.32)

Averaging these equations along the plate yields the mean wall temperature:

$$T_m(t) = \frac{q_0 L}{\lambda \text{Re}_L^{0.8}} (28.19 \tau^{0.125} - 2.94 \tau^{1.2}) \quad \tau = \frac{tU}{L} \le 1,$$

$$T_m(t) = 25.25 \frac{q_0 L}{\lambda \text{Re}_L^{0.8}} \quad \tau \ge 1$$

(2.33)

In the case of appreciable thermal capacity, part of the heat generated is stored within the wall, and the rest is transferred in the fluid. To determine these parts of heat, one should take into account that the corresponding heat transfer equations are coupled. Integrating along the plate an energy balance at the wall and assuming that heat at the wall is proportional to heat generated, $q = q_0 f(t)$, resulting in the expression for function $f(\tau)$:

$$q = q_0 - \hat{c} \frac{\partial T_w}{\partial t} \qquad f(\tau) = 1 - \frac{1}{X_L} \frac{d(T_m / T_{RS})}{d\tau}$$

(2.34)

Rearranging Equation (2.33) using Equation (2.34) leads to first-order differential equations [6]:

$$\frac{d(T_m / T_{RS})}{d\tau} + \frac{\text{Re}_L^{0.8} X_L \hat{C}}{F(\tau)} (T_m / T_{RS}) = X_L \quad \tau \le 1, \quad F(\tau) = 28.19 \tau^{0.125} - 2.94 \tau^{1.2}$$

(2.35)

$$\frac{d(T_m / T_{RS})}{d\tau} + \frac{\text{Re}_L^{0.8} X_L \hat{C}}{25.25} (T_m / T_{RS}) = X_L, \qquad \tau \ge 1$$

(2.36)

where $\bar{C} = B\sqrt{\alpha}/L = \hat{c}\alpha/\lambda$ is the dimensionless thermal capacity parameter. Equation (2.35) is solved for the mean wall temperature subject to initial condition $T_m(0)/T_{RS} = 0$. This solution at $\tau = 1$ is then used as the initial condition for Equation (2.36). The final results for the mean temperatures are

$$\frac{T_m}{T_{RS}} = X_L \exp\left(-\int_0^\tau \frac{Re_L^{0.8} X_L \hat{C}}{F(\tau)} d\eta\right) \int_0^\tau \exp\left(-\int_0^\tau \frac{Re_L^{0.8} X_L \hat{C}}{F(\tau)} d\eta\right) d\eta \quad \tau \le 1 \tag{2.37}$$

$$\frac{T_m}{T_{RS}} = X_L \exp\left[-\frac{Re_L^{0.8} X_L \hat{C}}{25.25}(\tau-1) - \int_0^1 \frac{Re_L^{0.8} X_L \hat{C}}{F(\tau)} d\eta\right] \int_0^1 \exp\left(\int_0^\tau \frac{Re_L^{0.8} X_L \hat{C}}{F(\tau)} d\eta\right) d\eta$$

$$+ \frac{25.25}{Re_L^{0.8} \hat{C}}\left\{1 - \exp\left[-\frac{Re_L^{0.8} X_L \hat{C}}{25.25}(\tau-1)\right]\right\} \quad \tau \ge 1 \tag{2.38}$$

The mean heat transfer coefficient for the plate of appreciable capacity is defined by relation, which after using Equations (2.37) and (2.38) yields

$$h_m = \frac{q_0 - \hat{c}(\partial T_m/\partial t)}{T_m}, \quad h_m = \frac{\lambda Re_L^{0.8}}{LF(\tau)}, \quad \tau \le 1 \quad h_m = 0.0396\frac{\lambda Re_L^{0.8}}{L} \quad \tau \ge 1 \tag{2.39}$$

Analytical results for the wall temperatures are compared with authors' experimental data for two values of the dimensionless wall capacities ($\hat{C} = 0.68 \cdot 10^{-3}$ and $2.39 \cdot 10^{-3}$) and for Reynolds number range of $5.3 \cdot 10^5 \le Re_L \le 2 \cdot !0^6$. Agreement of turbulent flow solution (Equations 2.37 and 2.38) with experimental data for all these parameters is observed. The slug flow solution (Equations 2.29 and 2.30) overestimates the wall temperatures. The experimental results show that the transient mean heat transfer coefficient for high wall thermal capacity and high flow velocities decreases gradually until it passes through a minimum and then starts rising, finally reaching its steady-state value. However, the values of mean heat transfer h_m obtained from the solution do not exhibit any minimum, so that the steady-state value of h_m is reached abruptly at $\tau = 1$.

Kawamura [22] used a similar approach to take into account the thermal capacity in the case of turbulent flow in an annulus.

Example 2.4: Unsteady Heat Transfer between a Fluid with Time-Varying Temperature and a Plate [17,18]

A thermally thin plate with lower surface temperature is exposed to a coolant at temperature T_c with heat transfer coefficient h_c, which is streamlined over its top by steady laminar flow with slug velocity distribution. Initially, the plate and fluid are both at the coolant temperature T_c, when suddenly the fluid temperature at $x = 0$ is changed to T_0 and subsequently held constant. Defining $\theta = T - T_0$, one gets the energy equation (2.3) and boundary and conjugate conditions (2.5) and (2.6) for the case of transient heat transfer and slug velocity in the boundary layer as follows:

$$\frac{\partial \theta}{\partial t} + U\frac{\partial \theta}{\partial x} = \alpha\frac{\partial^2 \theta}{\partial y^2} \tag{2.40}$$

$$t = 0, \ x > 0, \ y > 0, \ \theta = \theta_c = T_c - T_0, \qquad x = 0, \ t > 0, \ y > 0, \ \theta = 0, \tag{2.41}$$

$$y \to \infty \ t > 0, \ \theta \ \text{is finite}$$

$$y = 0, \ t > 0 \ x > 0 \quad \hat{c}\frac{\partial \theta}{\partial t} + \frac{h_c}{\lambda}(\theta - \theta_c) = \frac{\partial \theta}{\partial y} \tag{2.42}$$

where $\hat{c} = \rho_w c_w \Delta / \lambda$. The problem is solved using a Laplace transform with respect to both independent variables.

$$\tilde{\theta} = \int_0^\infty \theta \exp(-pt)dt \qquad \bar{\theta} = \int_0^\infty \theta \exp(-sx)dx \tag{2.43}$$

The system of Equations (2.40) through (2.42) becomes

$$\frac{d^2\tilde{\theta}}{dy^2} - \frac{Us+p}{\alpha}\tilde{\theta} = -\frac{\theta_c}{s\alpha} \tag{2.44}$$

$$y \to \infty, \ \tilde{\theta} \ \text{is finite} \ y = 0, \ \frac{d\theta}{dy} = \left(\frac{h_c}{\lambda} + \hat{c}p\right)\tilde{\theta} - \theta_c\left(\frac{\hat{c}}{s} + \frac{h_c}{sp\lambda}\right)$$

The solution of this ordinary linear equation and the inverse transform yields [18]

$$\theta/\theta_c = 1 - \sigma(\tau - 1)\{\operatorname{erf}(Y) + \exp(2\eta Y + \eta^2)\{\operatorname{erf}[\varepsilon(\tau-1) + \eta + Y] - \operatorname{erf}(\eta + Y)\}\}$$

$$\sigma(\tau - 1) = \begin{cases} 0 & \tau < 1 \\ 1 & \tau \geq 1 \end{cases} \tag{2.45}$$

Here, $\tau = Ut/x$, $Y = (y/2)\sqrt{U/\alpha x}$, $\eta = (h_c/\lambda)\sqrt{\alpha x/U}$, $\varepsilon = \sqrt{x/\alpha U}/2\hat{c}$ is the coupling parameter, and $\sigma(\tau - 1)$ is the unit step function. Putting $Y = 0$ in Equation (2.45) gives the plate temperature and then the dimensionless heat flux scaled by that for the isothermal surface:

$$\theta_w/\theta_c = 1 - \sigma(\tau - 1)\{\exp(\eta^2)\{\operatorname{erf}[\varepsilon(\tau - 1) + \eta] - \operatorname{erf}(\eta)\}\} \tag{2.46}$$

$$\frac{q_w}{\lambda\theta_c\sqrt{U/\pi\alpha x}} =$$

$$\begin{cases} 0 & \tau < 1 \\ \exp\{-[\varepsilon^2(\tau - 1)^2 + 2\eta\varepsilon(\tau - 1)]\} + \sqrt{\pi}\eta\exp(\eta^2)\{\operatorname{erf}[\varepsilon(\tau - 1) + \eta] - \operatorname{erf}(\eta)\} & \tau \geq 1 \end{cases}$$

$$\tag{2.47}$$

The surface temperature and the heat flux for the arbitrary inlet temperature $T_{x=0}(t)$ at $x = 0$ is found by applying the principle of superposition (Section 1.5):

$$T_w - T_c = \exp(\eta^2) \int_0^{\tau-1} \{\text{erf}[\varepsilon(\tau-1-\zeta)+\eta] - \text{erf}(\eta)\} \frac{d[T_{x=0}(\zeta) - T_c]}{d\zeta} a \qquad (2.48)$$

$$\frac{q_w}{\lambda\sqrt{U/\pi\alpha x}} = -\int_0^{\tau-1} \{\exp\{-[\varepsilon^2(\tau-1-\zeta)^2 + 2\eta\varepsilon(\tau-1-\zeta)]\}$$

$$+ \sqrt{\pi}\eta\exp(\eta^2)\{\text{erf}[\varepsilon(\tau-1-\zeta)+\eta] - \text{erf}(\eta)\}\} \frac{d[T_{x=0}(\zeta) - T_c]}{d\zeta} d\zeta \qquad (2.49)$$

The case when the fluid temperature at $x = 0$ varies linearly with time is of special interest, because experimental data show that the gas turbine inlet temperature varies approximately of that fashion during start-up and shutdown. The results obtained from Equations (2.48) and (2.49) for that case are given in [17].

Analysis of results leads to the following conclusions:

1. Equations (2.46) and (2.47) show that the plate at position x does not respond to the fluid temperature change at $x = 0$ until the front reaches the position x. This is a consequence of the neglect of axial convection in both plate and fluid and the slug velocity profile.

2. Because for laminar steady slug flow over an isothermal plate, the thermal boundary layer thickness is $\delta_t \sim \sqrt{\alpha x/U}$ the coupling parameter can be presented as $\varepsilon = \rho c_p \delta_t / \rho_w c_{pw}\Delta$. Hence, ε is a measure of the ratio of the thermal energy storage capacity per unit length of the boundary layer to that of the plate. Thus, for $\varepsilon \to 0$, viewed as being caused by a plate of very large thermal energy capacity, one would expect, on physical grounds, that the plate remains isothermal at T_c regardless of η, and according to Equation (2.47), the heat flux is $q_w/\lambda\theta_c\sqrt{U/\pi\alpha x} = 1$ for $\tau \geq 1$. For fixed η, it is seen that the larger the value of ε, the shorter the duration of the transient. In the limit as $\varepsilon \to \infty$, viewed as caused by a zero thermal energy storage capacity plate, Equation (2.47) yields

$$\frac{q_w}{\lambda\theta_c\sqrt{U/\pi\alpha x}} \to \sqrt{\pi}\eta\exp(\eta^2)\text{erfc}(\eta) \quad \text{for} \quad \tau \geq 1 \qquad (2.50)$$

That is, the plate responds instantly at $\tau = 1$ to the new fluid temperature, and there is only the simple transient due to the passage of the front fluid that was at $x = 0$ at $t = 0$.

3. In the case of the quasi-steady approach, the basic assumption is that the steady-state relations are valid at each instant of time. An expression for the quasi-steady surface heat flux in slug flow over a surface with an arbitrary temperature can be found using Lighthill's formula (Equation 1.42). The result for the surface temperature and heat flux for a step change in fluid temperature at $x = 0$ are [17]

$$\frac{\theta_{w.q.s.}}{\theta_c} = 1 - \exp(\eta^2)[\mathrm{erf}(\varepsilon\tau + \eta) - \mathrm{erf}(\eta)] \tag{2.51}$$

$$\frac{q_{w.q.s.}}{\lambda\sqrt{U/\pi\alpha x}} = \exp[-(\varepsilon^2\tau^2 + 2\eta\varepsilon\tau)] + \sqrt{\pi}\,\eta\exp(\eta^2)[\mathrm{erf}(\varepsilon\tau + \eta) - \mathrm{erf}(\eta)] \tag{2.52}$$

Comparisons show that the quasi-steady solution does not properly predict the nondimensional lag time, $\tau = 1$, which must expire before the plate can receive the information that the fluid temperature at $x = 0$ has changed. By examination of Equations (2.51) and (2.52), it is seen that if $\varepsilon\tau$ in the quasi-steady solution is replaced by $\varepsilon(\tau - 1)$, the quasi-steady solution becomes identical to the exact solution. Hence, in an attack on the more complicated problem involving a nonslug velocity profile, one might, as a first approximation, utilize a quasi-steady solution with the equivalent of τ replaced by $\tau - 1$, as long as axial conduction is still being neglected both in the plate and in the fluid. The error between the quasi-steady and exact solutions will be less than 10%, for both heat flux and temperature, for the step change in fluid temperature, if $\varepsilon \le 0.02$.

Later, Sucec [19] considered a similar problem for the case when the fluid temperature at the leading edge suddenly starts to vary sinusoidally. Using the method of complex temperature in conjunction with the Laplace transform, the exact solution is obtained for the case of slug velocity distribution.

Example 2.5: Heat Transfer between Two Fluids Separated by a Thin Plate [13]

In the case of slug flow, the governing system for both fluids simplifies to

$$U_i\frac{\partial T_i}{\partial x_i} = \alpha_i\frac{\partial^2 T_i}{\partial y_i} \quad y_i = 0, \ T_i = T_{i,w} \ \ y_i \to \infty, \ T_i \to T_\infty \ \ i = 1,2 \tag{2.53}$$

$$\lambda_2(\partial T_2/\partial y_2)_{y_2=0} = -\lambda_1(\partial T_1/\partial y_1)_{y_1=0} = (\lambda_w/\Delta)(T_{w2} - T_{w1}) \tag{2.54}$$

Introducing dimensionless variables, one presents the solution in the following form [13]:

$$\theta_1 = b\left\{1+\frac{\lambda_2\gamma_2}{\lambda_1\gamma_1}[erf(y_1/2\zeta_1)+\exp(\gamma_1 y_1+\gamma_1^2\zeta_1^2)erfc(y_1/2\zeta_1)+\gamma_1\zeta_1]\right\} \qquad (2.55)$$

$$\theta_2 = b\{erfc(y_2/2\zeta_2)-\exp(\gamma_2 y_2+\gamma_2^2\zeta_2^2)erfc[(y_2/2\zeta_2)+\gamma_2\zeta_2]\} \qquad (2.56)$$

$$\theta_i = \frac{T_i-T_{\infty 2}}{T_{\infty 1}-T_{\infty 2}}, \quad \gamma_1 = \frac{\lambda_w\left(\lambda_1 a_1^{-1/2}+\lambda_2 a_2^{-1/2}\right)}{\Delta\lambda_1\lambda_2 a_2^{-1/2}}, \quad \gamma_2 = \gamma_1\left(\frac{a_1}{a_2}\right)^{1/2}$$

$$b = \frac{\lambda_w\lambda_2\gamma_2}{\Delta}, \quad \zeta = \left(\frac{\alpha x}{U}\right)^{1/2}, \quad a = \frac{\alpha}{U}, \quad \sigma = \exp(\gamma^2\zeta^2)erfc(\gamma\zeta) \qquad (2.57)$$

Setting $y_1 = y_2 = 0$, $\gamma_1\zeta_1 = \gamma_2\zeta_2$, $q_1 = -q_2$ yields heat flux and heat transfer coefficients:

$$\frac{q_i\Delta}{\lambda_w(T_{\infty 2}-T_{\infty 1})} = \exp(\gamma^2\zeta^2)erfc(\gamma\zeta), \quad h_1 = \frac{(\lambda_w/\Delta)\sigma_1}{1-b[1+(\lambda_2\gamma_2/\lambda_1\gamma_1)\sigma_1]},$$

$$h_2 = \frac{(\lambda_w/\Delta)\sigma_2}{b(1-\sigma_2)} \qquad (2.58)$$

The other solution to this problem that provides the desired accuracy is also presented in the literature [13]. (See also Example 6.22.) A comparison shows that the assumption of slug velocity may be used with reasonable accuracy when the velocity boundary layer is relatively thin (i.e., in the vicinity of the leading edge and for the large Biot numbers). The other result is the seriousness of neglecting the actual wall temperature variation which leads to considerable errors in prediction of the heat transfer coefficients when the constant wall temperature or heat flux boundary condition is used.

2.4 Solutions of the Conjugate Convective Heat Transfer Problems in the Power Series

Series are used to solve conjugate problems in the early works. The first such solutions were obtained even before Perelman [2] coined the expression "conjugate heat transfer." For example, Sokolova [24] investigated the temperature of a radiated plate placed in supersonic gas flow using series expansion. Later, this method was applied to several problems and was grounded by proof of series convergence [25].

Example 2.6: A Thermally Thin Plate with Heat Sources in a Compressible Fluid [25]

The governing equations for compressible fluid and a thermally thin plate are as follows:

$$\frac{\partial \rho u}{\partial x} + \frac{\partial \rho v}{\partial y} = 0 \qquad \rho u \frac{\partial u}{\partial x} + \rho v \frac{\partial u}{\partial y} = \frac{\partial}{\partial y}\left(\mu \frac{\partial u}{\partial y}\right) \tag{2.59}$$

$$\rho u \frac{\partial T}{\partial x} + \rho v \frac{\partial T}{\partial y} = \frac{1}{\Pr}\frac{\partial}{\partial y}\left(\lambda \frac{\partial T}{\partial y}\right) + \frac{\mu}{c_p}\left(\frac{\partial u}{\partial y}\right)^2 \tag{2.60}$$

$$\frac{d^2 T_w}{dx^2} - 2\frac{q_w}{\lambda_w \Delta} = -\frac{w(x)}{\lambda_w} \qquad w(x) = \sum_{n=0}^{\infty} w_n x^n \tag{2.61}$$

where the heat source is presented in the power series. The boundary and conjugate conditions are Equations (2.5) and (2.6). For a symmetrically streamlined plate with insulated edges, the boundary conditions are

$$\left.\frac{\partial T_s}{\partial x}\right|_{x=0} = \left.\frac{\partial T_s}{\partial x}\right|_{x=L} = \left.\frac{\partial T_s}{\partial x}\right|_{(\Delta/2)} = 0 \tag{2.62}$$

The Chapman and Rubesin approximation is used for viscosity [26]:

$$\frac{\mu}{\mu_\infty} = C\frac{T}{T_\infty} \qquad C = \sqrt{\frac{T_{w.av}}{T_\infty}}\frac{T_\infty + S}{T_{w.av} + S} \qquad S = \text{constant} \tag{2.63}$$

In the dimensionless variables, the energy equation (2.60) becomes

$$\frac{\partial^2 \theta}{\partial \eta^2} + \Pr\varphi\frac{\partial \theta}{\partial \eta} - 2\Pr\varphi'x\frac{\partial \theta}{\partial x} = -\frac{\Pr}{4}(k-1)M_\infty^2(\varphi'')^2 \tag{2.64}$$

$$\varphi(\eta) = \psi/\sqrt{xv_\infty U L C}$$

Here, $\theta = T/T_\infty$ and dimensionless x are scaled by L. The Blasius velocity $\varphi'(\eta)$ is a self-similar solution for $\beta = 0$. The boundary conditions follow from Equation (2.5):

$$\theta(x,0) = \theta_w(x) = \theta_{\text{ad}} + \vartheta(x), \qquad \theta(x,\infty) = 1 \tag{2.65}$$

where θ_{ad} is an adiabatic surface temperature.

The partial solution of a linear inhomogeneous differential equation (2.64) is [27]

$$\theta(x,\eta) = N(\eta) = 1 + \frac{k-1}{2} M_\infty^2 r(\eta), \quad r(\eta) = \frac{\text{Pr}}{2} \int_\eta^\infty \left[\varphi''(\zeta)\right]^{\text{Pr}} \int_0^\zeta \left[\varphi''(\zeta)\right]^{2-\text{Pr}} d\zeta d\zeta \quad (2.66)$$

Here, $r(\eta) = 2c_p(T_{\text{ad}} - T_\infty)/U^2$ is a recovery factor, which is the ratio of the frictional temperature increase of the plate to that due to adiabatic compression. From this definition and Equation (2.66), one gets the surface adiabatic temperature:

$$T_{\text{ad}} = N(0) = T_\infty\left[1 + r(0)\frac{k-1}{2} M_\infty^2\right]$$

$$r(0) = \frac{\text{Pr}}{2} \int_0^\infty [\varphi''(\xi)]^{\text{Pr}} d\xi\left(\int_0^\xi [\varphi''(\eta)]^{2-\text{Pr}} d\eta\right)$$

$$(2.67)$$

The solution of the homogeneous equation (2.64) may be found by the method of separation of variables. Assuming $\theta = X(x)Y(\eta)$, one obtains

$$\frac{Y'' + \text{Pr}\varphi Y'}{\varphi'Y} = 2\text{Pr}\,x\frac{X'}{X} = \text{constant} \quad (2.68)$$

If the constant is chosen to be equal $2\text{Pr}\cdot m$, then $X'/X = m(1/x)$ and $X = x^m$. The function $Y(\eta)$ is determined from the following equation and boundary conditions:

$$Y'' + \text{Pr}\varphi Y' - 2\text{Pr}\varphi' mY = 0 \qquad Y(0) = 1 \qquad Y(\infty) = 0 \quad (2.69)$$

Assuming $m = \alpha n, \alpha n + \gamma_1, \alpha n + \gamma_2$ and using the principle of superposition lead to the solution of the energy equation:

$$\theta(x,\eta) = N(\eta) + \sum_{n=0}^\infty [a_n Y_{an}(\eta) + b_n Y_{an+\gamma_1}(\eta)x^{\gamma_1} + c_n Y_{an+\gamma_2}(\eta)x^{\gamma_2}]x^{an} \quad (2.70)$$

Then the surface temperature and the surface heat flux are obtained:

$$\theta_w(x,0) = \sum_{n=0}^\infty \left[a_n + b_n x^{\gamma_1} + c_n x^{\gamma_2}\right]x\alpha^{an} \quad (2.71)$$

$$q_w = -\frac{\lambda_\infty T_\infty[1 + (S/T_\infty)]}{2[\theta_w(x) + (S/T_\infty)]}\sqrt{\frac{U\theta_w(x)}{v_\infty x L C}}$$

$$\times\left[\sum_{n=0}^\infty a_n Y'_{an}(0) + b_n Y'_{an+\gamma_1}(0)x^{\gamma_1} + c_n Y'_{an+\gamma_2}(0)x^{\gamma_2}\right]x^{an}$$

$$(2.72)$$

After using this result, Equation (2.61) for plate becomes

$$\frac{\partial^2 \theta_w}{\partial x^2} + 2 \frac{\lambda_\infty L^2 [1 + (S/T_\infty)]}{\lambda_w \Delta [\theta_w(x) + (S/T_\infty)]} \sqrt{\frac{U \theta_w(x)}{\nu_\infty x L C}} \sum_{n=0}^{\infty} [a_n Y'_{an}(0) + b_n Y'_{an+\gamma_1}(0) x^{\gamma_1}$$

$$+ c_n Y'_{an+\gamma_2}(0) x^{\gamma_2}] x^{\alpha n} = -\frac{L^2}{\lambda_w T_\infty} \sum_{n=0}^{\infty} w_n(xL)^n \qquad (2.73)$$

When substituting Equation (2.71) into Equation (2.73), it is seen that in order to obtain the identity, the numbers defining constant m should be $\alpha = 3/2, \gamma_1 = 1/2, \gamma_2 = 1$. Then, the surface temperature distribution (Equation 2.71) takes the form

$$\theta_w(x,0) = \sum_{n=0}^{\infty} [a_n + b_n x^{1/2} + c_n x] x^{(3/2)n} \qquad (2.74)$$

Coefficients a_n, b_n, c_n are defining using a_0, c_0, coefficients of source, and values $Y'_{(3/2)n}(0), Y'_{(3/2)n+(1/2)}(0)$, and $Y'_{(3/2)n+1}(0)$. Condition (Equation 2.62) gives two equations defining a_0, c_0:

$$\frac{d\theta_w}{dx}\bigg|_{x=0} = c_0 = 0, \quad \frac{d\theta_w}{dx}\bigg|_{x=1} = \sum_{n=1}^{\infty} \left[\frac{3}{2} n a_0 + \left(\frac{3}{2} n + \frac{1}{2} \right) b_0 + \left(\frac{3}{2} n + 1 \right) c_0 \right] = 0 \qquad (2.75)$$

The first 12 coefficients of Equation (2.74) are calculated and given in Reference [25].

It is also demonstrated by the convergence of the series (Equation 2.74) and is presented as a discussion of the calculation of eigenfunctions $Y(\eta)$. The important result of this work is that the solution of the conjugate problem shows that the surface temperature is not an analytic function of x at $x = 0$. Thus, surface temperature cannot be presented in the form of the Taylor series of x.

The unsteady conjugate heat transfer problem for similar conditions and for flow of compressible fluid over a plate is solved by Perelman et al. The solution of this problem is published in three articles. In the first part [27], an integral relation is obtained using an approach similar to that described in this example. In the second part [28], the solution of this integral relation is given, and the function Ω_n that depends on dimensionless time and does not depend on the source is introduced. Finally, the solution of the problem is expressed through this function so that only this function Ω_n is required to calculate the temperature field. In the third article [29], the function Ω_n is calculated.

In this study, the dimensionless parameter $B = (1/\text{Re})(\alpha_w/\alpha_\infty C)$ is introduced. (C is the constant in Chapman-Rubesin's law [Equation 2.63].) This parameter is proportional to the ratio of two times $B \sim t/t_s$. The time t, in which heat impulse at some point of the plate appears at the point of the upper end of the boundary layer, has the order $t \sim \delta^2/C\alpha_\infty$, where $\delta \sim L/\sqrt{\text{Re}}$ and $C\alpha_\infty$ are the thermal diffusivity of the gas near a plate. The time t_s in which the same impulse appears at the other plate point at the distance L has the order $t_s \sim L^2/\alpha_s$. Then $B \sim \alpha_w/\mathrm{Re}C\alpha_\infty$. The case $B \ll 1 (t \ll t_s)$ means that physically, the heating of a boundary layer occurs during a period when heat impulse covers a distance in the plate, which is much less than L. Thus, conjugate unsteady heat transfer problems may be divided into two classes: if the number $E = \alpha_w/C\alpha_\infty$ for compressible fluid or number $E = \alpha_w/\alpha_\infty$ for incompressible fluid is introduced. At $\mathrm{Re} \leq E$, the problem may be solved as the unsteady one, both in the boundary layer and within the body, and at $\mathrm{Re} \gg E$, as the unsteady one within the body and the quasi-steady one in the boundary layer. The quasi-steady approach is used by many authors. For example, Sokolova [24] and Pomeransev [30] applied the quasi-steady method in early works, and later Sucec [5], Karvinen [7], and Dorfman [8] used this approach as well. (See also Example 2.2.)

Gdalevich and Khusid [31] used the power series to solve the system of Navier-Stokes and energy equations for an incompressible fluid over plate.

Example 2.7: Temperature of a Radiating Thin Plate Placed in Supersonic Gas Flow [24]

This problem is solved using Dorodnizin's [32] independent variables, which transform the equations for a compressible gas into the form of incompressible fluid (see Section 3.7):

$$u\frac{\partial u}{\partial \xi} + \tilde{v}\frac{\partial u}{\partial \eta} = v_0 \frac{\partial}{\partial \eta}\left[b(\phi)\frac{\partial u}{\partial \eta}\right] \qquad \tilde{v} = (1-m^2)^{\frac{k}{k-1}}u\frac{\partial \eta}{\partial x} + \frac{v}{\phi} \tag{2.76}$$

$$\frac{\partial u}{\partial \xi} + \frac{\partial \tilde{v}}{\partial \eta} = 0 \qquad u\frac{\partial \theta}{\partial \xi} + \tilde{v}\frac{\partial \theta}{\partial y} = \frac{v_0}{\mathrm{Pr}}\frac{\partial}{\partial \eta}\left[b(\phi)\frac{\partial \theta}{\partial \eta}\right] - \left(\frac{1}{\mathrm{Pr}} - 1\right)v_0\left[b(\phi)\frac{\partial m^2}{\partial \eta}\right] \tag{2.77}$$

$$\xi = (1-m_0^2)^{\frac{\kappa}{\kappa-1}}x, \quad \eta = \int_0^y \frac{\rho}{\rho_0}d\tilde{y} = \int_0^y \frac{(1-m_0^2)^{\frac{k}{k-1}}}{\phi}d\tilde{y}, \quad m_0^2 = \frac{\frac{k-1}{2}M^2}{1+\frac{k-1}{2}M^2},$$

$$\theta = \phi + m^2, \quad m^2 = \frac{u^2}{2c_p T_0}$$

Here, $\phi = T/T_0$ and 0 denote stagnation parameters. Side conditions for Equation (2.77) are

$$u(\xi,0) = \tilde{v}(\xi,0) = 0 \qquad u(\xi,\infty) = U, \quad \theta(\xi,\infty) = 1 \tag{2.78}$$

Three assumptions of the resistance of a plate are considered taking into account the energy balance of heat that a plate gets from gas and loses by radiation:

1. Zero thermal resistance

$$\frac{c_p \mu_0}{\mathrm{Pr}} \int_0^l b(\phi) \left(\frac{\partial \theta}{\partial \eta} \right)_{\eta=0} d\xi = \int_0^l \sigma T_0^3 (\phi_w^4 - \phi_\infty^4) \frac{d\xi}{(1-m_0^2)^{\frac{k}{k-1}}}$$

2. Infinite thermal resistance

$$\frac{c_p \mu_0}{\mathrm{Pr}} (1-m_0^2)^{\frac{k}{k-1}} b(\phi) \left(\frac{\partial \theta}{\partial \eta} \right)_{\eta=0} = \sigma T_0^3 (\phi_w^4 - \phi_\infty^4)$$

3. Finite thermal resistance

$$\frac{c_p \mu_0}{\mathrm{Pr}} (1-m_0^2)^{\frac{k}{k-1}} b(\phi) \left(\frac{\partial \theta}{\partial \eta} \right)_{\eta=0} = \sigma T_0^3 (\phi_w^4 - \phi_\infty^4) - \lambda_w (1-m^2)^{\frac{2k}{k-1}} \frac{\partial^2 \theta}{\partial \xi^2}$$

The function $b(\phi)$ for $\phi < 1$ is close to 1, then the law for viscosity takes the form $\mu = \mu_0 \phi$.

New variables reduce the hydrodynamic problem to Blasius Equation (1.7) at $\beta = 0$:

$$\zeta = \frac{1}{2} \eta \sqrt{\frac{U}{v_0 \xi}}, \qquad \psi = \sqrt{U v_0 \xi} \varphi(\zeta), \tag{2.79}$$

The energy equation (2.77) in new variables takes the form

$$\frac{\partial^2 \theta}{\partial \zeta^2} + \mathrm{Pr}\,\varphi(\zeta) \frac{\partial \theta}{\partial \zeta} = 2\,\mathrm{Pr}\,\xi \varphi'(\zeta) \frac{\partial \theta}{\partial \xi} + \frac{m_0^2}{2} (1-\mathrm{Pr})[\varphi'''(\zeta)\varphi'(\zeta) + \varphi''^2(\zeta)] \tag{2.80}$$

1. *Plate with zero thermal resistance* — In this case, the plate temperature is constant and hence $\partial \theta / \partial \xi = 0$. Equation (2.80) and the energy balance in the variables (Equation 2.79) becomes

$$\frac{\partial^2 \theta}{\partial \zeta^2} + \mathrm{Pr}\,\varphi(\zeta) \frac{\partial \theta}{\partial \zeta} = \frac{m_0^2}{2} (1-\mathrm{Pr})[\varphi'''(\zeta)\varphi'(\zeta) + \varphi''^2(\zeta)] \tag{2.81}$$

$$\frac{c_p \rho_0}{\mathrm{Pr}} U \sqrt{\frac{v_0}{U\xi}} \left(\frac{\partial \theta}{\partial \zeta} \right)_{\zeta=0} = \frac{\sigma T_0^3 (\phi_w^4 - \phi_\infty^4)}{(1-m_0^2)^{\frac{k}{k-1}}} \tag{2.82}$$

Assuming $\theta = 1 + Bm_0^2$, one gets

$$B'' + \mathrm{Pr}\,\varphi(\zeta)B' = \frac{1}{2}(1-\mathrm{Pr})[\varphi'''(\zeta)\varphi'(\zeta) + \varphi''^2(\zeta)] \tag{2.83}$$

with boundary conditions $B(\infty) = 0$ and $B(0) = (\phi_w - 1)/m_0^2$, which correspond to boundary conditions $\theta(\infty) = 1$ and $\theta(0) = \phi_w$. The solution of this equation gives for B and θ

$$\theta = m_0^2 \int_0^\zeta A(\tilde{\zeta})I(\tilde{\zeta})d\tilde{\zeta} - \frac{m_0^2 \int_0^\infty A(\zeta)I(\zeta)d\zeta + \phi_w - 1}{\int_0^\infty A(\zeta)d\zeta} \int_0^\zeta A(\tilde{\zeta})d\tilde{\zeta} + \phi_w \qquad (2.84)$$

$$A(\zeta) = \exp\left(-\Pr \int_0^\zeta \varphi(\tilde{\zeta})d\tilde{\zeta}\right), \quad I(\zeta) = \int_0^\zeta \frac{f(\tilde{\zeta})}{A(\tilde{\zeta})}d\tilde{\zeta},$$

$$f(\zeta) = \frac{1}{2}(1-\Pr)[\varphi'''(\zeta)\varphi'(\zeta) + \varphi''^2(\zeta)] \qquad (2.85)$$

Applying these results, one obtains $(\partial\theta/\partial\zeta)_{\zeta=0}$, and then after substitution into Equation (2.82), the equation for a desired plate temperature can be obtained:

$$\left(\frac{\partial\theta}{\partial\zeta}\right)_{\zeta=0} = -0.08498m_0^2 - \frac{\phi_w - 1}{1.6774} \qquad (\phi_w K_1)^4 + (\phi_w K_1) - K_2 = 0 \qquad (2.86)$$

$$K = 0.5962 \frac{c_p U \rho_\infty (1-m_0^2)^2}{\Pr \sigma T_\infty^3 \sqrt{Re}} \quad K_1 = K^{-1/3} \qquad (2.87)$$

$$K_2 = (1 - 0.14254m_0^2)K_1 + \phi_\infty^4 K^{4/3}$$

To determine the surface temperature, one should calculate from Equation (2.87) K, K_1 and K_2, then from Equation (2.86) determine $\phi_w K_1$, and finally estimate $\phi_w = (\phi_w K_1)/K_1$.

2. *Plate with infinite thermal resistance* — In this case, the energy balance in variables (Equation 2.79) can be presented as

$$\left(\frac{\partial\theta}{\partial\zeta}\right)_{\zeta=0} = \frac{2\Pr\sigma T_0^3}{c_p\mu_0(1-m_0^2)^{\frac{k}{k-1}}}\sqrt{\frac{v_0\xi}{U}}(\phi_w^4 - \phi_\infty^4) \qquad (2.88)$$

By introducing a new variable,

$$\sqrt{s} = \frac{2\Pr\sigma T_0^3}{c_p\mu_0(1-m_0^2)^{\frac{k}{k-1}}}\sqrt{v_0/U}\sqrt{\xi} = \frac{2\Pr\sigma T_\infty^3}{c_p\sqrt{\mu_\infty\rho_\infty U}(1-m_0^2)^2}\sqrt{x}, \qquad (2.89)$$

one gets Equation (2.80) with boundary conditions in the form

$$\frac{\partial^2\theta}{\partial\zeta^2} + \Pr\varphi(\zeta)\frac{\partial\theta}{\partial\zeta} = 2\Pr s\varphi'(\zeta)\frac{\partial\theta}{\partial s} + \frac{m_0^2}{2}(1-\Pr)[\varphi'''(\zeta)\varphi'(\zeta) + \varphi''^2(\zeta)]$$

$$\theta(s,\infty) = 1 \qquad \left(\frac{\partial\theta}{\partial\zeta}\right)_{\zeta=0} \quad \sqrt{s}(\phi_w^4 - \phi_\infty^4) \qquad (2.90)$$

For small s, solution of this equation and the plate temperature are presented by series

$$\theta(s,\zeta) = \sum_0^\infty \theta_n(\zeta)s^{n/2} \qquad \phi_w = \sum_0^\infty \theta_n(0)s^{n/2} \tag{2.91}$$

Substituting the series for θ into Equation (2.90) gives the equations for coefficients $\theta_n(0)$. The first six coefficients given in [24] provide sufficient result for $s \le 0.01$.

Calculation shows that the temperature distribution across the boundary layer for the plate with infinite resistances is very close to that for the plate with zero resistance. This fact is used to obtain an approximate solution of Equation (2.90), which can be applied for greater values of s. This solution is sought as a sum of two functions θ_0 and θ_1:

$$\theta = \theta_0 + \theta_1 \tag{2.92}$$

which satisfy the following equations and boundary conditions:

$$\frac{\partial^2 \theta_0}{\partial \zeta^2} + \Pr \varphi(\zeta)\frac{\partial \theta_0}{\partial \zeta} = \frac{m_0^2}{2}(1 - \Pr)[\varphi'''(\zeta)\varphi'(\zeta) + \varphi''^2(\zeta)] \tag{2.93}$$

$$\theta_0'(0) = 0 \quad \theta_0(\infty) = 1$$

$$\frac{\partial^2 \theta_1}{\partial \zeta^2} + \Pr \varphi(\zeta)\frac{\partial \theta_1}{\partial \zeta} = 2\Pr s\varphi'(\zeta)\frac{\partial \theta_1}{\partial s} \tag{2.94}$$

$$\theta_1'(0) = \sqrt{s}(\phi_w^4 - \phi_\infty^4) \qquad \theta_1(\infty) = 0$$

The system of three equations (Equations 2.92 through 2.94) contains three unknowns: θ_0, θ_1, and ϕ_w. Equation (2.93) is similar to Equation (2.80), and its solution satisfying conditions (Equation 2.93) is

$$\theta_0 = 1 - m_0^2 \int_\zeta^\infty A(\zeta)I(\zeta)d\zeta \tag{2.95}$$

Because it is assumed that θ is known and is given by solution (Equation 2.84), one determines from the sum (Equation 2.92) function θ_1 as difference:

$$\theta_1 = \theta - \theta_0 = \left[1 - \frac{\int_0^\zeta A(\tilde{\zeta})d\tilde{\zeta}}{\int_0^\infty A(\zeta)d\zeta}\right]\left[\phi_w - 1 + m_0^2 \int_0^\infty A(\zeta)I(\zeta)d\zeta\right] \tag{2.96}$$

Substituting this expression into Equation (2.93) after integrating the result and taking into account conditions (Equation 2.93), gives an equation that determines the average value of ϕ_w:

$$\frac{d\phi_w}{ds} + \frac{\phi_w}{2s} + 0.83800\frac{\phi_w^4}{\sqrt{s}} = 0.83800\frac{\phi_\infty^4}{\sqrt{s}} - \frac{0.14254m_0^2 - 1}{2s} \qquad (2.97)$$

This equation is numerically integrated giving ϕ_w as a function of s and m_0^2. Using these results from Table 3 in [24], one estimates the radiating plate temperature.

3. *Plate with finite thermal resistance* — In this case, the boundary conditions in variables (ζ, s) can be presented in the form

$$\theta(s,\infty) = 1, \quad \left(\frac{\partial\theta}{\partial\zeta}\right)_{\zeta=0} = \sqrt{s}(\phi_w^4 - \phi_\infty^4) - \frac{2\lambda_w \Pr(1 - m_0^2)^{\frac{k}{k-1}}}{\sigma T_0^3} A_1\sqrt{s}\frac{\partial^2\theta}{\partial s^2} \qquad (2.98)$$

Solving the basic equation (2.80) with these side conditions, one obtains

$$-\frac{a}{\sqrt{s}}\frac{d^2\phi_w}{ds^2} + \frac{d\phi_w}{ds} + \frac{\phi_w}{2s} + 0.83800\frac{\phi_w^4}{\sqrt{s}} = 0.83800\frac{\phi_\infty^4}{\sqrt{s}} - \frac{0.14254m_0^2 - 1}{2s} \qquad (2.99)$$

$$a = 0.83800\lambda_w A_1^2 \frac{(1 - m_0^2)^{10}}{\sigma T_\infty^3}, \qquad A_1 = \left[\frac{2\Pr\sigma T_0^3}{c_p\mu_0\sqrt{U/v_0}(1 - m_0^2)^{k/(k-1)}}\right]^2$$

Equation (2.99) differs from Equation (2.97) by the first term. The value of factor a of this term is not more than 10^{-4}. Thus, the first term of Equation (2.99) can be neglected in the case of large variable s. Physically, this means that the longitudinal conduction may be neglected far from the leading edge of the plate. Then, Equation (2.99) coincides with Equation (2.97), and its solution can be presented by series (Equation 2.91). Calculation shows that neglect of the plate conduction affects the results only near the leading edge where $s < 0.0015$, and the error due to that disregard is at the most 4%. Thus, estimation of the plate temperature can be done using the results obtained for the plate with infinite resistance. However, such an approximation is applicable only for plates without heat sources.

Calculation example [24] — A thin plate 5 m in length is flying on an altitude of 10 km with velocity 336 m/s, $c_p = 1kJ/kgK$, $\rho_\infty = 0.42\ kg/m^3$, $T_\infty = 223K$, $\mu = 0.151 \cdot 10^{-4} kg/ms$, $\sigma = 5.67 \cdot 10^{-11}\ kJ/m^2sK^4$, $\Pr = 0.74$, speed of sound $U_{sd} = 20.1\sqrt{T_\infty} = 300m/s$, $m_0^2 = 0.2$, $\mathrm{Re} = 4.68 \cdot 10^7$, $M = 1.12$.

The plate temperature with zero resistance is calculated using Equations (2.86) and (2.87): $K = 13.6, K_1 = K^{-1/3} = 0.419, K_2 = 0.42$. For small K_2, the dependency between $(\phi_w K_1)$ and K_2 is close to linear

$(\phi_w K_1) \approx 0.93 K_2$ [24]. Then $\phi_w \approx 0.932$, $T_0 = T_\infty/(1-m_0^2) = 223/0.8 = 279$, $T_w = 0.932 \cdot 279 \cdot = 260K$.

The plate temperature with infinite resistance is calculated using Equation (2.89), $s = 0.00769$. Because s is small, the series (Equation 2.91) can be used. Taken the coefficients from [24], one obtains $\phi_w = 0.933$, and then $T_w = 260K$. $M = 4.47$.

In the case of high Mach number $U = 1342 m/s$, the plate temperature with zero resistance calculated using Equations (2.86) and (2.87) is $T_w = 663K$, the plate temperature with infinite resistance is $T_w = 618K$, and the plate temperature without radiation is $T_w = 1028K$. Thus, the radiating significantly reduces the plate temperature, especially in the case of high Mach numbers.

Similar problems are considered in References [3] and [33]. In the first article, the series for small and large distance from leading edge are used, similar to the series developed by Perelman [1] (Example 2.1). The authors of the second work first used the integral transform to simplify the initial system of equations and then applied the series. Their results are in agreement with that obtained in this example from [24].

Example 2.8: Temperature of a Thin Plate after an Oblique Shock [30]

This problem is solved by Chapman and Rubesin's method [26]. The plate temperature distribution is described by polynomial

$$T_w = T_2 - B(t)\xi(2-\xi) \tag{2.100}$$

Here, $\xi = x/L$, subscript 2 denotes parameters after shock, and T_2 is the plate temperature at the leading edge, which is close to the temperature after shock:

$$T_2 = T_\infty + \frac{2kU_{sd}}{(k+1)^2 c_p} M^2 \left(1 - \frac{1}{1-M^2}\right)\left(1 + \frac{1}{kM^2}\right) \tag{2.101}$$

Here, U_{sd} is the speed of sound and Mach number is calculated using the free stream velocity component normal to the shock. Equation (2.100) also satisfies the condition $(\partial T_w/\partial\xi)_{\xi=1} = 0$, which follows from the fact that far from the leading edge the temperature is approximately constant. The energy balance for a thermally thin plate follows from Equation (2.7), which after using Equation (2.100) gives

$$\rho_w c_w \Delta\left[\frac{dT_w}{dt} - \frac{dB}{dt}(2\xi - \xi^2)\right] = q_w \tag{2.102}$$

Chapman and Rubesin's equation (2.64) is used to calculate the heat flux:

$$\frac{\partial^2 T}{\partial \eta^2} + \Pr \varphi(\eta)\frac{\partial T}{\partial \eta} - 2\Pr \xi \varphi'(\eta)\frac{\partial T}{\partial \xi} = -\frac{1}{4}\Pr(k-1)M^2\varphi''^2 \qquad (2.203)$$

In this case, the solution of Equation (2.103) is presented as

$$T = \theta - B[2Y_1(\eta)\xi - Y_2(\eta)\xi^2]$$

$$\theta = T_2\left\{1 + \frac{\Pr(k-1)}{4}M_2^2 r(\eta) - g(\eta)\left[1 + \frac{\Pr(k-1)}{4}M_2^2 r(0)\right] + T_{w0}g(\eta) + (T_2 - T_{w0})Y_0(\eta)\right\}$$

$$r(\eta) = \int_\eta^{\eta_2}\varphi''(\xi)^{\Pr}d\xi\int_{\eta_2}^{\xi}(\varphi'')^{2-\Pr}d\omega, \qquad g(\eta) = \left[\int_\eta^{\eta_2}\varphi''(\xi)^{\Pr}d\xi\right]\Big/\left[\int_0^{\eta_2}\varphi''(\xi)^{\Pr}d\xi\right]$$

$$(2.104)$$

where T_{w0} is the wall temperature at the first moment when the flow starts. Calculating heat flux by differentiating the first Equation (2.104) and substituting the result in Equation (2.102) gives the differential equation determining $B(t)$ in Equation (2.100):

$$\rho_w c_w \Delta\left[\frac{dT_2}{dt} - \frac{dB}{dt}(2\xi - \xi^2)\right] = \frac{\lambda_2}{2x}\sqrt{\Pr_2}\left[\frac{d\theta}{d\eta} - B[2Y_1'(\eta)\xi - Y_2'(\eta)\xi^2]\right]_{\eta\to 0} \qquad (2.105)$$

where $\Pr_2 = (c_p\rho U x/\lambda)_2$. Averaging Equation (2.105) and using $Y'(0)$ from Reference [26] yield

$$\frac{dB}{dt} + 0.62\frac{\lambda_2\sqrt{\Pr_2}}{c_w\rho_w L\Delta}B = -1.5\left[\frac{\lambda_2\sqrt{\Pr_2}}{c_w\rho_w L\Delta}\frac{d\theta}{d\eta}\bigg|_{\eta=0} + \frac{dT_2}{dt}\right] \qquad (2.106)$$

Using Equation (2.104) gives $(d\theta/d\eta)_{\eta=0} = 0.252 M_2^2 T_2$. Then Equation (2.106) and its solution become

$$\frac{dB}{dt} + m(t)B = 1.5\dot{T}_2(t) - N(t), \qquad m(t) = 0.62\frac{\lambda_2\sqrt{\Pr_2}}{c_w\rho_w L\Delta}, \qquad N(t) = 0.378\frac{\lambda_2\sqrt{\Pr_2}}{c_w\rho_w L\Delta}M_2^2 T_2$$

$$B = B_0\exp\left(-\int_0^t m(\tilde{t})d\tilde{t}\right) + \int_0^t\left[1.5\left(\frac{dT}{dt}\right)_2(\tilde{t}) - N(\tilde{t})\right]\exp\left(-\int_0^t m(\tilde{t})d\tilde{t}\right)d\tilde{t}$$

$$(2.107)$$

Calculation results obtained using this equation are compared with observations during the supersonic flight of a rocket V-2.

2.5 Solutions of the Conjugate Heat Transfer Problems by Integral Methods

Integral methods are used to study conjugate heat transfer significantly less than series. Yet, at least two early works are performed. Luikov [12] and Olsson [34] both considered the heat transfer from incompressible fluid flow to a plate and a wedge-shaped fin, respectively, applying integral methods.

Example 2.9: Heat Transfer between a Thin Plate and an Incompressible Flow Past It [12]

The integral heat transfer equation is used in the form (Section 1.4)

$$\frac{\partial}{\partial x}\int_0^{\delta_t}(T_\infty - T)u(y)dy = \alpha\left(\frac{\partial T}{\partial y}\right)_{y=0} \tag{2.108}$$

The approximate velocity and temperature distribution are presented as

$$\frac{u}{U} = \frac{3y}{2\delta} - \frac{1}{2}\left(\frac{y}{\delta}\right)^3, \quad \vartheta = T - T_* = a_1 + b_1 y + c_1 y^2 + d_1 y^3, \quad \vartheta_s = a_2 + b_2 y \tag{2.109}$$

where T_* is the isothermal wall temperature. Satisfying side conditions (2.5) and (2.6) leads to the temperature distributions across the thermal boundary layer and a plate:

$$\vartheta = \vartheta_w + \frac{3(\vartheta_\infty - \vartheta_w)}{2\delta_t}y - \frac{\vartheta_\infty - \vartheta_w}{2\delta_t^3}y^3, \quad \vartheta_s = \vartheta_w + \frac{\vartheta_w}{\Delta}y,$$

$$\vartheta_w = \frac{\vartheta_0 z}{1+z}, \quad z = \frac{3\lambda\Delta}{2\lambda_w\delta_t} \tag{2.110}$$

where ϑ_0 is the temperature of the plate with zero thermal resistance. Substituting Equation (2.110) into Equation (2.108) and taking into account that $\vartheta = \vartheta_\infty$ at $y > \delta$ (Example 1.2), one obtains

$$\frac{3}{20}\vartheta U\frac{d}{dx}\left[\frac{1}{1+z}\left(\frac{\xi^2}{z^2\delta} - \frac{\xi^4}{14z^4\delta^3}\right)\right] = \frac{3\alpha\vartheta_\infty z}{2z\xi\delta} \tag{2.111}$$

where $\xi = \delta_t/\delta$. Neglecting the small second term and putting $z \to 0$, one gets the equation and its solution:

$$\frac{d}{dx}\left(\frac{1}{z^2\delta}\right) = \frac{10\alpha}{z^2\xi^3\delta^3} \quad \xi = \frac{\delta_t}{\delta} = \sqrt[3]{\frac{13}{14\,\mathrm{Pr}}} \quad \mathrm{Nu}_{x*} = 0.332\,\mathrm{Pr}^{1/3}\,\mathrm{Re}_x^{1/2} \tag{2.112}$$

This is the same as the solution presented in [35].

Asymptotic solutions of Equation (2.111) for the case of conjugate heat transfer are obtained for small and large Brun numbers, which characterize nonuniformity of the plate temperature [12]:

$$\frac{Nu_x}{Nu_{x*}} = 1 + 0.33Br_x, \quad Br_x < 1.5, \qquad \frac{Nu_x}{Nu_{x*}} = 1.66 - \frac{0.34}{Nu_{x*}}$$

$$(2.113)$$

$$Br_x > 1.5, \qquad Br_x = \frac{\lambda\Delta}{\lambda_w x} Pr^m Re_x^n$$

Example 2.10: Heat Transfer between a Thin Radiating Wedge and an Incompressible Fluid Flow Past It [34]

The free stream velocity in such a case is given by the power function (Equation 1.5). Using polynomials and boundary and conjugate conditions (Equations 2.5 and 2.6), two functions describing the velocity and temperature distributions across the layers are obtained. Two cases are considered: when the thermal boundary layer is thinner ($\xi = \frac{\delta_t}{\delta} < 1$) and thicker ($\xi > 1$) than the velocity boundary layer:

$$G_1(\xi) = \frac{2}{15}\xi^2 - \frac{3}{140}\xi^4 + \frac{1}{180}\xi^5,$$

$$G_2(\xi) = \frac{1}{90}\xi^2 - \frac{1}{84}\xi^3 + \frac{3}{560}\xi^4 - \frac{1}{1080}\xi^5$$

$$(2.114)$$

$$G_1(\xi) = -\frac{3}{10} + \frac{3}{10}\xi + \frac{2}{15\xi} - \frac{3}{140\xi^3} + \frac{1}{180\xi^4},$$

$$G_2(\xi)\frac{1}{120} - \frac{1}{180}\xi + \frac{1}{840\xi^3} - \frac{1}{3024\xi^4}$$

$$(2.115)$$

Then the momentum and energy integral equations lead (Section 1.4) to an equation:

$$\left[G_1(\xi) + \frac{1260m}{37 + 300m}G_2(\xi) \right] = \frac{37 + 300m^B}{630 Pr\,\theta_w(L)L^{(m+1)/2}},$$

$$(2.116)$$

$$B = \int_0^L \theta_w(x) x^{\frac{m-1}{2}}\, dx$$

where $\theta_w = T_w - T_\infty$, $m = \beta/(2\pi - \beta)$ is the temperature head, and β is the leading edge angle. Because the wedge is assumed to be thin ($m \ll 1$), the temperature approximately may be considered as a function of x only $\theta_w(x)$. The balance for the wedge in the case of the linearized radiative law is

$$\lambda_w\beta\frac{d}{dx}\left(x\frac{d\theta_w}{dx}\right) = 2\theta_w\left[\frac{2\lambda}{\xi}\left(\frac{37 + 300mC}{1260v}\right)^{1/4} x^{(m-1)/2} + 4\varepsilon\sigma T_\infty^3 \right]$$

$$(2.117)$$

and $\theta_w(L) = \theta_0, \theta_w(0)$ – finite are boundary conditions, where θ_0 is the temperature of the wedge base. Here, C is constant in Equation (1.5) for power free stream flow. The system of two equations (2.116) and (2.117) determines two unknowns, ξ and $\theta_w(x)$. In the case of neglecting convection, Equation (2.117) becomes a Bessel differential equation that under the indicated conditions has the solution

$$\theta_w = \theta_0 \frac{I_0\left[\zeta_r(x/L)^{1/2}\right]}{I_0(\zeta_r)}, \qquad \zeta_r = \left(\frac{32\varepsilon L\sigma T_\infty^3}{\beta\lambda_w}\right)^{1/2} \qquad (2.118)$$

Similarly, if radiation is neglected,

$$\theta_w = \theta_0 \frac{I_0[\zeta_c(x/L)^{(m+1)/4}]}{I_0(\zeta_c)}, \quad \zeta_c = \frac{1}{m+1}\left(\frac{64\lambda}{\beta\lambda_w\xi}\right)^{1/2}\left(\frac{37+300mC}{1260v}\right)^{1/4}L^{(m+1)/4} \quad (2.119)$$

For combined convection and radiation, Equation (2.117) becomes

$$\frac{d}{dx}\left(z\frac{d\theta_w}{dz}\right) = \theta_w\left[\frac{\zeta_c^2}{4L^{(m+1)/2}} + \frac{\zeta_r^2}{L(m+1)^2}z^{\frac{1-m}{1+m}}\right], \qquad z = x^{(m+1)/2} \qquad (2.120)$$

Neglecting m with respect to 1 in the radiative term allows us to determine the solution in the series:

$$\theta_w = \theta_0\left[\frac{\displaystyle\sum_{n=0}^{\infty}C_n\left(\frac{x}{L}\right)^{n(m+1)/2}}{\displaystyle\sum_{n=0}^{\infty}C_n}\right], \qquad (2.121)$$

$$C_{n+2} = \frac{\zeta_c^2 C_{n+1} + 4\zeta_r^2 C_n}{4(n+2)^2}, C_0 = 1, C_1 = \frac{\zeta_c^2}{4} \quad n = 1,2\cdots$$

Introducing this solution into Equation (2.116) makes it possible to calculate ξ. Then the heat flux and temperature fields may easily be found by the usual methods. In order to estimate the accuracy of the approximate integral solution, the Blasius series is used to solve the same problem (Section 1.3). The system of Equations (2.1) to (2.6) for conjugate steady-state heat transfer is applied with two changes due to the specifics of the problem. The term $U(dU/dx)$ is added instead of $(1/\rho)(dp/dx)$ to take into account the gradient of the stream velocity, and Equation (2.4) for the body is used in cylindrical coordinates (r, ϕ):

$$u\frac{\partial u}{\partial x} + v\frac{\partial}{\partial y} = U\frac{dU}{dx} + v\frac{\partial^2 u}{\partial y^2}, \qquad \frac{\partial^2\theta_s}{\partial r^2} + \frac{1}{r}\frac{\partial\theta_s}{\partial r} + \frac{1}{r^2}\frac{\partial^2\theta_s}{\partial\phi^2} = 0 \qquad (2.122)$$

The free stream velocity and surface temperature distributions are presented as

$$U = x^m \sum_{j=0}^{\infty} u_j x^{jn}, \qquad \theta = T - T_{\infty} = \sum_{i=0}^{\infty} T_i x^{in}, \qquad T_0 \neq 0, u_0 \neq 0 \qquad (2.123)$$

The solution is sought in the form of the same series:

$$\psi = x^{(m+1)/2} (u_0 v)^{1/2} \sum_{j=0}^{\infty} \left(\frac{u_1}{u_0} \right)^{j} f_j(\eta) x^{jn},$$

$$\theta_i = T_i \sum_{j=0}^{\infty} \left(\frac{u_1}{u_0} \right)^{j-i} F_{ij}(\eta) x^{jn}, F_{ij} = 0, i > j \qquad (2.124)$$

Here, $\eta = (1/2) y x^{(m-1)/2} u_0^{1/2} v^{-1/2}$ is the Blasius variable. The expression for θ_i represents the solution for each term of the series (Equation 2.123) for determining θ. Introducing the series (Equation 2.124) in momentum and energy boundary layer equations and in side conditions and singling out coefficients for x give a recursive system of ordinary differential equations and boundary conditions for the functions $f_j(\eta)$ and $F_{ij}(\eta)$. Because the function $F_{ij}(\eta)$ is determined for each term of the series (Equation 2.123), the total solution for θ is obtained by a summation of these functions. Such a procedure is possible because the energy equation is linear.

The solution of Equation (2.122) for a wedge should satisfy the boundary condition of continuity in temperature across the body–fluid interface:

$$x = L, \theta_w = \theta_0 = \sum_{i=0}^{\infty} T_i L^{in} \qquad \phi = \beta/2, \theta_s = \theta_w = \sum_{i=0}^{\infty} T_i x^{in} \qquad (2.125)$$

Furthermore, the solution should be finite at $x = r = 0$ and symmetric with respect to $\phi = 0$. A solution that satisfies these conditions is as follows:

$$\theta_s = \sum_{i=0}^{\infty} \frac{T_i L^{in}}{\cos(in\beta/2)} \left(\frac{r}{L} \right)^{in} \cos in\phi + \sum_{i=0}^{\infty} B_i \left(\frac{r}{L} \right) \cos(2i+1) \frac{\pi\phi}{\beta}$$

$$B_i = \frac{4(-1)^i}{(2i+1)\pi} \sum_{j=1}^{\infty} \frac{j^2 n^2 T_j L^{jn}}{j^2 n^2 - [(2i+1)(\pi/\beta)]^2} \qquad (2.126)$$

The coefficient B_i was obtained through development of the first boundary condition (2.125) into a series of $\cos(2i+1)(\pi\phi/\beta)$.

The obtained expressions for the temperatures in the fluid and in the body contain yet unknown coefficients T_i. These coefficients may be found through satisfying the demand of continuity in the heat flux across the fluid–body interface. This is achieved by selecting the coefficients so that the heat flux in the fluid and in the body as obtained from Equations (2.124) and (2.126), respectively, match each other as well as possible according to the discrete least squares method.

Both results obtained by series and by integral approach are compared with surface temperature distribution determined experimentally on black-painted fins of copper and iron. Both results give a relatively exact description of the measured temperature distribution. The difference between the two results is less than the difference between either of them and experimental data.

The obtained results are used for discussion of difference between conjugate and usual approaches. Values of $Nu_x Re_x^{-1/2}$ calculated by conjugate approach for different values of wedge conductivity and different cooling media vary considerably from the values being obtained for infinite heat conductivity. The calculation indicates that these differences could lead to large errors in temperature and heat flux.

The values of heat flux obtained by $Nu_x Re_x^{-1/2}$, which corresponds to isothermal surface, are underestimated. For air, this approximation gives relatively good results, especially at high conductivities, but it leads to large errors for the other fluids. Studies indicate that the temperature distribution is more uniform at larger wedge angles, shorter fin lengths, and lower fluid velocities.

References

1. Perelman, T. L., 1961. Heat transfer between a laminar boundary layer and a thin plate with the inner sources, *J. Engn. Phys. Thermodyn.* 4 (5): 54–61.
2. Perelman, T. L., 1961. On conjugated problems of heat transfer, *Int. J. Heat Mass Transfer* 3: 293–303.
3. Kumar, I., and Bartman, A. B.,1968. Conjugate heat transfer from radiated plate to laminar boundary layer of compressible fluid (in Russian). In *Teplo-i Massoperenos*, Nauka i Technika, Minsk, 9: 181–198.
4. Kumar, I., 1968. Conjugate heat transfer problem in laminar boundary layer with blowing. *J. Engn. Phys. Thermodyn.* 14: 781–791.
5. Sucec, J., 1975. An improved quasi-steady approach for transient conjugated forced convection problems, *Int. J. Heat Mass Transfer* 24: 1711–1722.
6. Soliman, M., and Johnson, H. A., 1967. Transient heat transfer for turbulent flow over a flat plate of appreciable thermal capacity and containing time-dependent heat source, *ASME J. Heat Transfer* 89: 362–370.
7. Karvinen, R., 1976. Steady state and unsteady heat transfer between a fluid and a flat plate with coupled convection, conduction and radiation, *Acta Politech. Scand., Ser. E* 73: 5–36.
8. Dorfman, A. S., 1977. Solution of the external problem of unsteady state convection heat transfer with coupled boundary conditions, *Int. Chem. Engn.* 17: 505–510.
9. Tomlan, P. F., and Hudson, J. L., 1968. Transient response of countercurrent heat exchangers with short contact time, *Int. J. Heat Mass Transfer* 11: 1253–1265.

10. Perelman, T. L., 1961. A boundary value problem for equations of the mixed type in the theory of heat transfer, *J. Engn. Phys. Therodin.* 4 (8): 121–125.

11. Rizk, T. A., Kleinstreuer, C., and Ozisik, M. N., 1992. Analytic solution to the conjugate heat transfer problem of flow past a heated block, *Int. J. Heat Mass Transfer* 35: 1519–1525.

12. Luikov, A. V., 1974. Conjugate convective heat transfer problems, *Int. J. Heat Mass Transfer* 17: 257–265.

13. Viskanta, R., and Abrams, M., 1971. Thermal interaction of two streams in boundary layer flow separated by a plate. *Int. J. Heat Mass Transfer* 14: 1311–1321.

14. Sparrow, E. M., and DeFarias, F. N., 1968. Unsteady heat transfer in ducts with time varying inlet temperature and participating walls, *Int. J. Heat Mass Transfer* 11: 837–853.

15. Siegel, R., and Perlmutter, M., 1963. Laminar heat transfer in a channel with unsteady flow and heating varying with position and time, *ASME J. Heat Transfer* 85: 358–365.

16. Namatame, K., 1969. Transient temperature response of an annual flow with step change in heat generating rod, *J. Nucl. Sci.Tech.* 6: 591–600.

17. Sucec, J., 1975. Unsteady heat transfer between a fluid with time varying temperature and a plate, *Int. J. Heat Mass Transfer* 18: 25–36.

18. Sucec, J., 1973. Exact analytical solution to a transient conjugate heat transfer problem, *NASA TN D-7101*.

19. Sucec, J., 1980. Transient heat transfer between a plate and a fluid whose temperature varies periodically with time. *ASME J. Heat Transfer* 102: 126–131.

20. Chambre, P. L., 1964. Theoretical analysis of the transient heat transfer into a fluid flowing over a flat plate containing internal heat sources. In *Heat Transfer, Thermodynamics and Education*, edited by H. A. Johnson, pp. 59–69. McGraw-Hill, New York.

21. Travelho, J. S., and Santos, W. F. N., 1998. Unsteady conjugate heat transfer in a circular duct with convection from ambient and periodically varying inlet temperature. *ASME J. Heat Transfer* 120: 506–511.

22. Kawamura, H., 1974. Analysis of transient turbulent heat transfer in an annulus. Part I: Heating element with a finite (non zero) heat capacity and no thermal resistance. *Heat Transfer, Japan. Res.* 3: 45–68.

23. Carsllaw, H. S., and Jaeger, J. C., 1959. *Conduction of Heat in Solids,* 2nd ed. Oxford University Press, Oxford.

24. Sokolova, I. N., 1957. Temperature of a radiating thin plate placed in supersonic gas flow (in Russian), *Sbornik teoreticheskikh rabot po aerodinamike*, Sentralnyi Aerogidrodinamicheskii Institut, Gosoborongiz, Moscow.

25. Luikov, A. V., Perelman, T. L., Levitin, R. S., and Gdalevich, L. B., 1970. Heat transfer from a plate in a compressible gas flow, *Int. J. Heat Mass Transfer* 13: 1261–1270.

26. Chapman, D., and Rubesin, M., 1949. Temperature and velocity profiles in the compressible laminar boundary layer with arbitrary distribution of surface temperature, *J. Aeronaut. Sci.* 16: 547–565.

27. Perelman, T. L., Levitin, R. S., Gdalevich, L. B., and Khusid, B. M., 1972. Unsteady-state conjugated heat transfer between a semi-infinite surface and incoming flow of a compressible fluid — I. Reduction to the integral relation, *Int. J. Heat Mass Transfer* 15: 2551–2561.

28. Perelman, T. L., Levitin, R. S., Gdalevich, L. B., and Khusid, B. M., 1972. Unsteady state conjugated heat transfer between a semi-infinite surface and incoming flow of a compressible fluid-II. Determination of a temperature field and analysis of result, *Int. J. Heat Mass Transfer* 15: 2563–2573.

29. Kopeliovich, B. L., Perelman, T. L., Levitin, R. S., Khusid, B. M., and Gdalevich, L. B., 1976. Unsteady conjugate heat exchange between a semi-infinite surface and a stream of compressible fluid flowing over it, *J. Engn. Phys. Thermodyn.* 30: 337–339.

30. Pomeransev, A. A., 1960. Heating a wall by a supersonic gas flow. *J. Engn. Phys. Thermodyn.* 3 (8): 39–46.

31. Gdalevich, L. B., and Khusid, B. M., 1971. Conjugate nonstationary heat transfer for a thin plate in the flow of an incompressible fluid, *J. Engn. Phys Thermodyn.* 20: 748–753.

32. Dorodnizin, A. A., 1942. Laminar boundary layer in compressible gas, *Prikl. Math. Mech.* 6: 449–486.

33. Dunin, I. L., and Ivanov, V. V., 1974 Conjugate heat transfer problem with surface radiation taken into account, *Fluid Dynamic* 9: 667–670.

34. Olsson, U., 1973 Laminar flow heat transfer from wedge-shaped bodies with limited heat conductivity, *Int. J. Heat Mass Transfer* 16: 329–336.

35. Eckert, E. R. G., and Drake, R. M., 1959. *Heat and Mass Transfer*. McGraw-Hill, New York.

Part II

Theory and Methods

3

Heat Transfer from Arbitrary Nonisothermal Surfaces in a Laminar Flow

In essence, the conjugate heat transfer problem considers the thermal inter-action between a body and a fluid flowing over or inside it. As a result of such interaction, a particular temperature distribution establishes on the body–fluid interface. This temperature field determines the heat flux distri-bution on the interface and virtually the intensity of heat transfer. Hence, the properties of heat transfer of any conjugate problem are actually the same as those of some nonisothermal surface with the same temperature field, no matter how this field is established, as a result of conjugate heat transfer or given a priori. Thus, in general, a theory of conjugate heat transfer is in fact a theory of an arbitrary nonisothermal surface, because the temperature dis-tribution on the interface in a conjugate problem is unknown a priori.

Such theory applicable to both arbitrary nonisothermal and conjugate con-vective heat transfer presented in this and the next two chapters was devel-oped by the author as a result of studying this subject since 1970 [1]. The main part of this work was performed together with graduate students and colleagues from the Ukrainian Academy of Science and then by the author during his time as a visiting professor at the University of Michigan since 1996. Different parts of this theory have been published in many articles and in a book [2] that was the author's doctoral thesis. Although many of these publications were originally published in Russian, almost all are available in English.

3.1 The Exact Solution of the Thermal Boundary Layer Equation for an Arbitrary Surface Temperature Distribution [3]

There are only a few exact solutions of the thermal boundary layer equation. Most of them are derived for a specific surface temperature distribution. The first exact solution of the steady-state boundary layer equation was given by Pohlhausen [4] for a plate with constant surface temperature and free stream velocity. The same problem for a plate with a polynomial surface tempera-ture distribution was solved by Chapman and Rubesin [5]. Levy [6] gave the exact solution for the case of a power law distribution of both surface tem-perature and free stream velocity. Solution in the form of multiple series in terms of dynamic and thermal shape parameters was given by Bubnov and Grishmanovskaya [7] and Oka [8] (Section 1.6).

Here is given an exact solution to the thermal boundary layer equation for an arbitrary temperature distribution and a power law of the free stream velocity. The solution is constructed in the form of series containing consecutive derivatives of the temperature head distribution. Such series converge much more rapidly than multiple series of shape parameters.

The thermal boundary layer equation and the boundary conditions are used in the Prandtl-Mises form (Equation 1.3), which in Görtler variables (Equation 1.12) Φ, φ becomes

$$2\Phi\frac{\partial\theta}{\partial\Phi} - \varphi\frac{\partial\theta}{\partial\varphi} - \frac{1}{\Pr}\frac{\partial}{\partial\varphi}\left(\frac{u}{U}\frac{\partial\theta}{\partial\varphi}\right) = \frac{U^2}{c_p}\frac{u}{U}\left[\frac{\partial}{\partial\varphi}\left(\frac{u}{U}\right)\right]^2 \tag{3.1}$$

$$\varphi = 0, \ \theta = \theta_w(\Phi) \qquad \varphi \to \infty, \ \theta = 0 \tag{3.2}$$

$$\Phi = \frac{1}{\nu}\int_0^x U d\zeta \qquad \varphi = \frac{\psi}{\nu\sqrt{2\Phi}} \qquad \theta = T - T_\infty \tag{3.3}$$

The solution of Equation (3.1) is sought in the form

$$\theta = \sum_{k=0}^\infty G_k(\varphi)\Phi^k\frac{d^k\theta_w}{d\Phi^k} + G_d(\varphi)\frac{U^2}{c_p} \tag{3.4}$$

Substitution of Equation (3.4) into Equation (3.1) and replacement of the index $k+1$ by k in one of the sums yields the following two equations:

$$\sum_{k=0}^\infty \Phi^k\frac{d^k\theta_w}{d\Phi^k}\left[2kG_k + 2G_{k-1} - \varphi G_k' - \frac{1}{\Pr}\frac{\partial}{\partial\varphi}\left(\frac{u}{U}G_k'\right)\right] = 0 \tag{3.5}$$

$$2\frac{\Phi}{U^2}\frac{dU^2}{d\Phi}G_d - \varphi G_d' - \frac{1}{\Pr}\frac{\partial}{\partial\varphi}\left(\frac{u}{U}G_d'\right) - \frac{u}{U}\left[\frac{\partial}{\partial\varphi}\left(\frac{u}{U}\right)\right]^2 = 0 \tag{3.6}$$

Here, prime denotes a derivative with respect to variable φ.

If the free stream velocity has power law dependence $U = Cx^m$, the velocity distribution in the layer is self-similar (Section 1.2). In this case, the ratio of velocities in the layer and in the outer flow $u/U = \varphi'(\eta,\beta)$ depends only on variable η and parameter β, while the coefficient in the first term of the second equation equals 2β. The expressions in brackets in Equation (3.5) also turn out to depend only on φ and β.

Equating these expressions to zero leads to a system of ordinary equations:

$$(1/\Pr)[\omega(\varphi,\beta)G_k']' + \varphi G_k' - 2kG_k = 2G_{k-1} \quad (k = 0,1,2,\ldots) \tag{3.7}$$

determining the coefficients of series (3.4), where $\omega(\varphi, \beta)$ denotes the ratio u/U in self-similar flows. The boundary conditions for these equations follow from conditions (3.2):

$$\varphi = 0, \ G_0 = 1, \ G_k = 0 \ \ (k = 1, 2, 3 \ldots) \qquad \varphi \to \infty, \ G_k = 0 \ \ (k = 0, 1, 2 \ldots) \quad (3.8)$$

The equation and the boundary conditions for $G_d(\varphi)$ are found from Equation (3.6):

$$(1/\Pr)[\omega(\varphi, \beta)G_d']' + \varphi G_d' - 2\beta G_d = \omega(\varphi, \beta)[\omega'(\varphi, \beta)]^2$$

$$\varphi = 0, \ G_d = 0 \qquad \varphi \to \infty, \ G_d = 0 \tag{3.9}$$

The heat flux at the wall is defined via shear stress assuming that close to the wall $\tau \approx \tau_w$:

$$q_w = \lambda \left(\frac{\partial \theta}{\partial y}\right)_{y=0} = \frac{\lambda}{v\sqrt{2\Phi}}\left(-u\frac{\partial \theta}{\partial \varphi}\right)_{\varphi=0} \qquad \tau = \mu\left(\frac{\partial u}{\partial y}\right) = \frac{\rho}{\sqrt{2\Phi}}u\left(\frac{\partial u}{\partial \varphi}\right) \tag{3.10}$$

Integrating both sides of the last equation, one finds that close to the wall (except near the separation point where $\tau_w \approx 0$), the velocity $u_{\varphi \to 0}$ is determined via shear stress as

$$u = 2^{3/4}\Phi^{1/4}(\tau_w\varphi/\rho)^{1/2} \tag{3.11}$$

Using this result and determining the derivative $\partial\theta/\partial\varphi$ from Equation (3.4), one gets

$$q_w = h_*\left[\theta_w + \sum_{k=1}^{\infty} g_k \Phi^k \frac{d^k\theta_w}{d\Phi^k} - g_d\frac{U^2}{c_p}\right] \qquad St_* = \frac{g_0}{\Pr}\left(\frac{C_f}{2}\right)^{1/2}\Phi^{-1/4} \tag{3.12}$$

Here, $g_0 = -2^{1/4}(\varphi^{1/2}G_0')_{\varphi=0}$, $g_k = (G_k'/G_0')_{\varphi=0}$, $g_d = (G_d'/G_0')_{\varphi=0}$, h_* is the heat transfer coefficient for an isothermal surface in the case of negligible dissipation. In that case, the sum in series (Equation 3.12) becomes zero, and this equation reduces to the boundary condition of the third kind, and Equation (3.12) for the Stanton number gives for $\Pr = 1$ and $\beta = 0$ (Figure 3.1)

$$St_* = 0.576 \, Re_x^{-1/4}(C_f/2)^{1/2} = C_f/2 \tag{3.13}$$

as it should be according to the Reynolds analogy.

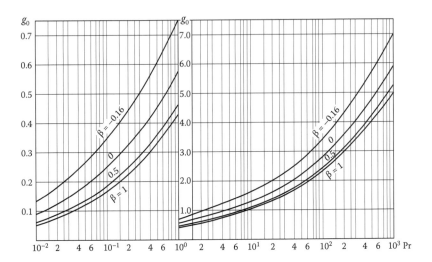

FIGURE 3.1
Dependence of coefficient g_0 on Pr and β for a laminar boundary layer.

Series (3.12) is an exact solution for a power law free stream velocity. For gradientless flow, $\Phi = Ux./\nu$, and this series becomes

$$q_w = h_* \left[\theta_w + \sum_{k=1}^{\infty} g_k x^k \frac{d^k \theta_w}{dx^k} - g_d \frac{U^2}{c_p} \right] \qquad (3.14)$$

It is seen that in this case, the heat flux is governed by the derivatives of temperature head with respect to the coordinate x. It follows from Equation (3.12) that when there is a pressure gradient, the role of the longitudinal coordinate plays the variable Φ, and the heat flux is determined by the derivatives of the temperature head with respect to Φ. Thus, the function $\theta_w(\Phi)$ plays the same role in the case of a pressure gradient flow as the function $\theta_w(x)$ does for gradientless flow. This situation is plausible physically and reflects the fact that the flow characteristics at a given point are governed not only by local quantities but also by the prehistory of the flow, which is taken into account by the variable Φ.

The coefficients G_k are determined by inhomogeneous Equation (3.7). Presenting these coefficients by sum gives for G_k and kG_k the following expressions:

$$G_k = \sum_{i=0}^{i=k} \frac{(-1)^{k+i}}{(k-i)!} F_i,$$

$$kG_k = \sum_{i=0}^{i=k} \frac{(-1)^{k+i}}{(k-i)!} iF_i = \sum_{i=0}^{i=k-1} \frac{(-1)^{k+i}}{(k-i)!} kF_i + kF_k - \sum_{i=0}^{i=k-1} \frac{(-1)^{k+i-1}}{(k-i)!} (k-i)F_i \quad (3.15)$$

Substituting the sums (Equation 3.15) into Equations (3.7) and (3.8) transforms these into homogeneous equations and the following boundary conditions:

$$(1/\operatorname{Pr})[\omega(\varphi,\beta)F_i']' + \varphi F_i' - 2iF_i = 0 \quad (i = 0,1,2...k) \tag{3.16}$$

$$F_0(0) = 1, \quad \sum_{i=0}^{i=k} \frac{(-1)^{k+i}}{(k-i)!} F_i(0) = 0, \quad \sum_{i=0}^{i=k} \frac{(-1)^{k+i}}{(k-i)!} F_i(\infty) = 0 \tag{3.17}$$

From the last sum, one gets $F_i(\infty) = 0$. Because $F_0(0) \neq 0$, the other values of $F_i(0)$ in the second sum cannot be equal to zero. In this case, one should take $F_i(0) = 1/i!$. Then, the second sum becomes zero as a sum of binomial coefficients with alternate signs. Thus, the boundary conditions for the functions $F_i(\varphi)$ are

$$\varphi = 0, F_i = 1/i!, \quad \varphi \to \infty, \quad F_i \to 0 \tag{3.18}$$

Equation (3.16) at the wall ($\varphi = 0$) in the case when $i \neq 0$ has a singularity as well as the Prandtl-Mises equation (Equation 1.3). This singularity can be removed by using a new variable $z = \varphi^{1/2}$, which transforms Equation (3.16) and boundary conditions (Equation 3.18) into the form

$$(1/\operatorname{Pr})[\omega(z,\beta)F_i']'z + [2z^4 - \omega(z,\beta)]F_i' - 8iz^3F_i = 0, \quad F_i(0) = 1/i!, \quad F_i(\infty) \to 0 \tag{3.19}$$

The solution of this linear problem can be presented as a sum of two others:

$$F_i(z) = \frac{1}{i!}\left[V_i(z) - \frac{V_i(\infty)}{W_i(\infty)}W_i(z)\right], \quad V_i(0) = 1, W_i(0) = 0, \quad V_i'(0) = 0, W_i'(0) = 1 \tag{3.20}$$

Using the Runge-Kutta method for numerical integration, one gets $V_i(z)$ and $W_i(z)$ and then $F_i(\varphi)$, coefficients $G_k(\varphi)$ from sum (Equation 3.4), and finally, coefficients of the series (3.12):

$$g_0 = \frac{V_0(\infty)}{W_0(\infty)} \qquad g_k = \sum_{i=0}^{i=k} \frac{(-1)^{k+i}}{i!(k-i)!} \frac{V_i(\infty)/V_0(\infty)}{W_i(\infty)/W_0(\infty)} \tag{3.21}$$

Equation (3.16) simplifies in the limiting cases $\operatorname{Pr} \to 0$ and $\operatorname{Pr} \to \infty$. In the first case, a slug velocity profile may be used (Section 2.3). Then, Equations (3.16) and (3.18) become

$$F_i'' + \operatorname{Pr}(\varphi F_i' - 2iF_i) = 0, \qquad \varphi = 0, F_i = 1/i! \qquad \varphi \to \infty, F_i \to 0 \tag{3.22}$$

The functions giving the solution of this problem and its first derivative at $\varphi = 0$ are [9]

$$F_i = \frac{1}{2^i i!(2i-1)!}\exp(-\xi^2)\frac{d^{2i}}{d\xi^{2i}}\left[\exp(\xi^2)\left(1-\frac{2}{\sqrt{\pi}}\int_0^\xi \exp(-\zeta^2)d\zeta\right)\right] \quad \xi = \varphi\left(\frac{\Pr}{2}\right)^{1/2}$$

(3.23)

$$F_0'(0) = -\frac{2}{\sqrt{\pi}}, \quad F_i'(0) = \frac{2(-1)^{i+1}}{\sqrt{\pi}i!}\left\{1+(2i+1)\left[(-1)^i + \sum_{m=1}^{m=i-1}\frac{(-1)^m i!}{(2i-2m+1)m!(1-m)!}\right]\right\}$$

(3.24)

Calculation shows that the following simple formulae are valid:

$$\frac{\sqrt{\pi}}{2}F_i'(0) = -\frac{1}{1\cdot3\cdot5\cdot\ldots\cdot(2i-1)} = -\frac{2^i}{(2i-1)!!}, \quad g_k = \frac{(-1)^{k+1}}{k!(2k-1)} \quad (3.25)$$

The coefficients decrease rapidly with increasing number:

$$g_1 = 1, g_2 = -1/6, g_3 = -1/30, g_4 = -1/168, g_5 = 1/1080, g_6 = -1/7920 \quad (3.26)$$

In the other limiting case, when $\Pr \to \infty$, the linear velocity profile $c\varphi^{1/2}$ may be used (Section 2.2), and Equation (3.16) and boundary conditions (Equation 3.18) take the form

$$\left(c\varphi^{1/2}F_i'\right)' + \Pr(\varphi F_i' - 2iF_i) = 0, \quad \varphi = 0, F_i = 1/i! \quad \varphi \to \infty, F_i \to 0 \quad (3.27)$$

The new variables $H_i = \exp(z)(z/3)^{-1/3}F_i$ and $z = (2\Pr/3c)\varphi^{1/2}$ transform Equation (3.27) into a confluence hypogeometric equation:

$$zH_i'' + [(4/3-z)]H_i' - [(4i/3)+1]H_i = 0 \quad (3.28)$$

To satisfy the corresponding boundary conditions, one should use the asymptotic expression for the confluence hypogeometric function $H(a,b,z)$. Then, the function $F_i(z)$ and the coefficients of series (3.12) are obtained as follows:

$$F_i = \frac{1}{i!}\exp(z)\left\{H\left[\frac{2}{3}(2i+1),\frac{2}{3},z\right]\right\}$$

$$-\frac{\Gamma(2/3)\Gamma[(4i/3)+1]}{\Gamma(4/3)\Gamma[(4/3)i+2/3]}z^{1/3}H[(4i/3)+1,4/3,z] \quad (3.29)$$

$$g_k = \frac{1}{3}\Gamma\left(\frac{2}{3}\right)\sum_{i=0}^{i=k}\frac{(-1)^{k+i}\Gamma(4i/3)}{(k-i)!(i-1)!\Gamma[(4i/3)+2/3]} \quad (3.30)$$

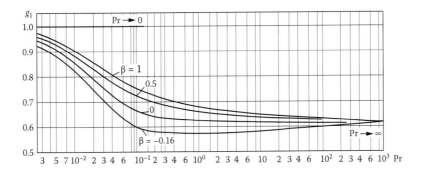

FIGURE 3.2
Dependence of coefficient g_1 on Pr and β for a laminar boundary layer.

The calculation gives

$$g_1 = 0.6123, \quad g_2 = -0.1345, \quad g_3 = 0.0298, \quad g_4 = -0.0057 \tag{3.31}$$

Coefficients g_k calculated by Equation (3.16) for different free stream velocity gradients (different β) and various Prandtl numbers are plotted in Figures 3.1, 3.2, and 3.3. These are compared with limiting values (Equations 3.26 and 3.31). It follows from Figures 3.2 and 3.3 that

1. Coefficient g_1 depends on the velocity gradient and on the Prandtl number; this dependence is more significant for small Prandtl numbers (Pr < 0.5); for Pr $\rightarrow$ 0 and Pr $\rightarrow$ ∞, the values of g_1 for all β tend to the greatest $g_1 = 1$ and to the lowest $g_1 = 0.6123$ values, respectively.

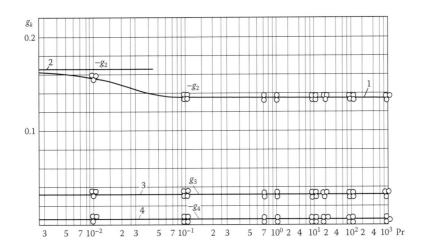

FIGURE 3.3
Dependence of coefficients g_k on Pr and β for a laminar boundary layer. $\circ$ – Numerical solution of Equation (3.16), $1 - Pr \rightarrow \infty$, $2, 3, 4 - Pr \rightarrow 0$.

2. Coefficient g_2 is practically independent of the velocity gradient and depends slightly on the Prandtl number in the region of small Prandtl numbers; for $\text{Pr} \to 0$ and $\text{Pr} \to \infty$, the values of g_2 also tend to the greatest absolute value $|g_2| = 1/6$ and to the lowest absolute value $|g_2| = 0.1345$, respectively.

3. Coefficients g_3 and g_4 are independent of both the velocity gradient and the Prandtl number. Therefore, the values obtained by numerical integration of Equation (3.16) practically coincide with the limiting values $g_3 = 1/30$ and $g_4 = -1/168$.

4. All subsequent coefficients will apparently be more independent of the velocity gradient and of the Prandtl. This follows from Equation (3.7) because the role of the first term, which depends only on Pr and β, reduces as the number k increases.

The independence of the coefficients of both β and Pr for large k makes it possible to use for all Prandtl numbers and for arbitrary pressure gradient the limiting values of g_k (3.26) for $\text{Pr} \to 0$.

3.2 Generalization for an Arbitrary Velocity Gradient in a Free Stream Flow

The results obtained above, in particular, series (Equations 3.12 and 3.14), are exact solutions for a power law velocity in a free stream flow. These expressions are also a highly accurate approximate solution for an arbitrary free stream velocity. The following considerations show this fact. If the coefficients g_k were independent of the pressure gradient (i.e., of parameter β), Equations (3.12) and (3.14) would be exact solutions for an arbitrary pressure gradient. Actually, all coefficients g_k are practically independent of β, except g_1. So, the exactness of series (Equations 3.12 and 3.14) for an arbitrary pressure gradient depends only on variation of coefficients g_1 via parameter β.

According to Figure 3.2, coefficient g_1 slightly depends on β in the range of mean Prandtl numbers, but for limiting cases $\text{Pr} \to 0$ and $\text{Pr} \to \infty$, this dependence deteriorates. The greatest effect of the pressure gradient on g_1 is in the range of Prandtl numbers close to $\text{Pr} = 0.1$, where the deviation from the average value of the first coefficient $g_1 = 0.675$ reaches $\pm 12\%$ when the velocity gradient in the self-similar flow changes from $\beta = 1$ (accelerated flow) to $\beta = -0.16$ (close to separation flow).

Thus, the relations (Equations 3.12 and 3.14)

$$q_w = h_* \left[\theta_w + \sum_{k=1}^{\infty} g_k \Phi^k \frac{d^k \theta_w}{d\Phi^k} \right], \quad q_w = h_* \left[\theta_w + \sum_{k=1}^{\infty} g_k x^k \frac{d^k \theta_w}{dx^k} \right] \qquad (3.32)$$

give the high-accuracy expression for heat flux from nonisothermal surfaces in the case of an arbitrary free stream velocity gradient. Dividing both sides of these equations by $q_{w*} = h_*\theta_w$, one obtains corresponding relations for the coefficient of nonisothermicity:

$$\chi_t = \frac{q_w}{q_{w*}} = 1 + \sum_{k=1}^{\infty} g_k \frac{\Phi^k}{\theta_w} \frac{d^k\theta_w}{d\Phi^k}, \qquad \chi_t = \frac{q_w}{q_{w*}} = 1 + \sum_{k=1}^{\infty} g_k \frac{x^k}{\theta_w} \frac{d^k\theta_w}{dx^k}, \quad (3.33)$$

which shows how much the heat transfer intensity from a particular nonisothermal surface is more or less than that from an isothermal surface.

The value of the parameter β in a general case should be determined satisfying an integral equation that can be obtained by integrating Equation (3.1) across the boundary layer. Omitting the dissipative term, one gets

$$\int_0^\infty \left[2\Phi \frac{\partial\theta}{\partial\Phi} - \varphi \frac{\partial\theta}{\partial\varphi} - \frac{1}{\text{Pr}} \frac{\partial}{\partial\varphi}\left(\frac{u}{U}\frac{\partial\theta}{\partial\varphi}\right)\right] d\varphi = 0 \qquad (3.34)$$

Substituting the self-similar solution $\omega(\varphi,\beta)$ for the velocity ratio u/U, yields

$$\int_0^\infty \left[2\Phi \frac{\partial\theta}{\partial\Phi} - \varphi \frac{\partial\theta}{\partial\varphi} - \frac{1}{\text{Pr}} \frac{\partial}{\partial\varphi}\left(\omega(\varphi,\beta)\frac{\partial\theta}{\partial\varphi}\right)\right] d\varphi = 0 \qquad (3.35)$$

The parameter β should be determined from this equation. However, Equation (3.35) is satisfied by any value of β. This is clearly seen by subtracting Equations (3.34) and (3.35):

$$\int_0^\infty \frac{\partial}{\partial\varphi}\left\{\left[\frac{u}{U} - \omega(\varphi,\beta)\right]\frac{\partial\theta}{\partial\varphi}\right\} d\varphi = \left[\frac{u}{U} - \omega(\varphi,\beta)\right]\frac{\partial\theta}{\partial\varphi}\Big|_0^\infty = 0 \qquad (3.36)$$

This equation is satisfied by any value of β because both exact and self-similar velocity profiles satisfy the same boundary conditions at the end points of interval $(0,\infty)$. This means that solutions (Equation 3.32) satisfy the exact integral equation (Equation 3.34), in other words it means that series (Equation 3.32) in the case of an arbitrary free stream velocity are on the average (across the boundary layer) exact solutions.

Because the integral equation is satisfied, one can increase the exactness of solutions (Equation 3.32) by using some additional condition to estimate the parameter β. The simplest such condition is obtained by equating the average values of the given $U(x)$ and power law Cx^m velocity distributions on each interval $(0,x)$:

$$\int_0^x C\zeta^m d\zeta = \int_0^x U(\zeta)d\zeta, \qquad \beta = 2\left(1 - \frac{\Phi}{\text{Re}_x}\right) \qquad (3.37)$$

As the coefficient g_1 slightly depends on parameter β, the method of determining the latter is not important, and so β may be estimated in some other way, for instance, by approximating $U(x)$ by the power function $U = Cx^m$, as in [10]. In fact, different estimations of β lead to practically the same final values of coefficient g_1.

The heat transfer coefficient for isothermal surface h_* in Equation (3.32) is determined using Equation (3.12) for the Stanton number and Figure 3.1. For laminar flow, the heat transfer coefficient can also be calculated by methods such as those reviewed by Spalding and Pun [11].

3.3 General Form of the Influence Function of the Unheated Zone: Convergence of the Series [12]

The superposition principle leads to Equation (1.38) for the heat flux on a surface with an arbitrary temperature distribution. In the case of a continuous temperature head distribution without breaks for gradientless flow over a plate, this equation has the form

$$q_w = h_* \left[\theta_w(0) + \int_0^x f(\xi/x) \frac{d\theta_w}{d\xi} d\xi \right] \tag{3.38}$$

The influence function of an unheated zone $f(\xi/x)$ is calculated for some simple cases using approximate methods. All of these functions have a form (Section 1.5):

$$f(\xi/x) = \left[1 - (\xi/x)^{C_1} \right]^{-C_2} \tag{3.39}$$

Because series (3.32) is an exact solution for the case of power law free stream velocity, one can estimate exactness of the function (Equation 3.39) by comparing two forms of solution defining heat flux: differential (Equation 3.32) and integral (Equation 3.38). This can be done by determining the relation between these two expressions [12].

It is shown above that in the case of the gradient free stream flow, the variable x in Equation (3.14) should be substituted for by Φ. To compare both differential and integral forms for heat flux in the general case, the same substitution should be done in Equation (3.38) to get

$$q_w = h_* \left[\theta_w(0) + \int_0^\Phi f(\xi/\Phi) \frac{d\theta_w}{d\xi} d\xi \right] \tag{3.40}$$

This equation is transformed using integration of parts assuming for the k-transform:

$$u_k = \frac{d^k \theta_w}{d\xi^k}, \qquad v_k = \int_0^\zeta v_{k-1} d\gamma + \frac{(-1)^k \Phi}{k!}, \qquad \zeta = \frac{\xi}{\Phi} \tag{3.41}$$

Taking $v_0 = \Phi f(\zeta)$, one obtains for $k=1$, $v_1 = \Phi \int_0^\zeta f(\gamma)d\gamma - \Phi$ and then

$$q_w = h_* \left\{ \Phi \frac{d\theta_w}{d\Phi} \left[\int_0^1 f(\zeta)d\zeta - 1 \right] + \Phi \frac{d\theta_w}{d\Phi} \bigg|_{\Phi=0} - \Phi \int_0^\Phi \left[\int_0^\zeta f(\gamma)d\gamma - 1 \right] \frac{d^2\theta_w}{d\xi^2} d\xi + \theta_w(0) \right\} \tag{3.42}$$

Repeating integration by parts yields

$$q_w = h_* \left[\Phi \frac{d\theta_w}{d\Phi} f_1(1) - \Phi^2 \frac{d^2\theta_w}{d\Phi^2} f_2(1) + \ldots (-1)^{k+1}\Phi^k \frac{d^k\theta_w}{d\Phi^k} f_k(1) \ldots + \theta_w(0) \right.$$

$$\left. + \frac{\Phi}{1!} \frac{d\theta_w}{d\Phi} \bigg|_{\Phi=0} + \frac{\Phi^2}{2!} \frac{d^2\theta_w}{d\Phi^2} \bigg|_{\Phi=0} + \ldots + \frac{\Phi^k}{k!} \frac{d^k\theta_w}{d\Phi^k} \bigg|_{\Phi=0} + (-1)^k \Phi^k \int_0^\Phi \frac{d^{k+1}\theta_w}{d\Phi^{k+1}} f_k(\xi/\Phi)d\xi \right] \tag{3.43}$$

$$f_k(\zeta) = \int_0^\zeta d\zeta \int_0^\zeta d\zeta \ldots \int_0^\zeta f(\zeta)d\zeta + \sum_{n=1}^{n=k} \frac{(-1)^n \zeta^{k-n}}{n!(k-n)!} \tag{3.44}$$

In the last expression, the integral is repeated k times. The second sum in Equation (3.43) beginning from $\theta_w(0)$ represents an expansion of function $\theta_w(\Phi)$ as a Taylor series at $\Phi = 0$. Therefore, one concludes that the expression (Equation 3.43) coincides with Equation (3.32) if the remainder in the form of the integral in Equation (3.43) goes to zero when $k \to \infty$ and if it is set

$$\lim_{k\to\infty} \Phi^k \int_0^\Phi f_k\left(\frac{\xi}{\Phi}\right) \frac{d^{k+1}\theta_w}{d\xi^{k+1}} d\xi \to 0 \qquad g_k = (-1)^{k+1} f_k(1) \tag{3.45}$$

Then, because

$$\int_0^1 d\zeta \int_0^\zeta d\zeta \ldots \int_0^\zeta f(\zeta)d\zeta = \frac{1}{(k-1)!} \int_0^1 (1-\zeta)^{k-1} f(\zeta)d\zeta \tag{3.46}$$

$$\sum_{n=1}^{n=k} \frac{(-1)^n \zeta^{k-n}}{n!(n-k)!} \frac{1}{k!} = \frac{1}{k!}[(\zeta-1)^k - \zeta^k]$$

one obtains, according to Equations (3.45) and (3.46), the relation between the coefficients g_k and the influence function of an unheated zone in the form

$$g_k = \frac{(-1)^{k+1}}{k!} \left[k \int_0^1 (1-\zeta)^{k-1} f(\zeta) d\zeta - 1 \right] \qquad (3.47)$$

Thus, expressions (Equations 3.38 and 3.40) determining the heat flux are integral sums of series (Equation 3.32), which present the same heat flux in equivalent differential forms.

Corresponding integral forms for a nonisothermicity coefficient equivalent to (Equation 3.33) are

$$\chi_t = \frac{1}{\theta_w} \left[\int_0^\Phi f\left(\frac{\xi}{\Phi}\right) \frac{d\theta_w}{d\xi} d\xi + \theta_w(0) \right] \qquad \chi_t = \frac{1}{\theta_w} \left[\int_0^x f\left(\frac{\xi}{x}\right) \frac{d\theta_w}{dx} d\xi + \theta_w(0) \right] \qquad (3.48)$$

Results just obtained make it possible to find the relation between exponents C_1 and C_2 in the influence function (Equation 3.39) and coefficients g_k. Substituting Equation (3.39) into Equation (3.47), expanding $(1-\zeta)^{k-1}$ via a binomial formula, and introducing a new variable $r = \zeta^{C_1}$, one reduces the integral to the sum of beta functions, and Equation (3.47) becomes

$$g_k = \frac{(-1)^{k+1}}{k!} \left[\frac{k}{C_1} \sum_{n=0}^{k-1} (-1)^n \frac{(k-1)!}{n!(k-n-1)!} B\left(\frac{n+1}{C_1}, 1-C_2\right) - 1 \right]$$
$$(3.49)$$

$$B(i,j) = \int_0^1 r^{i-1}(1-r)^{j-1} dr$$

According to Equation (1.40), the exponents in Equation (3.39) in the case of the gradientless free stream flow are $C_1 = 3/4$ and $C_2 = 1/3$. The values of coefficients $g_1 = 0.612, g_2 = -0.131, g_3 = 0.030, g_4 = -0.0056$ obtained from Equation (3.49) using these exponents differ only little from the results (Equation 3.31) obtained from the exact solution for $Pr \to \infty$.

Thus, although Equation (3.39) is found by an approximate integral method, it is reasonably accurate, so that the calculation by Equation (3.38) with function (Equation 3.39) and $C_1 = 3/4$ and $C_2 = 1/3$ gives virtually the same result as that obtained by the exact solution (Equation 3.32).

Another result derived from comparing the solutions in two forms, of the series and of the integral, is the influence function of the unheated zone in the general case. A comparison of Equations (3.32) and (3.38) shows that if one neglects the slight dependence $g_1(\beta)$, Equation (3.39) can be used for the influence function in the general case after substituting variable ξ/Φ for ξ/x:

$$f(\xi/\Phi) = [1-(\xi/\Phi)^{C_1}]^{-C_2} \tag{3.50}$$

In such a case, the exponents C_1 and C_2 should be determined so that the values of g_1 and g_2 obtained by applying Equation (3.50) would be the same as those given by Figures 3.2 and 3.3.

First, it should be considered how formula (Equation 3.50) conforms to known cases. In particular, for the case of self-similar gradient flows and for $\mathrm{Pr} > 0.5$, for which the coefficients g_k are practically independent on Pr (Figure 3.2), Equation (3.50) should yield formula (1.44) derived by Lighthill. Because for the self-similar flows $U = Cx^m$ and, consequently, $\Phi = [C/\nu(m+1)]x^{m+1}$, it is evident that in this case Equation (3.50) gives Lighthill's formula. It is also readily seen that the result obtained by Equation (3.50) with exponents $C_1 = 1$ and $C_2 = 1/2$ corresponds to the exact solution in the limiting case $\mathrm{Pr} \to 0$. This follows from the fact that for $C_1 = 1$ and $C_2 = 1/2$, the beta function and the expression (3.49) for g_k can be presented in the following form:

$$B(n+1,1/2) = \frac{\Gamma(n+1)\Gamma(1/2)}{\Gamma[n+1+(1/2)]} = \frac{n!}{[n+(1/2)][n-(1/2)][n-(3/2)]\dots(1/2)} \tag{3.51}$$

$$g_k = (-1)^{k+1}\left[\sum_{n=0}^{k-1} \frac{(-1)^n 2^n}{(2n+1)!!(k-n-1)!} - \frac{1}{k} \right] \tag{3.52}$$

Direct calculations show that g_k given by this formula are in agreement with the values (Equation 3.26) obtained from the exact solution for $\mathrm{Pr} \to 0$.

Thus, the influence function (Equation 3.50) describes quite accurately the effect of an unheated zone in the known cases for self-similar flows with pressure gradient and for practically all relevant values of Prandtl numbers. In these cases, the exponents C_1 and C_2 vary relatively little from 3/4 and 1/3 at medium and high Prandtl numbers to 1 and 1/2 for $\mathrm{Pr} \to 0$. Figure 3.4 shows the exponents C_1 and C_2 for the general case as the functions of Pr for various β. These are found by solving a system of two equations obtained by substituting g_1 and g_2 into Equation (3.49).

To estimate β for determining exponents C_1 and C_2 in the general case, Equation (3.37) can be used in the same way as for estimating coefficients g_1 and g_2.

Considering function (Equation 3.50) and expression (Equation 3.45), the convergence of series (Equation 3.32) is studied. According to the theorem of the mean value, one gets

$$J_k = \Phi^k \int_0^\Phi f_k\left(\frac{\xi}{\Phi}\right) \frac{d^{k+1}\theta_w}{d\xi^{k+1}}\, d\xi \le \Phi^k \int_0^\Phi f_k\left(\frac{\xi}{\Phi}\right)\left|\frac{d^{k+1}\theta_w}{d\xi^{k+1}}\right|\, d\xi = \Phi^{k+1}\left|\frac{d^{k+1}\theta_w}{d\xi^{k+1}}\right|_{\Phi=\bar{\Phi}} \int_0^1 |f_k(\zeta)|\, d\zeta \tag{3.53}$$

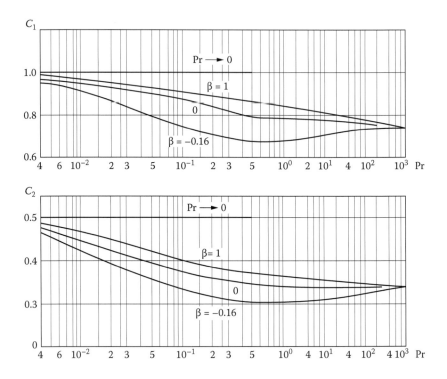

FIGURE 3.4
Dependence of exponents C_1 and C_2 on Pr and β for a laminar boundary layer.

where $\bar{\Phi}$ is a suitable value of Φ from segment $[0, \Phi]$. Applying Equation (3.47) gives an estimation for the integral in Equation (3.53):

$$\int_0^1 |f_k(\zeta)|d\zeta \le \frac{1}{k!}\left[\int_0^1 (1-\zeta)^k |f_k(\zeta)|d\zeta + \int_0^1 |(1-\zeta)^k - \zeta^k|d\zeta\right]$$

$$\le \frac{1}{k!}\left[\int_0^1 (1-\zeta)^k |f_k(\zeta)|\,d\zeta + \int_0^1 |(1-\zeta)^k + \zeta^k|d\zeta\right]$$

$$= \frac{1}{k!}\left[\int_0^1 (1-\zeta)^k |f_k(\zeta)|\,d\zeta + \frac{2}{k+1}\right] \qquad (3.54)$$

The influence function (Equation 3.50) increases as the exponent C_1 decreases and as the exponent C_2 increases. Because the range of the values of these exponents is $3/4 \leq C_1 \leq 1/2$ and $1/3 \leq C_2 \leq 1/2$, the maximum value of the function (Equation 3.50) is when $C_1 = C_2 = 1/2$. Hence, the integrand in Equation (3.54) can be estimated using these values of the exponents:

$$(1-\zeta)^k |f_k(\zeta)| < (1-\zeta)(1-\zeta^{1/2})^{-1/2} < 8\sqrt{3}/9\sqrt{2} < 10/9 \tag{3.55}$$

where $8\sqrt{3}/9\sqrt{2}$ is the maximum value of function (Equation 3.55) in interval [0,1]. Thus,

$$\int_0^1 |f_k(\zeta)| d\zeta \leq \frac{1}{k!}\left(\frac{10}{9k}+\frac{2}{k+1}\right) = \frac{1}{(k+1)!}\left(\frac{10}{9}\frac{k+1}{k}+2\right) < \frac{5}{(k+1)!} \tag{3.56}$$

and finally, estimation of the remainder (3.45) is obtained as follows:

$$J_k = \Phi^k \int_0^\Phi f_k\left(\frac{\xi}{\Phi}\right)\frac{d^{k+1}\theta_w}{d\xi^{k+1}}d\xi < \frac{5}{(k+1)!}\Phi^{k+1}\left|\frac{d^{k+1}\theta_w}{d\Phi^{k+1}}\right|_{\Phi=\bar{\Phi}} \tag{3.57}$$

This expression differs from the estimation of the Taylor series remainder only by factor 5. Therefore, the known results for Taylor series convergence are valid for series (3.32) as well. In particular, if the function $\theta_w(\Phi)$ has derivatives of all orders in an interval $[0, \Phi]$, the series (3.32) converges to integral (3.40).

3.4 The Exact Solution of the Thermal Boundary Layer Equation for an Arbitrary Surface Heat Flux Distribution [13]

This section considers the inverse problem when surface heat flux distribution is specified and the corresponding temperature head distribution needs to be established. Accordingly, Equation (3.32) is solved for the heat flux to obtain

$$\theta_w + \sum_{k=1}^\infty g_k \Phi^k \frac{d^k\theta_w}{d\Phi^k} = \frac{q_w}{h_*} = \theta_{w*}(\Phi), \tag{3.58}$$

where $\theta_{w*}(\Phi)$ determines the temperature head on an isothermal surface with the same heat flux distribution and is a known function of x and, hence, of Φ, because $q_w(x)$ is given. Equation (3.58) can be considered as a differential equation defining the unknown function $\theta_w(\Phi)$. The solution of this equation is sought in a form similar to that of series (Equation 3.32):

$$\theta_w = \theta_{w*} + \sum_{n=1}^{\infty} h_n \Phi^n \frac{d^n \theta_{w*}}{d\Phi^n} \qquad (3.59)$$

where h_k are coefficients similar to g_k. Substituting Equation (3.59) into Equation (3.58), one gets

$$\sum_{n=1}^{\infty} h_n \Phi^n \frac{d^n \theta_{w*}}{d\Phi^n} + \sum_{k=1}^{\infty} g_k \Phi^k \frac{d^k \theta_{w*}}{d\Phi^k} + \sum_{k=1}^{\infty} g_k \Phi^k \frac{d^k}{d\Phi^k} \sum_{n=1}^{\infty} h_n \Phi^n \frac{d^n \theta_{w*}}{d\Phi^n} = 0 \qquad (3.60)$$

Performing the differentiation in the last term and assembling terms containing like expressions of $\Phi^k (d^k \theta_{w*}/d\Phi^k)$ and $\Phi^n (d^n \theta_{w*}/d\Phi^n)$ yields the relation

$$h_k + g_1(kh_k + h_{k-1}) + g_2[k(k-1)h_k + 2(k-1)h_{k-1} + h_{k-2}]$$

$$+ g_3[k(k-1)(k-2)h_k + 3(k-1)(k-2)h_{k-1} + 3(k-2)h_{k-2} + h_{k-3}]$$

$$+ g_4[k(k-1)(k-2)(k-3)h_k + 4(k-1)(k-2)(k-3)h_{k-1}$$

$$+ 6(k-2)(k-3)h_{k-2} + 4(k-3)h_{k-3} + h_{k-4}] + \ldots = 0 \quad k = 1, 2, 3 \ldots, h_0 = 1 \quad (3.61)$$

Then, the expressions determining coefficients h_k via known g_k are obtained:

$$h_1 + g_1(h_1 + 1) = 0, h_2 + g_1(2h_2 + h_1) + g_2(2h_2 + 2h_1 + 1) = 0,$$

$$h_3 + g_1(3h_3 + h_2) + g_2(6h_3 + 4h_2 + h_1) + g_3(6h_3 + 6h_2 + 3h_1 + 1) = 0,$$

$$h_4 + g_1(4h_4 + h_3) + g_2(12h_4 + 6h_3 + h_2) + g_3(24h_4 + 18h_3 + 6h_2 + h_1)$$

$$+ g_4(24h_4 + 24h_3 + 12h_2 + 4h_1 + 1) = 0 + \ldots \qquad (3.62)$$

Figure 3.5 presents the values of the first four coefficients h_k as functions of the Prandtl number and β. For the limiting cases $\mathrm{Pr} \to 0$ and $\mathrm{Pr} \to \infty$, the coefficients h_k are

$$h_1 = -1/2, \; h_2 = 3/16, \; h_3 = 5/96, \; h_4 = 35/1968$$

$$h_1 = -0.38, \; h_2 = 0.135, \; h_3 = -0.037, \; h_4 = 0.00795 \qquad (3.63)$$

Like the coefficients g_k, the first few coefficients h_k are weak functions of β, and the rest are practically independent of β and Pr.

The weak dependence of the coefficients h_k on β means that Equation (3.59), like Equation (3.32), can be used with high accuracy for arbitrary free stream flows. In this case, β can be found as before using Equation (3.37).

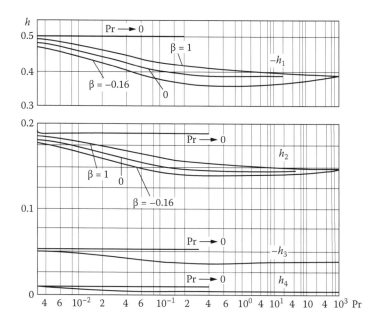

FIGURE 3.5
Dependence of coefficients h_k on Pr and β for a laminar boundary layer.

It was shown in the preceding section that differential form (3.32) for the heat flux corresponds to the integral presentation (Equation 3.40). To obtain a solution of the inverse problem in an integral form corresponding to differential form (Equation 3.59), the temperature head from Equation (3.40) should be found, taking $q_w(\Phi)$ as a given function. For the case of gradientless free stream flow, such a problem was considered in Chapter 1, and the solution for the temperature head was obtained in the form (Equation 1.50). Because the expressions for gradient and gradientless flows differ only by variables Φ and x, the solution of the inverse problem for gradient flows is found by substituting variable Φ for x in Equation (1.50) to get

$$\theta_w = \frac{C_1}{\Gamma(C_2)\Gamma(1-C_2)} \int_0^\Phi \left[1 - \left(\frac{\xi}{\Phi}\right)^{C_1}\right]^{C_2-1} \left(\frac{\xi}{\Phi}\right)^{C_1(1-C_2)} \frac{q_w(\xi)}{h_*(\xi)\xi} d\xi \qquad (3.64)$$

In the simplest case of gradientless flow and $\mathrm{Pr} \approx 1$ for which $\Phi = x, C_1 = 3/4$, $C_2 = 1/3$, and $h_* = 0.332\lambda(U/xv)^{1/2}\,\mathrm{Pr}^{1/3}$, Equation (3.64) reduces to the following well-known relation [14]:

$$\theta_w = 0.623\,\mathrm{Re}_x^{-1/2}\,\mathrm{Pr}^{-1/3} \int_0^x \left[1 - \left(\frac{\xi}{x}\right)^{3/4}\right]^{-2/3} d\xi \qquad (3.65)$$

For the case of a constant heat flux, Equations (3.59) and (3.64) can be used to obtain a relation between heat transfer coefficients for an isothermal ($T_W =$ const.) and for the surface with $q_w =$ const. Dividing each term of Equations (3.59) and (3.64) by q_w, one gets

$$\frac{1}{h_q} = \frac{1}{h_*} + \sum_{k=1}^{\infty} h_k \Phi^k \frac{d^k(1/h_*)}{d\Phi^k} = \frac{C_1}{\Gamma(C_2)\Gamma(1-C_2)} \int_0^{\Phi} \left[1 - \left(\frac{\xi}{\Phi}\right)^{C_1} \right]^{C_2-1}$$

$$\times \left(\frac{\xi}{\Phi}\right)^{C_1(1-C_2)} \frac{d\xi}{h_*(\xi)\xi} \qquad\qquad (3.66)$$

If the customary power law relation is used for the isothermal heat transfer coefficient, $h_* = C\Phi^{-n}$, the ratio h_*/h_q is independent of x and is determined by one of the relations

$$\frac{h_*}{h_q} = 1 + \sum_{k=1}^{\infty} n(n-1)(n-2)\ldots(n-k+1),$$

$$\frac{h_*}{h_q} = \Gamma\left(1 - C_2 + \frac{n}{C_1}\right) / \Gamma(1-C_2)\Gamma\left(1 + \frac{n}{C_1}\right) \qquad\qquad (3.67)$$

These expressions give the ratio h_*/h_q for arbitrary gradient flow and arbitrary Prandtl number. For gradientless flow, the well-known results follow from Equation (3.67). For $\text{Pr} \approx 1$ and $\text{Pr} \to 0$, one gets $h_*/h_q = 0.74$ and $h_*/h_q = 2/\pi$, respectively. For the stagnation point, the isothermal heat transfer coefficient is independent of x, so $n = 0$, and $h_*/h_q = 1$.

To derive the influence function of an unheated zone for the case of the heat flux jump, Equation (3.64) is integrated by parts setting

$$u(\xi) = q_w, \quad v(\xi, \Phi) = \frac{C_1}{\Gamma(C_2)\Gamma(1-C_2)} \int_{\xi}^{\Phi} \left[1 - \left(\frac{\zeta}{\Phi}\right)^{C_1} \right]^{C_2-1} \left(\frac{\zeta}{\Phi}\right)^{C_1(1-C_2)} \frac{d\zeta}{h_*(\zeta)\zeta} \quad (3.68)$$

Because according to Equations (3.66) and (3.68) $v(\Phi, \Phi) = 0$ and $v(0, \Phi) = 1/h_q$, one has

$$\theta_w = \frac{q_w(0)}{h_q} + \int_0^{\Phi} v(\xi, \Phi) \frac{dq_w}{d\xi} d\xi \qquad\qquad (3.69)$$

Because ratio $q_w(0)/h_q$ determines the temperature head $(\theta_w)_q$ on the surface with $q_w =$ const, Equation (3.69) has the same form as Equation (3.40).

This implies that quantity $v(\xi, \Phi)$ in front of the derivative in the integrand, like $f(\xi/\Phi)$ in Equation (3.40), gives the influence function of the unheated zone. The integrand in Equation (3.69) is the increment of the temperature head, and the value $(dq_w/d\xi)d\xi$ is the corresponding increment of the heat flux. Consequently, the function $v(\xi, \Phi)$ is equal to the reciprocal of the heat transfer coefficient $(h_q)_\xi$ after the heat flux jump at the point $\Phi = \xi$, and the influence function $f_q(\xi/\Phi)$ in this case is determined as follows:

$$\frac{1}{f_q(\xi, \Phi)} = \frac{h_*}{(h_q)_\xi} = \frac{C_1 h_*}{\Gamma(C_2)\Gamma(1-C_2)} \int_{\xi/\Phi}^{1} (1-\zeta^{C_1})^{C_2-1}\zeta^{C_1(1-C_2)-1} \frac{d\zeta}{h_*(\Phi\zeta)} \quad (3.70)$$

Because in the problem at hand the heat flux is given and $(h_q)_\xi = q_w/\theta_w$, Equation (3.70) in fact determines the distribution of the temperature head after the heat flux jump for the arbitrary gradient flows. To obtain the well-known expression for the gradientless flow from Equation (3.70), consider that in this case, $\Phi = \mathrm{Re}_x, h_* \sim \mathrm{Re}^{-1/2}$, and hence, $h_*(\Phi\zeta) = h_*(x)\zeta^{-1/2}$. Then, using variable $\sigma = 1 - \zeta^{C_1}$ and beta function (3.49), Equation (3.70) is transformed to the following form:

$$f_q\left(\frac{\xi}{x}\right) = \frac{(h_q)_\xi}{h_*} = \frac{B(C_2, 1-C_2)}{B_\sigma\{C_2, [C_1(1-C_2)+1/2]/C_1\}}, \quad B_\sigma(i, j) = \int_0^\sigma r^{i-1}(1-r)^{j-1} dr$$

$$(3.71)$$

where $B_\sigma(i, j)$ is an incomplete beta function. The well-known formula

$$\theta_w = 0.276(q_w/h_*)B_\sigma(1/3, 4/3) \quad (3.72)$$

determining the temperature head after the heat flux jump for gradientless flow follows from Equation (3.71) for $\mathrm{Pr} \approx 1$ and $C_1 = 3/4$ and $C_2 = 1/3$.

It is evident from Equation (3.70) that influence function $f_q(\xi, \Phi)$ for the case of known heat flux, unlike the influence function $f(\xi, \Phi)$ for the case of known temperature, depends not only on the ratio ξ/Φ but also on the function $h_*(\Phi)$. Therefore, the function $f_q(\xi, \Phi)$ in contrast to the function $f(\xi/\Phi)$ depends on each of the variables ξ and Φ.

3.5 Temperature Distribution on an Adiabatic Surface in an Impingent Flow

This problem is considered as an example when the heat flux naturally is prescribed and temperature head distribution needs to be found.

Let a thermally insulated section of a surface be preceded by an isothermal section. At the entrance to the adiabatic section, the heat flux drops abruptly to zero, whereas the temperature head decreases gradually, becoming practically equal to zero only at a certain distance from entrance point. To determine the variation of the temperature head in this case, Equation (3.64) is again integrated by parts putting

$$u(\xi) = \frac{q_w}{h_*} = \theta_{w*}, \quad v\left(\frac{\xi}{\Phi}\right) = \frac{C_1}{\Gamma(C_2)\Gamma(1-C_2)} \int_\xi^\Phi \left[1-(\zeta)^{C_1}\right]^{C_2-1} (\zeta)^{C_1(1-C_2)} \frac{d\zeta}{\zeta} \qquad (3.73)$$

Because in this case $v(1) = 0$, $v(0) = 1$, and $q_w(0)/h_*(0) = \theta_w(0)$, Equation (3.64) becomes

$$\theta_w = \theta_{w*}(0) + \int_0^\Phi v\left(\frac{\xi}{\Phi}\right) \frac{d\theta_{w*}}{d\xi} d\xi \qquad (3.74)$$

where $\theta_{w*}(0)$ is the temperature head on the isothermal section at the entrance to the adiabatic section. It is seen that the function $v(\xi/\Phi)$ of the integrand in Equation (3.74) is sought. Reasoning as in the derivation of Equation (3.70), one arrives at the conclusion that the function $v(\xi/\Phi)$ describes the variation of the temperature head after the heat flux jump, referring to the corresponding variation of the temperature head that would occur on an isothermal surface for the same heat flux. In the studied case of flow impingent on the adiabatic section, the temperature head changes from the temperature head $\theta_{w*}(0)$ on the preceding isothermal surface at the entrance to the adiabatic section to some value θ_w after the heat flux jump (i.e., on $\theta_{w*}(0) - \theta_w$). Hence, in the problem in question, the function in Equation (3.74) is $v(\xi/\Phi) = [\theta_{w*} - \theta_w]/\theta_w$. Solving this equation for θ_w and using Equations (3.73) and (3.49), one finds the temperature head variation on the adiabatic section in an impingement flow in the general case:

$$\theta_w = \theta_{w*}(0)\left[1 - \frac{B_\sigma(C_2, 1-C_2)}{B(C_2, 1-C_2)}\right], \qquad \sigma = 1 - \left(\frac{\xi}{\Phi}\right)^{C_1} \qquad (3.75)$$

For gradientless flow and $\mathrm{Pr} \approx 1$, the familiar relation follows from Equation (3.75):

$$\theta_w = \theta_{w*}(0)\left[1 - \frac{B_\sigma(1/3, 2/3)}{B(1/3, 2/3)}\right], \qquad \sigma = 1 - \left(\frac{\xi}{x}\right)^{3/4} \qquad (3.76)$$

3.6 The Exact Solution of an Unsteady Thermal Boundary Layer Equation for Arbitrary Surface Temperature Distribution

Consider an incompressible laminar steady-state flow with the free stream velocity $U(x)$ and temperature T_∞ past a body with surface temperature $T_w(x)$. At the moment $t = 0$, the surface temperature starts to change with time according to a function $T_w(t, x)$. The problem is to determine the temperature field and the surface heat flux distribution for $t \geq 0$ [15].

The unsteady boundary layer Equation (2.3) in dimensionless variables (Equation 3.3) becomes

$$\left[\frac{2\Phi v}{Ux}\left(\frac{U}{u} - z\right) + \frac{2\Phi v}{U^2}\frac{dU}{dx}z\right]\frac{\partial\theta}{\partial z} + 2\Phi\frac{\partial\theta}{\partial\Phi} - \varphi\frac{\partial\theta}{\partial\varphi} - \frac{1}{Pr}\frac{\partial}{\partial\varphi}\left(\frac{u}{U}\frac{\partial\theta}{\partial\varphi}\right) = 0 \quad z = \frac{Ut}{x} \tag{3.77}$$

The initial and boundary conditions are as follows:

$$z \geq 0, \; \varphi = 0, \; \theta = \theta_w(x, z) \quad \varphi \to \infty, \; \theta \to 0 \tag{3.78}$$

In the case of power law free stream velocity $U = Cx^m$, the terms $2\Phi v/Ux$ and $(2\Phi v/U^2)(dU/dx)$ depend only on exponent m, and the term $\Phi(\partial\theta/\partial\Phi)$ is equal to $x(\partial\theta/\partial x)/(m+1)$. Then, Equation (3.77) takes the following form:

$$\frac{2}{m+1}\left[\frac{U}{u} + (m-1)z\right]\frac{\partial\theta}{\partial z} + \frac{2}{m+1}x\frac{\partial\theta}{\partial x} - \varphi\frac{\partial\theta}{\partial\varphi} - \frac{1}{Pr}\frac{\partial}{\partial\varphi}\left(\frac{u}{U}\frac{\partial\theta}{\partial\varphi}\right) = 0 \tag{3.79}$$

The solution of this equation subjected to initial and boundary conditions (3.78) can be presented in a series similar to series (Equation 3.4):

$$\theta = \sum_{k=0}^{\infty}\sum_{i=0}^{\infty} G_{ki}(z, \varphi)\frac{x^{k+i}}{U^i}\frac{\partial^{k+i}\theta_w}{\partial x^k \partial t^i} = G_{00}\theta_w + G_{10}x\frac{\partial\theta_w}{\partial x} + G_{01}\frac{x}{U}\frac{\partial\theta_w}{\partial t}$$

$$+ G_{20}x^2\frac{\partial^2\theta_w}{\partial x^2} + G_{02}\left(\frac{x}{U}\right)^2\frac{\partial^2\theta}{\partial x^2} + G_{11}\frac{x^2}{U}\frac{\partial^2\theta_w}{\partial x\partial t} + \cdots \tag{3.80}$$

Substituting this series into Equations (3.78) and (3.79), one obtains a set of equations and initial and boundary conditions that determine the coefficients of the series (3.80):

$$\frac{2}{m+1}\left[\frac{U}{u}+(m-1)z\right]\left[\frac{\partial G_{ki}}{\partial z}+G_{k(i-1)}\right]+\frac{2}{m+1}\left[(k+i)G_{ki}+G_{(k-1)i}\right]-\varphi\frac{\partial G_{ki}}{\partial \varphi}$$

$$-\frac{1}{\text{Pr}}\frac{\partial}{\partial \varphi}\left(\frac{u}{U}\frac{\partial G_{ki}}{\partial \varphi}\right)=0 \qquad (3.81)$$

$$z\geq 0, \varphi=0, G_{00}=1, G_{ki}=0, (i=0, k>0, \ k=0, i>0, \ i>0., k>0), \ \varphi\rightarrow\infty, G_{ki}\rightarrow 0$$
$$(3.82)$$

For a power law free stream velocity, the ratio u/U depends only on variable φ. Hence, the coefficients G_{ki} are a function only of variables z, φ and parameters m and Pr.

The surface heat flux is determined by differentiating Equation (3.80) to obtain

$$q_w=-\lambda\left(\frac{\partial\theta}{\partial y}\right)_{y=0}=\sum_{k=0}^{\infty}\sum_{i=0}^{\infty}\left(\frac{\partial G_{ki}}{\partial\varphi}\right)_{\varphi=0}\frac{x^{k+i}}{U^i}\frac{\partial^{k+i}\theta_w}{\partial x^k\partial t^i}=h_*\left(\theta_w+g_{10}x\frac{\partial\theta_w}{\partial x}\right.$$

$$\left.+g_{01}\frac{x}{U}\frac{\partial\theta_w}{\partial t}+g_{20}x^2\frac{\partial^2\theta_w}{\partial x^2}+g_{02}\left(\frac{x}{U}\right)^2\frac{\partial^2\theta_w}{\partial t^2}+g_{11}\frac{x^2}{U}\frac{\partial^2\theta_w}{\partial x\partial t}+...\right) \qquad (3.83)$$

$$g_{ki}=[(\partial G_{ki}/\partial\varphi)/(\partial G_{00}/\partial\varphi)]_{\varphi=0} \ (i=0, k>0, \ k=0, i>0 \ i>0, k>0) \qquad (3.84)$$

Equation (3.81) subjected to conditions (Equation 3.82) are solved numerically for a plate and $\text{Pr}=1$ using the finite difference method. The coefficients of the first four terms containing the derivatives with respect to time ($i\neq 0$) are given in Figure 3.6. If the surface temperature head depends on the coordinate only, one puts $i=0$, and Equations (3.80) and (3.83) become the proper form of the steady-state solutions (3.4) and (3.32) with coefficients g_k (Equation 3.31). The coefficients $g_{ki}(z)$ gradually grow with time and finally attain the values of $(g_{ki})_{t\rightarrow\infty}$ that coincide with those obtained by Sparrow without initial conditions [16]. To obtain a satisfactory result, one can use only several terms of the series (Equation 3.83), because coefficients g_{ki} decrease rapidly with growing the value (ki). The ratio $g_{ki}(z)/(g_{ki})_{t\rightarrow\infty}$ is about 0.99 when $z=Ut/x=2.4$. Hence, for $z>2.4$, coefficients g_{ki} are practically independent of time and become the values $(g_{ki})_{t\rightarrow\infty}$.

Applying the same technique of repeated integration by part as in the case of steady-state heat transfer (Section 3.3), one can show that differential form

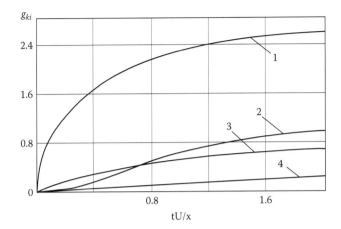

FIGURE 3.6
Dependence of coefficients g_{ki} on dimensionless time $z = tU/x$ for an unsteady gradientless laminar boundary layer. $\mathrm{Pr} = 1. 1 - g_{01}, 2 - g_{02}, 3 - g_{11}, 4 - g_{21}$.

(Equation 3.83) for heat flux is identical to the following integral expression:

$$q_w = h_* \left[\theta_w(t,0) + \int_0^x f(\xi/x,0,z)\frac{\partial \theta_w}{\partial \xi}\,d\xi + \int_0^t f(0,\eta/t,z)\frac{\partial \theta_w}{\partial \eta}\,d\eta \right.$$

$$\left. + \int_0^t d\eta \int_0^x f(\xi/x,\eta/t,z)\frac{\partial^2 \theta_w}{\partial \xi \partial \eta}\,d\xi \right] \tag{3.85}$$

Here $f(\xi/x,\eta/t,z)$ is an influence function of an unheated zone for the unsteady heat transfer that depends on the integration variables ξ/x and η/t and dimensionless time z. The relation between the function of the unheated zone and the coefficients of series is

$$g_{ki}(z) = \frac{(-1)^{k+i+1}}{(k-1)!(i-1)!}\int_0^z (z-\sigma)^{i-1}\int_0^1 (1-\zeta)^{k-1} f(\zeta,\sigma/z,z)\,d\sigma\,d\zeta \qquad \zeta = \frac{\xi}{x},\ \sigma = \frac{\eta}{t}$$

$$\tag{3.86}$$

The expression for the nonisothermicity coefficient follows from Equation (3.83):

$$\chi_t = 1 + g_{10}\frac{x}{\theta_w}\frac{\partial \theta_w}{\partial x} + g_{01}\frac{x}{U\theta_w}\frac{\partial \theta_w}{\partial t} + g_{20}\frac{x^2}{\theta_w}\frac{\partial^2 \theta_w}{\partial x^2}$$

$$+ g_{02}\frac{x^2}{U^2\theta_w}\frac{\partial^2 \theta_w}{\partial t^2} + g_{11}\frac{x^2}{U\theta_w}\frac{\partial^2 \theta_w}{\partial x \partial t} + \cdots \tag{3.87}$$

3.7 The Exact Solution of a Thermal Boundary Layer Equation for a Surface with Arbitrary Temperature in a Compressible Flow

In this section, the Dorodnizin (or Illingworth-Stewartson) independent variables [17] (or [18]) are used to show that solution (Equation 3.4) for incompressible fluid is valid in the case of gradientless compressible flow past the plate [2]. Substituting a Dorodnizin's variable η for the variable y and transforming energy equation (Equation 2.60) for compressible fluid to the variables x and φ,

$$\eta = \sqrt{\frac{U}{2v_\infty Cx}} \int_0^y \frac{\rho}{\rho_\infty} d\xi, \quad \varphi = \frac{\psi}{\rho_\infty \sqrt{2Cv_\infty xU}}, \tag{3.88}$$

one arrives at the following thermal boundary equation of Prandtl-Mises-Görtler's type:

$$2x\frac{\partial i}{\partial x} - \varphi\frac{\partial i}{\partial \varphi} - \frac{1}{\Pr}\frac{\partial}{\partial \varphi}\left(\frac{u}{U}\frac{\partial i}{\partial \varphi}\right) - (k-1)M_\infty^2 \frac{u}{U}\left[\frac{\partial}{\partial \varphi}\left(\frac{u}{U}\right)\right]^2 = 0 \tag{3.89}$$

Here, C is a coefficient in Chapman-Rubesin's law (Equation 2.63) for viscosity, $i = (J - J_\infty)/J_\infty$ is the dimensionless difference of a gas enthalpy, and k is the specific heat ratio. This equation is in agreement with Equation (3.1) if one writes the latter for the gradientless flow and takes into account that in the case of gradientless incompressible flow $\Phi = \mathrm{Re}_x$, and terms $(k-1)M_\infty^2$ and enthalpy difference i become U^2/c_p and θ, respectively. Consequently, the corresponding changes in expression (3.4) transform it into a solution of Equation (3.89):

$$i = \sum_{k=0}^\infty G_k(\varphi)x\frac{\partial^k i_w}{\partial x^k} + G_d(\varphi)(k-1)M_\infty^2 \tag{3.90}$$

The heat flux and shear stress are determined in the case of compressible fluid as

$$q_w = \frac{\rho_w \lambda_w T_\infty}{\rho_\infty \sqrt{2Cv_\infty xU}}\left(u\frac{\partial i}{\partial \varphi}\right)_{\varphi=0}, \quad \tau_w = \frac{\rho_w \mu_w T_\infty}{\rho_\infty \sqrt{2Cv_\infty xU}}\left(u\frac{\partial u}{\partial \varphi}\right)_{\varphi=0} \tag{3.91}$$

Integrating both sides of the second equation, substituting the result obtained for u in the first equation, and using the relation for shear stress [10]

$\tau_w = 0.332\rho_\infty C_x \sqrt{U^3 v_\infty / Cx}$ yields the expression for heat flux in two forms:

$$q_w = 0.576 g_0 \lambda T_\infty C_x \sqrt{\frac{U}{Cxv_\infty}} \left[i_w + \sum_{k=1}^{\infty} g_k x^k \frac{d^k i_w}{dx^k} - g_d(k-1)M_\infty^2 \right]$$

$$C_x = \left(\frac{T_w}{T_\infty}\right)^{1/2} \frac{T_\infty + S}{T_w + S} \tag{3.92}$$

$$q_w = \frac{q_{w\bullet} C_x}{C}\left(i_{ad} + \sum_{k=1}^{\infty} g_k x^k \frac{d^k i_{ad}}{dx^k} \right) \qquad i_{ad} = \frac{J_w - J_{ad}}{J_\infty} = i_w - \frac{r}{2}(k-1)M_\infty^2$$

The first part of Equation (3.92) is written using the dimensionless gas enthalpy i_w, but in the second form, the dimensionless stagnation gas enthalpy i_{0w} is employed, where $J_{ad.}$ is an adiabatic wall enthalpy and r is a recovery factor (Section 3.9.2). Coefficient $C_x = \lambda_w \rho_w / \lambda_\infty \rho_\infty = \mu_w \rho_w / \mu_\infty \rho_\infty$ in contrast to coefficient C is determined using the local wall temperature (see Equation 2.63), and q_{w*} is the heat flux on an isothermal surface with the average temperature head of the studied nonisothermal surface. The coefficients g_k and g_d are the same as in the case of incompressible fluid given in Section 3.1.

Equations (3.90) and (3.92) are the exact solutions of the thermal boundary layer equation for an arbitrary plate temperature distribution. Chapman-Rubesin's solution [5] for a polynomial plate temperature distribution follows from Equations (3.90) and (3.92).

3.8 The Exact Solution of a Thermal Boundary Layer Equation for a Moving Continuous Surface with Arbitrary Temperature Distribution

A number of industrial processes, like a forming of synthetic films and fibers, the rolling of metals, glass production, and so forth, are based on the systems, in which a continuous material goes out of a slot and moves through a surrounding coolant with a constant velocity U_w. As a result of a coolant viscosity, a boundary layer is formed on such a surface (Figure 3.7).

Although this boundary layer is similar to that on a stationary or moving plate, it differs. In this case, the boundary layer grows in the direction of the motion, as opposed to flow over the plate, on which it grows in the opposite direction of that in which it is moving. It can be shown that in a coordinate system attached to the moving surface, the boundary layer equations differ from the equations for the case of flow over a plate, but the boundary

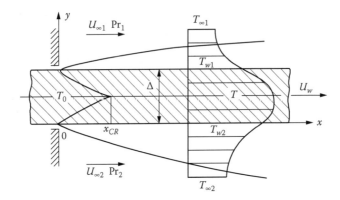

FIGURE 3.7
Schematic of a boundary layer on a moving plate for symmetrical and asymmetrical (Example 8.2) flows.

conditions are identical. These equations of a moving surface in the moving frame are unsteady, but if the coordinate system is fixed and attached to the slot, the problem becomes steady, and both boundary layer equations coincide; however, the boundary conditions differ because the flow velocity on the moving surface is not zero.

Exact solutions of the dynamic and thermal boundary layer problems analogous to Blasius and Pohlhausen solutions for a streamlined semi-infinite plate are given in References [19] and [20]. The friction coefficient on the moving surface is greater by 34%, and the heat transfer coefficient for isothermal surface and $Pr = 0.7$ is greater by 20% than for a plate.

The exact solution for a nonisothermal surface is obtained in Reference [21] for stationary and moving coolants with different ratios $\varepsilon = U_\infty/U_w$ of

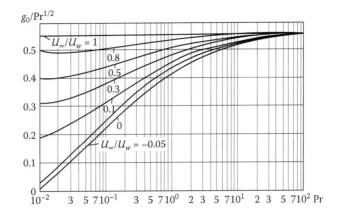

FIGURE 3.8
Dependency of $g_0 \, Pr^{1/2}$ on the Prandtl number and ratio of velocities $\varepsilon = U_\infty/U_w$ for a plate moving through surrounding medium.

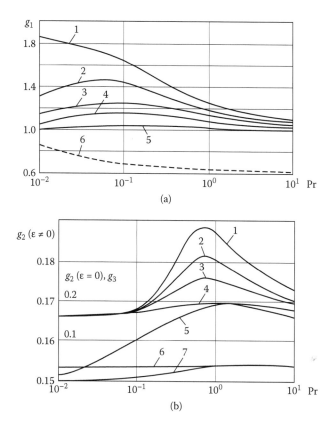

FIGURE 3.9
Dependence of coefficients $g_1(a)$ and $g_2(b)$ on the Prandtl number and ratio of velocities $\varepsilon = U_\infty/U_w$ for a plate moving through surrounding medium. $1-\varepsilon = 0$, $2-0.1$, $3-0.3$, $4-0.5$, $5-0.8$, $6-$ streamlined plate, $7(b)-g_3$.

the velocities of a surface U_w and fluid U_∞. The expressions for the heat flux and nonisothermicity coefficient are the same Equations (3.32) and (3.40) and Equations (3.33) and (3.48), respectively, but only in variable x, because the coolant flow is gradientless. The heat transfer coefficient for an isothermal surface is

$$\text{Nu}_* = g_0\, \text{Re}^{1/2}\, x^{-1/2} \quad \text{or} \quad h_* = g_0\sqrt{U_w/vx} \tag{3.93}$$

where Nu, Re, and dimensionless x in the first formula may be defined by any length. Coefficients g_0 and g_k are given in Figures 3.8 and 3.9. The corresponding exponents C_1 and C_2 are plotted in Figure 3.10.

As in the case of a streamlined plate, the coefficients g_k for $k \geq 3$ are practically independent of the Prandtl number and the parameter ε and

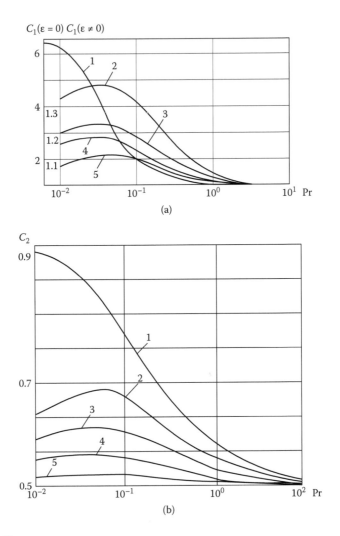

FIGURE 3.10
Dependence of exponent C_1 (a) and C_2 (b) on Prandtl number and ratio of velocities $\varepsilon = U_\infty/U_w$ for a plate moving through surrounding medium. $1 - \varepsilon = 0$, $2 - 0.1$, $3 - 0.3$, $4 - 0.5$, $5 - 0.8$.

can be calculated by formula (Equation 3.25). The coefficient g_1 for flow over a stationary plate in the intervals of large and medium values of Pr is considerably smaller than in the case of a continuous moving surface. For Pr = 0.7, for example, the coefficient g_1 is 1.7 times for $\varepsilon = 0.8$ and twice for $\varepsilon = 0$ as great as in the case of a flow over a fixed surface. This means that the influence of nonisothermicity is substantially greater for a continuous moving surface than for the case of a flow over a stationary plate.

3.9 The Other Solution of a Thermal Boundary Layer Equation for an Arbitrary Surface Temperature Distribution

Some other solutions of the thermal boundary layer equation are obtained for arbitrary nonisothermal surfaces. Because the approach used in all cases presented here is the same as in the above-discussed problems, these solutions are described briefly, indicating only the distinctions of each problem.

3.9.1 Non-Newtonian Fluid with a Power Law Rheology [2]

The power rheology means that fluid obeys basic power laws [22]:

$$\hat{\tau} = K_\tau \left(\frac{1}{2}I_2\right)^{\frac{n-1}{2}} \hat{e}, \quad \vec{q} = K_q \left(\frac{1}{2}I_2\right)^{\frac{s}{2}} gradT, \quad I_2 = 4\left(\frac{\partial u}{\partial x}\right)^2 + 4\left(\frac{\partial v}{\partial y}\right)^2 + 2\left(\frac{\partial u}{\partial y} + \frac{\partial v}{\partial x}\right)^2$$

$$(3.94)$$

Here, $\hat{\tau}$ and $\hat{e}$ are the stress and rate of deformation tensors, $\vec{q}$ is the heat flux vector, and I_2 is the second invariant of a rate of deformation tensor. Power laws (Equation 3.94) adequately describe the behavior of such atypical fluids as suspensions, polymer solutions and melts, starch pastes, clay mortars, and so forth.

Newton's friction and Fourier's heat conduction laws follow from Equation (3.94) when $n = 1$ and $s = 0$. Therefore, the deviation of n and s from these values can be a measure of fluid anomaly. The boundary layer equations for power law fluids are as follows:

$$\frac{\partial u}{\partial x} + \frac{\partial v}{\partial y} = 0, \quad u\frac{\partial u}{\partial x} + v\frac{\partial u}{\partial y} - U\frac{dU}{dx} - \frac{K_\tau}{\rho}\frac{\partial}{\partial y}\left[\left|\frac{\partial u}{\partial y}\right|^n\right] = 0$$

$$u\frac{\partial T}{\partial x} + v\frac{\partial T}{\partial y} - \frac{K_q}{\rho c_p}\frac{\partial}{\partial y}\left[\left|\frac{\partial u}{\partial y}\right|^s \frac{\partial T}{\partial y}\right] - \frac{K_\tau}{\rho c_p}\left|\frac{\partial u}{\partial y}\right|^{n+1} = 0 \qquad (3.95)$$

Self-similar solutions of system (Equation 3.95) exist [22] in the same cases as for Newtonian fluids — that is, when the exponents in laws (Equation 3.94) are equal ($s = n - 1$) and free stream velocity and temperature head distributions obey the power laws (1.5). The equality $s = n - 1$ means that viscosity and heat conductivity defined by expression $[(1/2)I_2]$ in laws (Equation 3.94) are proportional to each other. In such a case, the thermal boundary layer in Equation (3.95) has an exact solution for the case of power law free stream distribution.

Transforming the third part of Equation (3.95) to Prandtl-Mises-Görtler's variables, one gets

$$\Phi = \frac{\rho}{K_\tau}\left(\frac{L}{U^3}\right)^{n-1}\int_0^x U^{2n-1}(\xi)d\xi, \quad \varphi = \frac{\psi}{[n(n+1)(K_\tau/\rho)^2(U^3/L)^{n-1}\Phi]^{\frac{1}{n+1}}} \tag{3.96}$$

$$n(n+1)\Phi\frac{\partial\theta}{\partial\Phi} - n\varphi\frac{\partial\theta}{\partial\varphi} - \frac{1}{Pr}\frac{\partial}{\partial\varphi}\left[\left(\frac{u}{U}\right)^n\left|\frac{\partial}{\partial\varphi}\left(\frac{u}{U}\right)\right|^{n-1}\frac{\partial\theta}{\partial\varphi}\right]$$

$$-\frac{U^2}{c_p}\left(\frac{u}{U}\right)^n\left|\frac{\partial}{\partial\varphi}\left(\frac{u}{U}\right)\right|^{n+1} = 0 \tag{3.97}$$

Substituting solution (Equation 3.4) into Equation (3.97) leads to ordinary differential equations similar to Equations (3.7) and (3.9):

$$(1/\,Pr)\{[\omega(\varphi,\beta,n)]^n \mid \omega'(\varphi,\beta,n)\mid^{n-1}G_k'\}' + n\varphi G_k' - n(n+1)kG_k = n(n+1)G_{k-1} \tag{3.98}$$

$$(1/\,Pr)\{[\omega(\varphi,\beta,n)]^n \mid \omega'(\varphi,\beta,n)\mid^{n-1}G_d'\}' + n\varphi G_d'$$

$$-n(n+1)\beta G_d = -[\omega(\varphi,\beta,n)]^n \mid \omega'(\varphi,\beta,n)\mid^{n-1} \tag{3.99}$$

The gradient pressure parameter is dependent in this case not only on exponent *m* as in the case of Newtonian fluids, but also on exponent *n*:

$$\beta = \frac{(n+1)m}{(2n-1)m+1} \tag{3.100}$$

Because of that, the pressure gradient is characterized not using β, which in this case depends also on *n*, but by using exponent *m*. Boundary conditions for Equations (3.98) and (3.99) remain the same conditions (3.8) and (3.9) as well as Equations (3.32) and (3.33) for the heat flux and nonisothermicity coefficient. For an isothermal surface, one obtains

$$\frac{Nu_*}{Re^{\frac{n}{n+1}}} = g_0\left(\frac{C_f}{2}Re^{\frac{1}{n+1}}\right)^{\frac{2n-1}{2n}}\left(\frac{\Phi}{Re}\right)^{-\frac{1}{2(n+1)}}, \quad g_0 = -2^{1/2}[n(n+1)]^{-\frac{1}{2(n+1)}}(\varphi G_0')_{\varphi=0} \tag{3.101}$$

where Nu and Re are generalized numbers as given in the Nomenclature.

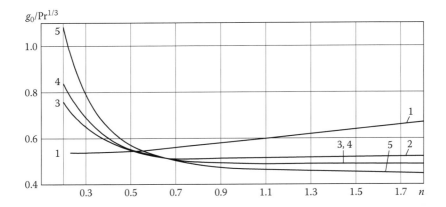

FIGURE 3.11
Dependence of $g_0 \, Pr^{1/3}$ on Prandtl number and exponents n and m for non-Newtonian fluid $s = n - 1$, $1 - m = 0$, $Pr > 10$, $2 - m = 1/3$, $Pr > 10$, $3 - m = 1$, $Pr = 1000$, $4 - m = 1$, $Pr = 100$, $5 - m = 1$, $Pr = 10$.

Calculations are performed for large Prandtl numbers $Pr = 10, 100, 1000$ typical for non-Newtonian fluids, exponents n from 0.2 to 1.8 and $m = 0, 1/3$ and 1. The results are given in Figures 3.11 and 3.12.

It is seen from Figure 3.11 that for large Pr, the value $g_0 / Pr^{1/3}$ slightly depends on the Prandtl number. This indicates that the heat transfer coefficient for an isothermal surface for non-Newtonian fluids is proportional to $Pr^{1/3}$ as well as for usual Newtonian fluids. The dependencies of coefficients g_k on Prandtl number and pressure gradient (Figure 3.12) are similar to those for Newtonian fluids. In particular, for $Pr > 10$ and small pressure gradients $m = 0$ and 1/3, the coefficients g_k are practically independent on Pr. As the pressure gradient increases, this dependence becomes more marked. Functions $g_1(n)$ and $g_2(n)$ for $m = 1/3, m = 1$ and $Pr = 100$ practically merge in one curve.

Because in the case under consideration $s = n - 1$, the basic relation (3.32) remains the same as for Newtonian fluids, all other formulae derived above remain valid as well, in particular, integral form (Equation 3.40) and Equation (3.48). Of course, the exponents C_1 and C_2 of the influence function in integral form, coefficients h_k in formula (3.58) and others should be determined according to coefficients g_k given in Figures 3.11 and 3.12. This can be done similar to that performed in previous sections. As in the case of Newtonian fluids, the relation obtained for self-similar free stream velocity distributions can be used with high accuracy for an arbitrary pressure gradient, but in this case instead of expression (3.37) for β, a similar formula for m should be used:

$$m = \left[Ux / \int_0^x U(\xi) d\xi \right] - 1 \qquad (3.102)$$

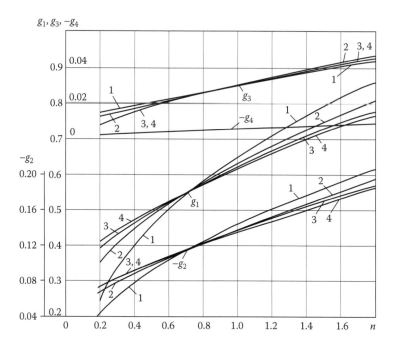

FIGURE 3.12

Dependence of coefficient g_k on the Prandtl number and exponents n and m for non-Newtonian fluids $= n - 1, 1 - m = 0,$ Pr $= 10, 2 - m = 1,$ Pr $= 100, m = 1/3,$ Pr $> 10, 3 - m = 1,$Pr $> 1000, 4 - m = 0,$ Pr > 10.

For the general case of arbitrary exponents n and s in laws (3.94), only approximate solutions for arbitrary nonisothermal surface have been obtained [23–25].

3.9.2 The Effect of Mechanical Energy Dissipation [2]

The effect of dissipation is minor for incompressible fluids because this effect is proportional to the square of velocity, which in this case is typically relatively small. Therefore, the effect of dissipation in the case of an incompressible fluid can be significant only for large Prandtl numbers. The effect of dissipation is determined for both Newtonian and non-Newtonian fluids by the second term of Equation (3.12). The coefficient g_d in this equation can be calculated by integration of ordinary differential equation (3.9), similar to computing coefficients g_k. Some results are given in Figure 3.13.

Computing a heat of dissipation is important for recovery factor and for determining adiabatic wall temperature. To obtain these quantities, the case of known heat flux distribution should be considered. This problem can be solved using the same approach as in Section 3.4, where this problem is considered ignoring dissipation. Again, using relation

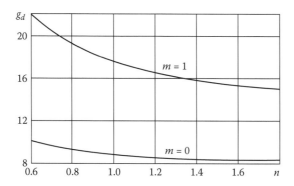

FIGURE 3.13
Dependence of coefficient g_d in determining the dissipative term on exponents n and m for non-Newtonian and Newtonian ($n = 1$) fluids.

(Equation 3.58) as an ordinary differential equation defining the temperature head as a sum,

$$\theta_{w*} = \frac{q_w}{h_*} + g_d \frac{U^2}{c_p},$$ (3.103)

the same differential (Equation 3.59) and integral (Equation 3.64) relations are obtained. Assuming in sums (Equations 3.59 and 3.64) for the case of adiabatic walls $q_w = 0$ yields the expression for recovery factor:

$$r = \frac{T_{ad.} - T_\infty}{U^2/2c_p} = 2g_d \left(1 + \sum_{k=1}^{\infty} h_k \frac{\Phi^k}{U^2} \frac{d^k U^2}{d\Phi^k} \right)$$ (3.104)

The corresponding integral form follows from Equation (3.64):

$$r = \frac{2g_d C_1}{\Gamma(C_2)\Gamma(1-C_2)U^2} \int_0^{\Phi} \left[1 - \left(\frac{\xi}{\Phi} \right)^{C_1} \right]^{C_2-1} \left(\frac{\xi}{\Phi} \right)^{C_1(1-C_2)} \frac{U^2(\xi)}{\xi} d\xi$$ (3.105)

Coefficients h_k in Equation (3.104) for Newtonian fluids are given in Section 3.4. They can be similarly calculated for non-Newtonian fluids using Equation (3.62) and known g_k (Figure 3.12).

3.9.3 Axisymmetric Streamlined and Rotating Bodies [2]

Stepanov [26] and Mangler [27] suggested variables that transform a problem for an axisymmetric streamlined body to an equivalent two-dimensional problem. For the general case of non-Newtonian fluids including Newtonian fluids, those variables are as follows [22]:

$$\tilde{x} = \int_0^x R^{n+1}(\xi)d\xi, \quad \tilde{y} = Ry \tag{3.106}$$

where R is a cross-sectional radius. Variables (Equation 3.106) transform an axisymmetric problem to an equivalent two-dimensional problem, so the relations obtained in previous sections are also valid for this case if one substitutes in expression (3.96) $R^{n+1}dx$ for dx and determines the heat flux and stress using $\tilde{q}_w$ and $\tilde{\tau}_w$ for the two-dimensional problem, according to relations

$$\Phi = \frac{\rho}{K_\tau}\left(\frac{L}{U^3}\right)^{n-1}\int_0^x U^{2n-1}(\xi)R^{n+1}(\xi)d\xi. \quad q_w = R^n\tilde{q}_w \quad \tau_w = R^n\tilde{\tau}_w \tag{3.107}$$

Analogous variables for the turbulent boundary layer are given in Reference [28].

An exact solution is also obtained for the case of rotating axisymmetric bodies [29]. The self-similar solutions are obtained for bodies with power-law radius and surface temperature distributions. These are presented in the form [30].

$$\frac{u}{R\omega} = \tilde{u}(\eta, \beta) \quad \eta = y\sqrt{Cm} \quad \beta = \frac{1+3m}{4m}, \quad R = Cx^m \tag{3.108}$$

Here, ω and R are angular velocity and body radius, respectively. The exact solution is presented using independent variables similar to those in other cases:

$$\Phi = \left(\frac{\omega}{\nu}\right)^2\int_0^x R^3(\xi)d\xi \quad \varphi = \frac{\psi}{\sqrt{2(\nu^3/\omega)\Phi}} \tag{3.109}$$

Using these variables, one gets the energy boundary layer equation in the following form [29]:

$$2\Phi\frac{\partial\theta}{\partial\Phi} - \varphi\frac{\partial\theta}{\partial\varphi} - \frac{1}{Pr}\frac{\partial}{\partial\varphi}\left(\frac{u}{\omega R}\frac{\partial\theta}{\partial\varphi}\right) = 0 \tag{3.110}$$

Substituting self-similar velocity distribution (3.108) for $u/\omega R = \tilde{u}$ in Equation (3.110) leads to an equation similar to Equation (3.7). Then, coefficients G_k, g_k and other quantities can be calculated.

3.9.4   Thin Cylindrical Bodies [2]

Results obtained in the previous subsection are only valid when the body radius is large in comparison with the boundary layer thickness. This relation is usually violated for a streamlined thin cylindrical body when the thickness of the growing boundary layer becomes equal and even exceeds the body radius. The problem of boundary layer for thin cylinder is complicated and cannot be reduced to a two-dimensional case.

The asymptotic solutions of this problem in series are known for small and large values of the curvature parameter [10]:

$$X = \frac{\nu x}{U_\infty R^2} \tag{3.111}$$

Approximate solutions for an isothermal body, which are valid for a whole range of values of parameter (Equation 3.111), are also known (see References [31] and [32]). An approximate solution for the arbitrary nonisothermal thin cylinder obtained here is presented in the same form as the other exact solutions. The thermal boundary equation in the Prandtl-Mises form for the cylindrical body is

$$\frac{\partial \theta}{\partial x} = \frac{\partial}{\partial \psi}\left(r^2 u \frac{\partial \theta}{\partial \psi} \right) \tag{3.112}$$

To express $r^2 u$ as a function of ψ, the approximate solution for velocity distribution across the boundary layer obtained by integral method [31] is employed:

$$u/U_\infty = [\ln(r/R)]/\gamma(X) \tag{3.113}$$

Applying the definition of the stream function and performing integration, one finds two relations:

$$u = \frac{1}{r}\frac{\partial \psi}{\partial r}, \quad \psi = \frac{U_\infty R^2}{4\gamma(X)}\left[1 + \left(\frac{r}{R}\right)^2\left(2\ln\frac{r}{r} - 1\right)\right], \quad r^2 u = \frac{U_\infty}{\gamma(X)}r^2\ln\frac{r}{R} \tag{3.114}$$

The desired relation $r^2 u = f(\psi)$ can be obtained by solving the second equation for (r/R). Because this equation is transcendental relative to this ratio, such a relation is approximated by the power function:

$$\left(\frac{r}{R}\right)^2\frac{u}{U_\infty}\gamma(X) = A(X)\left[\frac{\psi}{U_\infty R^2}\gamma(X)\right]^{\varepsilon(X)} \tag{3.115}$$

Expending functions in the last two equations of Equation (3.114) in the series at $(r/R) = 1$ and taking only the first nonvanishing terms, one finds that close to the surface, the exponent in Equation (3.115) is $\varepsilon = 1/2$. As the distance from the surface increases, the value of ε grows, and as follows from Equation (3.114), $\varepsilon \to 1$ as $(r/R) \to \infty$. Taking into account that ε changes relatively slightly across the boundary layer and that the characteristics at the surface are most important, it is assumed that the exponent is constant and equal to 1/2. For the same reason, Equation (3.115) is multiplied by derivative $\partial\theta/\partial r$, which has the maximum at the wall. Then, integrating Equation (3.115) across the boundary layer, solving the resulting equation for $A(X)$, and using Equations (3.113) and (3.114), yields an approximation function:

$$A(X) = \frac{2 \int\limits_1^{1+\delta/R} \frac{r}{R} \ln\frac{r}{R} d\left(\frac{r}{R}\right)}{\int\limits_1^{1+\delta/R} \left(\frac{r}{R}\right)^{-1}\left[\left(\frac{r}{R}\right)^2\left(2\ln\frac{r}{R}-1\right)+1\right]^{1/2} d\left(\frac{r}{R}\right)}$$

(3.116)

Now, introducing Görtler's variables,

$$\Phi = \int\limits_0^X \frac{A(\xi)}{\gamma^{1/2}(\xi)} d\xi, \qquad \varphi = \frac{\psi^{3/2}}{U_\infty^3 R^3 \Phi},$$

(3.117)

and applying Equation (3.115), transforms Equation (3.112) to the form similar to others:

$$\Phi \frac{\partial\theta}{\partial\Phi} - \varphi \frac{\partial\theta}{\partial\varphi} - \frac{1}{\Pr}\frac{9}{4}\varphi^{1/3}\frac{\partial}{\partial\varphi}\left(\varphi^{2/3}\frac{\partial\theta}{\partial\varphi}\right) = 0$$

(3.118)

The solution of this equation can be presented by the same series (3.4) without a dissipative term and with coefficients g_k determined by the following equation:

$$(9/4)(1/\Pr)\varphi^{1/3}\left(\varphi^{2/3}G_k'\right)' + \varphi G_k' - kG_k = G_{k-1}$$

(3.119)

The heat flow per unit surface of a thin cylinder is usually calculated instead of the heat flux:

$$Q_w = 2\pi R\lambda \left(\frac{\partial\theta}{\partial r}\right)_{r=R} = g_0\lambda\theta_w A(X)\gamma^{1/2}(X)\Phi^{1/3}\chi_t$$

(3.120)

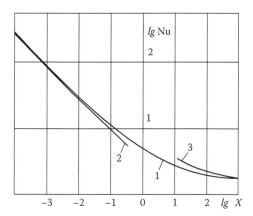

FIGURE 3.14
Dependence of the Nusselt number on curvature parameter X for a thin cylinder. 1 — Equation (3.120); 2,3 — asymptotic solutions for small and large values of X [10].

The nonisothermicity coefficient is determined by the same equations (Equation 3.33), but also using other values of coefficients g_k given by similar relations. Solutions of Equation (3.119) can be expressed using special functions. In particular, for large Prandtl numbers, the following formula can be obtained: $g_k = (-1)^{k+1} / [k!(3k-1)]$

In Figure 3.14, the results of calculation according to formula (Equation 3.120) are given in the form $Nu = Q_w / \lambda\, \theta_w = f(X)$. These agree with asymptotic solutions in series for small and large values of parameter X.

Examples of using results obtained in this chapter are given in Chapters 5 and 6.

References

1. Dorfman, A. S., 1970. Heat transfer from liquid to liquid in a flow past two sides of a plate, *High Temperature* 8: 515–520.
2. Dorfman, A. S., 1982. *Heat Transfer in Flow around Nonisothermal Bodies* (in Russian). Mashinostroenie, Moscow.
3. Dorfman, A. S., 1971. Exact solution of the equation for a thermal boundary layer for an arbitrary surface temperature distribution, *High Temperature* 9: 870–878.
4. Schlichting, H., 1979. *Boundary Layer Theory.* McGraw-Hill, New York.
5. Chapman, D., and Rubesin, M., 1949. Temperature and velocity profiles in the compressible laminar boundary layer with arbitrary distribution of surface temperature, *J. Aeronaut Sci.* 16: 547–565.
6. Levy, S., 1952. Heat transfer to constant property laminar boundary layer flows with power-function free stream velocity and surface temperature, *J. Aeronaut Sci.* 19: 341–348.

7. Bubnov, V. A., and Grishmanovskaya, K. N., 1964. Concerning exact solutions of problems of a nonisothermal boundary layer in an incompressible liquid (in Russian). *Trudy Leningrad. Politekhn. Inst.* 230: 77–83.

8. Oka, S., 1968. Calculation of the thermal laminar boundary layer of the incompressible fluid on a flat plate with specified variable surface temperature. In *Heat and Mass Transfer, Proceedings of the 3rd All-Union Conference on Heat and Mass Transfer,* 9: 74–91. Minsk.

9. Dorfman, A. S., 1973. Exact solution of equations for a thermal boundary layer with an arbitrary temperature distribution on a streamlined surface and a Prandl number $Pr \to 0$. *Int. Chem. Engn.* 13: 118–121.

10. Loytsyanskiy, L. G., 1962. *The Laminar Boundary Layer* (in Russian). Fizmatgiz Press, Moscow [Engl. trans., 1966].

11. Spalding, D. B., and Pun, W. M., 1962. A review of methods for predicting heat transfer coefficients for laminar uniform-property boundary layer flows. *Int. J. Heat Mass Transfer* 5: 239–244.

12. Dorfman, A. S., 1973. Influence function for an unheated section and relation between the superposition method and series expansion with respect to form parameters, *High Temperature* 11: 84–89.

13. Dorfman, A. S., 1982. Exact solution of the thermal boundary layer equation for an arbitrary heat flux distribution on a surface, *High Temperature* 20: 567–574.

14. Kays, W. M., 1969 and 1980. *Convective Heat and Mass Transfer.* McGraw-Hill, New York.

15. Dorfman, A. S., 1995. Exact solution of nonsteady thermal boundary layer equation, *ASME J. Heat Transfer* 117: 770–772.

16. Sparrow, E. M., 1958. Combined effects of unsteady flight velocity and surface temperature on heat transfer, *Jet Propulsion* 28: 403–405.

17. Dorodnizin, A. A., 1942. Laminar boundary layer in compressible gas, *Prikl. Math. Mech.* 6: 449–486.

18. Stewartson, K., 1949. Correlated compressible and incompressible boundary layers, *Proc. Roy. Soc. A* 200: 84–100.

19. Sakiadis, B. C., 1961. Boundary layer behavior on a continuous solid surface. *A. J. Ch. E. J.* 7 (Pt. 1): 26–28, (Pt. 2): 221–225.

20. Tsou, F. K., Sparrow, E. M., and Goldstein, R. J., 1967. Flow and heat transfer in the boundary layer on a continuous moving surface, *Int. J. Heat Mass Transfer* 10: 219–235.

21. Dorfman, A. S., and Novikov, V. G., 1980. Heat transfer from a continuously moving surface to surroundings, *High Temperature* 18: 898–901.

22. Shulman, Z. P., and Berkovskii, B. M., 1966, *Boundary Layer of Non-Newtonian Fluids* (in Russian). Nauka i Technika, Minsk.

23. Dorfman, A. S., 1967. Application of Prandtl-Mises transformation in boundary layer theory. *High Temperature* 5: 761–768.

24. Dorfman, A. S., and Vishnevskii, V. K., 1971. Approximate solution of dynamic and thermal boundary layer equations for non-Newtonian fluids and arbitrary pressure gradients, *Int. Chem. Engn.* 11: 377–383.

25. Dorfman, A. S., and Vishnevskii, V. K., 1972. Approximate solution of dynamic and thermal boundary layer equations for non-Newtonian fluids with arbitrary pressure gradients and surface temperature, *Int. Chem. Engn.* 12: 288–294.

26. Stepanov, E. I., 1947. About integration of laminar boundary equation in the case of axial symmetry, *Prikl. Mat. I Mech.* 11: 9–15.

27. Mangler, W., 1948. Zusammenhang zwichen ebenen und rotationssymmetrischen Grenzschichten in kompressiblen Flüssigkeiten, *ZAMM* 28: 97–103.

28. Fedyaevskii, K. K., Ginevskii, A. S., and Kolesnikov, A. V., 1973. Calculation of turbulent boundary layer in incompressible fluid (in Russian), *Sudostrenie*, Moscow.

29. Dofman, A. S., and Selyavin, G. F., 1977. Exact solution of equation of the thermal boundary layer for axisymmetric rotating bodies with an arbitrary surface temperature distribution, *Heat Transfer — Soviet Research* 9: 105–113.

30. Geis, Th., 1955. Änlishe Grenzschichten an Rotationskörpern. In *50 Jahre Grenzschichten forschung*, edited by H. Görtler and W. Tolmien, pp. 294–303. Vieweg, Braunschwieg.

31. Glauert, M. B., and Lighthill, M. J., 1955. The axisymmetric boundary layer on a long thin cylinder, *Proc. Roy. Soc. A* 230: 188–203.

32. Borovskii, V. R., Dorfman, A. S., Shelimanov, V. A., and Grechannyy, O. A., 1973. Analytical study of the heat transfer in a longitudinal flow past cylindrical bodies of small radius at constant temperature. *J Eng. Physic.* 25: 1344–1349.

4

Heat Transfer from Arbitrary Nonisothermal Surfaces in Turbulent Flow

In this chapter, the dependence between heat flux and temperature head for turbulent boundary layer in the differential (Equation 3.32) and integral (Equation 3.40) forms are derived. The approach used here is the same as that used in Chapter 3. First, the thermal boundary layer equation is solved for the equilibrium turbulent boundary layers as has been done for the self-similar laminar boundary layers. Then, the solution is generalized for the arbitrary free stream velocity using a slight dependency of coefficients g_k on the pressure gradient.

4.1 Basis Relations for the Equilibrium Boundary Layer

Equilibrium turbulent boundary layers studied in the literature [1–5] are flows similar to the self-similar laminar boundary layers with constant dimensionless pressure gradient:

$$\beta = \frac{\delta^*}{\tau_w} \frac{dp}{dx} \tag{4.1}$$

In the case of laminar self-similar flows, the parameter β depends by simple relation (1.5) on the exponent in the free stream velocity power law $U = Cx^m$. In the case of the equilibrium turbulent boundary layer, such dependency is more complicated but can be obtained from Equation (4.1).

Although there are more physically grounded one- and two-equation turbulent models, algebraic models including the models based on equilibrium turbulent boundary layer laws are still widely used. The reason for this is that as indicated by Wilcox [6, pp. 163– 164], "Certainly the $k - \varepsilon$ model is the most widely used two-equation model.... While the model can be fine tuned for a given application, it is not clear that this represents an improvement over algebraic models."

Mellor and Gibson's turbulent model [4,5] is one of the most often used algebraic models. In the Science Citation Index, these two papers are referenced over 154 times by the authors of theoretical and experimental studies. Mellor and Gibson's model is based on the modern velocity profile structure in the turbulent layer which consists of three distinct regions: the viscous sublayer, the log layer, and the defect layer. The viscous sublayer is the region between the surface and the log layer. The defect layer lies between the log layer and the edge of the boundary layer. The log layer is an overlap region in which both the viscous sublayer and defect layer laws are valid. Such a turbulent velocity profile structure was obtained first experimentally and later proved theoretically using the perturbation analysis [6].

The results obtained by Mellor and Gibson are adopted in this study to determine the velocity and the eddy-viscosity distribution across the boundary layer. The composite function describing the kinematic effective viscosity $v_e = v + v_{tb}$ across the boundary layer according to velocity profile structure consists of three parts [4]. In the inner part of the boundary layer, this function is presented in the form

$$\frac{v_e}{v} = \varepsilon_1 \left(\frac{y^2}{v} \frac{du}{dy} \right) \qquad (4.2)$$

This dependency is found from similarity considerations, and the function ε_1 is established on the basis of measurements made by Laufer [7]. The results can be approximated by composite function using two variables, ω and r (see Equation 4.6):

$$\left. \begin{array}{l} 0 < r < 2, \ \varpi = r^2 - r, \ \varepsilon_1(\omega) = 1 \\[2mm] 2 < r < 4.5, \ \omega = 2.75r - 1.5, \ \varepsilon_1(\omega) = \dfrac{r^2}{\varpi} \\[2mm] 4.5 < r < 11, \ \omega = 11, \ \varepsilon_1(\omega) = \dfrac{r^2}{11} \end{array} \right\} \qquad (4.3)$$

In the outer part of the layer, where the molecular viscosity is negligible, turbulent viscosity is usually assumed to be independent on the transverse coordinate:

$$v_e = \kappa_1 U \delta^* = 0.016 U \delta^* \qquad (4.4)$$

Prandtl's formula with mixing length

$$l = \kappa y = 0.4y \qquad v_e = \kappa^2 y^2 \left| \frac{du}{dy} \right| \qquad (4.5)$$

is used to describe the effective viscosity in the overlap part of the boundary layer, where both the wall and the defect laws are valid. The whole effective viscosity function is used to calculate the velocity profile by integrating boundary layer equations. For the inner part, it is found that

$$u^+ = u_v^+ + \frac{2}{\kappa}[(1+\beta_v y^+)^{1/2} - 1] + \frac{1}{\kappa}\ln\left[\frac{4}{\beta_v}\frac{(1+\beta_v y^+)^{1/2} - 1}{(1+\beta_v y^+)^{1/2} + 1}\right],$$

$$u_v^+ = \lim_{\zeta \to 0}\left[\int_\zeta^{y^+} \frac{\sigma(r)}{\kappa^2 y^{+2}}dy^+ - \frac{\ln\zeta}{\kappa}\right]$$

$$u^+ = \frac{u}{u_\tau}, \quad y^+ = \frac{yu_\tau}{v}, \quad \beta_v = \frac{\beta}{Re_{\delta^*}\sqrt{C_f/2}},$$

$$\sigma(r) = \omega - r, \omega = \kappa^2 y^{+2}\frac{du^+}{dy^+}, \quad r = \kappa y^+(1+\beta_v y^+)^{1/2}$$

(4.6)

Outside the wall layer ($\omega > 11$), the function $\sigma(r)$ vanishes, and the first term in Equation (4.6) becomes a constant $u_v^+ = D^+(\beta_v)$ (Table 4.1).

Equation (4.6) differs from the common profile of the wall law by taking into account the effect of the pressure gradient. In the case of gradientless flow ($\beta_v = 0$), Equation (4.6) has a limit and approaches the usual simple form $u^+ = (1/\kappa)\ln y^+ + D^+$, when $y^+ \ll 1/\beta_v$. For unfavorable pressure gradients, the result obtained by Equation (4.6) differs significantly from that given by this simple equation, especially for the flows near separation.

Because at the point of separation $u_\tau \to 0$ and $\beta_v \to \infty$, the new variables [4] are introduced for large values of β_v, and Equation (4.6) take the following form:

$$u^{++} = \frac{u}{u_{pv}}, \quad y^{++} = \frac{u_{pv}y}{v}, \quad u_{pv} = \frac{v}{\rho}\frac{dp}{dx}$$

(4.7)

TABLE 4.1 (a)

Values of D^+ and $D^+/\beta_v^{1/3}$ (a), $D, D/\beta^{1/2}, B, B/\beta^{1/2}$, and $A, A/\beta^{1/2}$

β_v	D^+	$1/\beta_v$	$D^+/\beta_v^{1/3}$
−0.01	4.92	0.0	1.33
0.0	4.90	0.1	5.63
0.02	4.94	0.5	6.74
0.05	5.06	1.0	7.34
0.10	5.26	2.0	8.12
0.20	5.63	3.0	8.70
		4.0	9.18
		5.0	9.62

TABLE 4.1(B)

β	D	B	A	$1/\beta$	$D/\beta^{1/2}$	$B/\beta^{1/2}$	$A/\beta^{1/2}$
-0.5	-2.54	2.3	4.74	0.75	2.97	7.2	8.56
0.0	-0.59	4.3	6.58	0.5	3.54	7.0	7.87
0.5	1.07	6.0	8.01	0.25	4.58	6.0	7.08
1	2.53	7.4	9.18	0.1	5.84	7.5	6.49
				0.0	10.0	10.3	5.90

$$u^{++} = u_v^{++} + \frac{2}{\kappa}[(\beta_v^{-2/3} + y^{++})^{1/2} - \beta_v^{-1/3}] + \frac{1}{\kappa\beta_v^{1/3}} \ln\left[\frac{4}{\beta_v} \frac{(\beta_v^{-2/3} + y^{++})^{1/2} - \beta_v^{-1/3}}{(\beta_v^{-2/3} + y^{++})^{1/2} + \beta_v^{-1/3}}\right]$$

(4.8)

$$u_v^{++} = \lim_{\zeta \to 0}\left[\int_\zeta^{y^{++}} \frac{\sigma(r)}{\kappa^2 y^{++2}} dy^{++} - \frac{1}{\kappa\beta_v^{1/3}} \ln\frac{\zeta}{\beta_v^{1/3}}\right], \quad r = \kappa(\beta_v^{-2/3} + y^{++})^{1/2} y^{++} \quad (4.9)$$

Outside the wall layer, the first term in Equation (4.8) becomes a constant $u_v^{++} = D^+/\beta_v^{1/3}$ (Table 4.1). Equation (4.8) has a limit at $1/\beta_v = 0$ and approaches $(2/\kappa)y^{++1/2} + (1/\kappa\beta_v^{1/3})[\ln(4/\beta_v) - 2] + D^+/\beta_v^{1/3}$ when $y^{++} \gg 1/\beta_v$.

The friction coefficient is determined as follows:

$$\left(\frac{2}{C_f}\right)^{1/2} = \frac{1}{\kappa}\ln Re_{\delta^*} + B, \ (\beta \leq 1) \quad \left(\frac{2}{C_f\beta}\right)^{1/2} = \frac{1}{\kappa}\ln Re_{\delta^*} + B/\beta^{1/2}, \ (\beta \geq 1) \quad (4.10)$$

The velocity defect profiles for the outer part of the layer for different values of the parameter β and Reynolds number $Re_{\delta^*} = U\delta^*/\nu$ are calculated by integrating the complete boundary layer equation for the entire range $-0.5 \leq \beta \leq \infty$. Using variables

$$\eta = \frac{y}{\delta_\tau}, \ \delta_\tau = \frac{\delta^* U}{u_\tau}, \ \frac{U - u}{u_\tau} = f'(\eta) \ and \ \xi = \eta\beta^{1/2}, \ u_p = u_\tau\beta^{1/2}, \ \frac{U - u}{u_p} = F'(\xi)$$

(4.11)

for small and large values of β, respectively, yields an ordinary differential equation that is solved numerically. Using the definition of β (Equation 4.1), one can see that u_p is determined by a pressure gradient and has a scale of velocity. Therefore, u_p is named the pressure velocity similar to the friction velocity u_τ.

It is shown [5] that functions (Equation 4.11) $f'(\eta, \beta)$ and $F'(\eta, \beta)$ only slightly depend on Re_{δ^*}. The maximum variation in the range $10^3 \leq \mathrm{Re}_{\delta^*} \leq 10^9$ from the values for $\mathrm{Re}_{\delta^*} = 10^5$ is less than 2% of the main stream velocity. Therefore, the values of functions $f'(\eta, \beta)$ and $F'(\eta, \beta)$ are tabulated and given in Reference [5] for $\mathrm{Re}_{\delta^*} = 10^5$.

The velocity distribution across the entire boundary layer is obtained by using Equations (4.6) or (4.8) for inner parts and data from Table 1 in [5] for outer parts of the profile, which coincide in the overlap area.

4.2 Solution of the Thermal Turbulent Boundary Layer Equation for an Arbitrary Surface Temperature Distribution [8,9]

The thermal boundary layer equation and the boundary conditions for the turbulent flow are used in the Prandtl-Mises-Görtler's form similar to that in Equations (3.1), (3.2), and (3.3) for laminar flow. In this case, using Görtler's independent variables in the form

$$\Phi = \frac{1}{\nu} \int_0^x U\, d\zeta \qquad \varphi = \frac{\psi}{\nu \Phi} \tag{4.12}$$

leads to the following equations:

$$2\Phi \frac{\partial \theta}{\partial \Phi} - \varphi \frac{\partial \theta}{\partial \varphi} - \frac{1}{\mathrm{Pr}} \frac{\partial}{\partial \varphi} \left(\frac{u}{U} \varepsilon_\alpha \frac{\partial \theta}{\partial \varphi} \right) = \frac{U^2}{c_p} \varepsilon_u \frac{u}{U} \left[\frac{\partial}{\partial \varphi} \left(\frac{u}{U} \right) \right] \tag{4.13}$$

$$\varepsilon_\alpha = \frac{\alpha_e}{\nu \Phi} = \frac{1}{\Phi} \left(\frac{1}{\mathrm{Pr}} + \frac{\varepsilon - 1}{\mathrm{Pr}_{tb}} \right), \quad \varepsilon_u = \frac{\nu_e}{\nu \Phi} \quad \varepsilon = \frac{\nu_e}{\nu} \tag{4.14}$$

The boundary conditions (Equation 3.2) remain the same as in the case of laminar flow.

The kinematic effective viscosity ν_e and the effective thermal diffusivity $\alpha_e = \nu_e/\mathrm{Pr}_{tb}$ are determined using Mellor and Gibson's composite function for the effective viscosity presented in the previous section.

In computations involving heat transfer in turbulent flows, it is necessary to have turbulent Prandtl number, Pr_{tb}, distribution across the boundary layer. A review and comparison of results obtained by different authors are given in the literature [10–14]. Unfortunately, the existing data are so contradictory that, at present, it does not appear to be possible to establish reliably a function that determines the turbulent Prandtl number distribution across the boundary layer. The absence of sufficiently reliable and general data for the distribution of the turbulent Prandtl number led to the result that in the majority of modern investigations as well as in the early research, its value was assumed constant and either close to or equal to one. The calculations show that results obtained using such an assumption are in agreement with experimental data [6,10,14,15].

The effect of the turbulent Prandtl number on the intensity of heat transfer from a flat plate is numerically investigated in the present study. The thermal turbulent boundary layer equation is integrated at different values of turbulent Prandtl number, and the results are compared with the experimental data. The details and analysis given in Section 4.4 lead us to the conclusion that the numerical results with $Pr_{tb} = 1$ in the majority of cases better correlate with experimental data than the other studied assumptions. On the basis of this and other mentioned research, the assumption $Pr_{tb} = 1$ is used in the following.

The solution of Equation (4.13) can be formulated in the form of Equation (3.4) as for laminar flow:

$$\theta = \sum_{k=0}^{\infty} G_k(\varphi)\Phi^k \frac{d^k\theta_w}{d\Phi^k} + G_d(\varphi)\frac{U^2}{c_p} \tag{4.15}$$

Substituting Equation (4.15) into Equation (4.13) and replacing the index k for $k+1$ in one of the sums yields the following two equations:

$$\sum_{k=0}^{\infty} \Phi^k \frac{d^k\theta_w}{d\Phi^k}\left[kG_k + G_{k-1} - \varphi G_k' - \frac{\partial}{\partial\varphi}\left(\frac{u}{U}\varepsilon_\alpha G_k'\right)\right] = 0 \tag{4.16}$$

$$\frac{\Phi}{U^2}\frac{dU^2}{d\Phi}G_d - \varphi G_d' - \frac{\partial}{\partial\varphi}\left(\frac{u}{U}\varepsilon_\alpha G_d'\right) - \varepsilon_u \frac{u}{U}\left[\frac{\partial}{\partial\varphi}\left(\frac{u}{U}\right)\right]^2 = 0 \tag{4.17}$$

For equilibrium flows, when the distribution of the relative velocity u/U in the boundary layer can be described by a function of one variable $\eta = y/\delta_\tau$, depending on the parameters β and Re_{δ^*}, the variable φ is also a function only of η and depends on the same parameters. To get the function defining variable φ, the relations from Reference [5] are used:

$$-\frac{U}{\delta_\tau}\frac{d\delta_\tau}{dU} = 1 + \frac{\beta_m}{\beta}, \quad \beta_m = \frac{(\kappa + \sqrt{C_f/2})[\beta + H(1+\beta)]}{\kappa + \sqrt{C_f/2}(H-1)}, \quad H = \frac{\delta^*}{\delta^{**}} = \frac{1}{1 - A\sqrt{C_f/2}} \tag{4.18}$$

where δ_τ, δ^* and δ^{**} are defined by Equations (4.11) and (1.17), respectively. The values of A are given in Table 4.1 as a function of β.

Using relations (4.1) for β and (4.11) for δ_τ and taking into account that $U dU = -dp/\rho$, $\tau_w = \rho u_\tau^2$, one finds from Equation (4.18) an expression $d(\delta_\tau U)/dx = u_\tau \beta_m$. Substituting this result into Equation (4.12), using the relation $u = \partial \psi / \partial y$, and ignoring, as usual, the weak dependence of C_f, H, u_τ and β_m on x yields the basic variables as

$$\Phi = \frac{1}{v} \int_0^x U d\xi = \frac{1}{v} \int_0^{(U\delta_\tau)} \frac{U}{u_\tau \beta_m} d(U\delta_\tau) = \frac{2\,\mathrm{Re}_{\delta^*}}{\beta_m C_f}, \quad \varphi = \beta_m \sqrt{\frac{C_f}{2}} \int_0^x \frac{u}{U} d\eta \qquad (4.19)$$

It follows from the last equation that for equilibrium flows and fixed Reynolds number Re_{δ^*}, variable φ is a unique function of η, and so the relative velocity u/U and effective viscosity $\varepsilon = v_e/v$, which in such flows depends only on η, can be written as functions of φ only. Then ε_a and ε_u defined by Equation (4.14) for given Prandtl number are also functions only of φ, if the turbulent Prandtl number Pr_{tb} is assumed to be a constant. It follows also from Equations (4.19) and (4.1) that the coefficient at G_d in Equation (4.17) depends only on β and Re_{δ^*}:

$$\frac{\Phi}{U^2} \frac{dU^2}{d\Phi} = \frac{2\Phi}{U^2} \frac{dU}{dx} = \frac{2\beta}{\beta_m} \qquad (4.20)$$

Under these conditions, the expressions in brackets in Equation (4.16) and the left-hand side of Equation (4.17) are functions only of φ and the parameters β and Re_{δ^*}. Hence, equating the expressions in brackets to zero gives a system of ordinary equations for the coefficients $G_k(\varphi)$ of the series (Equation 4.15):

$$(\omega \varepsilon_a G_k')' + \varphi G_k' - k G_k = G_{k-1}, \qquad (k = 0, 1, \ldots) \qquad (4.21)$$

where $\omega(\varphi, \beta, \mathrm{Re}_{\delta^*})$ denotes the ratio u/U in equilibrium flows.

Further calculations are similar to those performed in the case of laminar flow. The boundary conditions for these equations are the same condition (Equation 3.8):

$$\varphi = 0, \ G_0 = 1, \ G_k = 0 \ (k = 1, 2, 3 \ldots) \ \varphi \to \infty, \ G_k = 0 \ (k = 0, 1, 2 \ldots) \qquad (4.22)$$

Substituting velocity distribution in equilibrium layer $\omega(\varphi, \beta, \mathrm{Re}_{\delta^*})$ for u/U and using Equation (4.20), one gets from Equation (4.17) the analogous to (3.9) ordinary equation for $G_d(\varphi)$, the analogous to (3.9) ordinary equation for $G_d(\varphi)$:

$$[\omega(\varphi, \beta, \mathrm{Re}_{\delta^*}) \varepsilon_a G_d']' + \varphi G_d' - 2\beta/\beta_m = \varepsilon_u \omega(\varphi, \beta, \mathrm{Re}_{\delta^*}) [\omega(\varphi, \beta, \mathrm{Re}_{\delta^*})]^2$$

$$\varphi = 0, G_d = 0 \qquad \varphi \to \infty, G_d = 0 \qquad (4.23)$$

To get the expression for the heat flux, the same approach as in the case of laminar flow is used. Similar calculations lead to the connection between the velocity in the vicinity to the wall and the shear stress $u = u_\tau (2\varphi\Phi)^{1/2}$. Then differentiating Equation (4.15) with respect to variable φ gives

$$q_w = -\frac{\lambda}{\nu} u_\tau \Phi^{-1/2} \left[\sum_0^\infty (\sqrt{2\varphi}G_k')_{\varphi=0} \Phi^k \frac{d^k\theta_w}{d\Phi^k} + \frac{U^2}{c_p} (\sqrt{2\varphi}G_d')_{\varphi=0} \right] \qquad (4.24)$$

Dividing both sides of this equation by $\rho c_p U\theta_w$ and using Equation (4.19) for Φ yields

$$St = St_* \left(\chi_t + g_d \frac{U^2}{c_p\theta_w} \right), \qquad (4.25)$$

The coefficient of nonisothermicity χ_t, the Stanton number for isothermal surface St_*, and coefficients g_k in Equation (4.25) are defined as

$$\chi_t = 1 + \sum_1^\infty g_k \frac{\Phi^k}{\theta_w} \frac{d^k\theta_w}{d\Phi^k} \qquad St_* = g_0 \frac{C_f}{2} \qquad (4.26)$$

$$g_0 = -\left(\frac{2\beta_m}{Re_{\delta^*}} \right)^{1/2} \frac{(\varphi^{1/2}G_0')_{\varphi=0}}{Pr}, \quad g_k = \left(\frac{G_k'}{G_0'} \right)_{\varphi=0} \quad g_d = -\frac{G_d'}{G_0'} \qquad (4.27)$$

The relation $2St_*/C_f = g_0$ is the Reynolds analogy coefficient. For gradientless flow and $Pr = 1$ on the plate, as it should be according to Reynolds analogy, Equation (4.27) (Figure 4.1), $g_0 = 1$, while in other cases this relation shows how much the heat transfer coefficient differs from the friction coefficient. Equation (4.26) represents the nonisothermicity coefficient in the differential form. The corresponding integral form is the same as in laminar flow (Equation 3.48):

$$\chi_t = \frac{1}{\theta_w} \left[\int_0^\Phi f\left(\frac{\xi}{\Phi} \right) \frac{d\theta_w}{d\xi} d\xi + \theta_w(0) \right] \qquad (4.28)$$

Equations (4.21) and (4.23) are solved by the same approach as used for similar equations for coefficients G_k and G_d in the case of laminar flow. Inhomogeneous Equation (4.21) is reduced to a homogeneous one using the sums of Equation (3.15). Then, using a new variable $z = \varphi^{1/2}$, the latter is transformed into an equation without singularity:

$$[\omega\varepsilon_a F_i']'z + [2z^4 - \omega\varepsilon_a] F_i' - 4iz^3 F_i = 0 \quad z = 0, F_i = 1/i! \quad z \to \infty F_i \to 0 \qquad (4.29)$$

The solution of this equation is found by numerical calculation of two functions $V(z)$ and $W(z)$ satisfying Equation (4.29) under known initial conditions (Equation 3.20). Then, Equation (3.21) is used to get the coefficients g_k. The expression for coefficient g_0 follows from Equation (4.27):

$$g_0 = \left(\frac{1}{\mathrm{Pr}}\right)\sqrt{\frac{2\beta_m}{\mathrm{Re}_{\delta^*}}}\frac{W_0(\infty)}{V_0(\infty)} \tag{4.30}$$

There are some special effects in the heat transfer calculation at high and low Prandtl numbers. At high Prandtl numbers, the thermal boundary layer is thin and is located inside the laminar sublayer. In such a case, the damping of turbulent pulsating in the vicinity of the wall significantly affects the heat transfer. It can be shown that the turbulent viscosity decreases as the third or fourth power of the distance from the wall [16]. However, there is no evidence as to which is the first term of the series describing the turbulence damping. Therefore, some investigators think that close to the wall the turbulence decays as y^3, but the others assume that the term with y^3 vanishes, and hence, the turbulence decays as y^4. The latter assumption is used here, and the effective viscosity close to the wall for high Prandtl numbers is presented as in Reference [17]:

$$v_e/v = 1 + 0.0092(y^+)^4 \tag{4.31}$$

In the case of low Prandtl numbers, the thermal boundary layer is much thicker than the dynamic layer, and one should take into account the thermal conductivity outside of the dynamic boundary layer. It is known that each point of the real outer border of the turbulent boundary layer oscillates within the limits from $(0.3 - 0.4)\delta$ to 1.2δ. Thus, the intensity of the turbulent pulsating is practically the same near the outer edge of the layer at both sides of the line $y = \delta$. On the other hand, there is no turbulent pulsating far away from this line if the free stream is initially laminar without turbulent pulsating. On the basis of such considerations, one of two assumptions is usually applied for the calculation of heat transfer in the case of small Prandtl numbers [18]: the turbulent viscosity outside the dynamic boundary layer is the same as that on the outer edge of the dynamic turbulent boundary layer, or the turbulent viscosity outside of the dynamic boundary layer is equal to zero. The first assumption is adopted in this study. Some calculations performed using the other assumption for $\mathrm{Pr} = 0.01$ and $\mathrm{Re}_{\delta^*} = 10^3$ and 10^5 indicate that differences of both final results are about 5%.

The results obtained by the formulae derived above are given in the next two sections.

4.3 Intensity of Heat Transfer from an Isothermal Surface: Comparison with Experimental Data [19]

Calculations have been performed for $\beta = -0.3$ (flow at stagnation point), $\beta = 0$ (gradientless flow), $\beta - 1$ and $\beta = 10$ (flows with weak and strong adverse pressure gradients), and following Prandtl and Reynolds numbers:

$$\text{Pr} = 0.01, 0.1, 1, 10, 100, 1000 \qquad \text{Re}_{\delta^*} = 10^3, 10^5, 10^9 \qquad (4.32)$$

4.3.1 Reynolds Analogy

Figure 4.1 shows results of calculations in the form of the Reynolds analogy coefficient $2St/C_f$ as a function of the Prandtl number, for various values of Re_{δ^*}. For comparatively low values of Re, and for Pr near unity, the results of the computations agree with value $1/0.863$ [20] obtained for air (Pr = 0.7) and two formulae:

$$2St/C_f = \text{Pr}^{-0.6} \qquad 2St/C_f = 1 + 16\sqrt{C_f/2}\,(\text{Pr}^{0.55} - 1) \qquad (4.33)$$

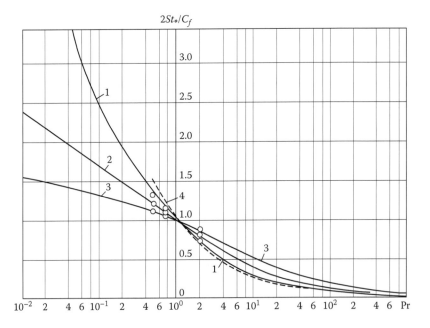

FIGURE 4.1

Reynolds analogy coefficient as a function of Prandtl and Reynolds numbers for gradientless ($\beta = 0$) turbulent flow. $1 - \text{Re}_{\delta^*} = 10^3$, $2 - 10^5$, $3 - 10^9$, 4 — first Equation (4.33); $\circ$ — second Equation (4.33).

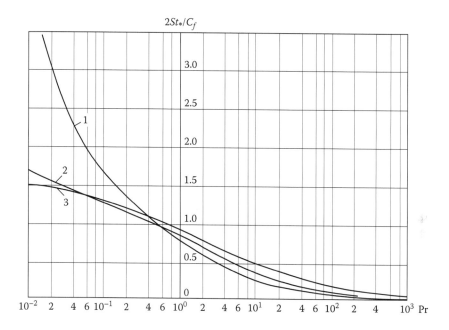

FIGURE 4.2
Reynolds analogy coefficient as a function of Prandtl and Reynolds numbers for favorable gradient ($\beta = -0.3$) turbulent flow near the stagnation point ($1 - \text{Re}_{\delta^*} = 10^3, 2 - 10^5, 3 - 10^9$).

The first formula is based on experimental data obtained in Reference [21] for $0.5 < \text{Pr} < 50$ and $10^5 < \text{Re}_x < 10^7$. The second relation is derived in Reference [22] by numerical integration of the system of turbulent boundary layer equations for $0.5 < \text{Pr} < 2$ and $1.2 \cdot 10^5 < \text{Re}_x < 1.1 \cdot 10^9$.

It follows from Figure 4.1 that for large values of Re the analogy coefficient differs substantially from the corresponding values at low Re. The increase in Re leads to a growth for $\text{Pr} > 1$ and to reduction for $\text{Pr} < 1$ in the analogy coefficient. Similar results are obtained for gradient flows (Figures 4.2, 4.3, and 4.4).

However, in contrast to gradientless flow, for which the effect of the Reynolds number changes at $\text{Pr} = 1$ (Figure 4.1), for the gradient flows the effect of the Reynolds number changes, at $\text{Pr} > 1$ for favorable gradients (Figure 4.2) and at $\text{Pr} < 1$ for adverse gradients (Figures 4.3 and 4.4). Comparison between data for gradient and gradientless flows shows that adverse gradients lead to growth and favorable gradients lead to reduction of the Reynolds analogy coefficients.

4.3.2 Relations for Heat Transfer from an Isothermal Surface in Gradientless Flow

To obtain approximation relations for calculating the heat transfer coefficients, the following considerations are taken into account. It has been shown in

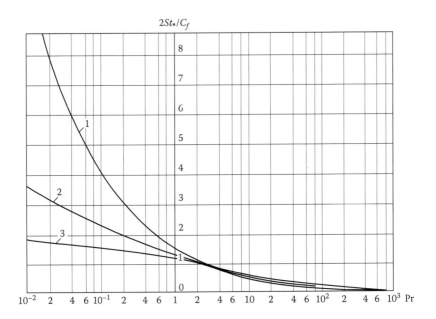

FIGURE 4.3

Reynolds analogy coefficient as a function of Prandtl and Reynolds numbers for slightly adverse gradient ($\beta = 1$) turbulent flow ($1 - \mathrm{Re}_{\delta^*} = 10^3,\ 2 - 10^5,\ 3 - 10^9$).

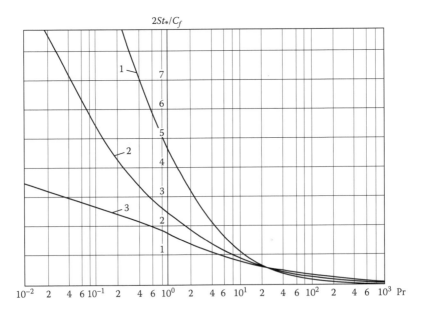

FIGURE 4.4

Reynolds analogy coefficient as a function of Prandtl and Reynolds numbers for highly adverse gradient ($\beta = 10$) turbulent flow ($1 - \mathrm{Re}_{\delta^*} = 10^3,\ 2 - 10^5,\ 3 - 10^9$).

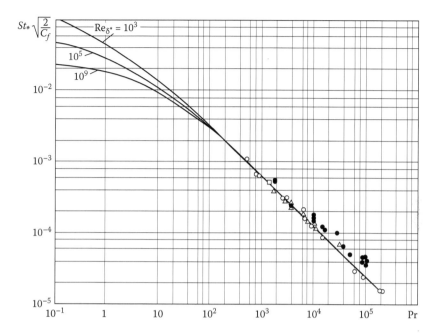

FIGURE 4.5
Comparison between calculation (second Equation 4.34) and experimental data [12] for gradientless flow and large Prandtl numbers.

Reference [12] that for Pr → ∞, the Stanton number is proportional to $\sqrt{C_f/2}$. It is also well known that for Pr = 1, the Stanton number is proportional to $C_f/2$. From this, one can expect that for other values of Prandtl numbers there is a proportionality between St and $(C_f/2)^n$, where the exponent decreases with increase of Prandtl number, from 1 at Pr = 1 to 1/2 for Pr → ∞. It follows from calculations that this proportionality actually exists for Pr > 1. Defining the coefficients of proportionality between St and $C_f/2$, one finds

$$St_* = \mathrm{Pr}^{-1.35}(C_f/2)^{1-0.3\lg\mathrm{Pr}}(1 < \mathrm{Pr} < 50)$$

$$St_* = 0.113\,\mathrm{Pr}^{-3/4}(C_f/2)^{1/2}(\mathrm{Pr} > 50)$$

(4.34)

Figure 4.5 compares the results of the calculation using the last formula with experimental data from Reference [12]. The calculated curves $St\sqrt{2/C_f} = f(\mathrm{Pr})$ for different Reynolds numbers, which merge into one for large Prandtl numbers, are continued into region Pr > 10³ by calculating the slope of the tangent at the point Pr = 10³. Good agreement is observed between the computed and experimental data: the coefficient 0.113 in Equation (4.34) determined by calculation practically coincides with the value 0.115 determined from experimental data in Reference [12]. The

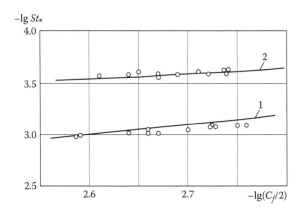

FIGURE 4.6
Comparison between calculation (first Equation 4.34) and experimental data [14] for gradient-less flow and large Prandtl numbers. 1 — water (Pr = 5.5), 2 — oil (Pr = 55).

value 0.113 is obtained using Pr_{tb} = 1. The calculations performed using Pr_{tb} = 1.5 and Pr_{tb} = 0.5 give for the coefficient in Equation (4.34) the values 0.136 and 0.096, respectively, which differ from 0.115 markedly. Comparison between the results obtained by Equation (4.34) for region $1 < Pr < 50$ and experimental data [14], given in Figure 4.6, also shows agreement.

Analysis of calculation data for $Pr < 1$ indicates that these cannot be approximated by functions of the type in Equation (4.34) because the dependence $\ln St = f[\ln(C_f/2)]$ in this case is nonlinear. However, it turns out, as indicated in Reference [18], that for $Pr < 1$ there is unique relation $Nu_x = f(Pe_x)$. In Figure 4.7, such a relation derived using calculations is compared with the results of experiments obtained in Reference [21] for air and in Reference [23] for liquid metals. These data also agree with other experimental results [14,24]. This dependence can be approximated by the following formula:

$$(Nu_x)_*^{-0.023} = 1.04 - 0.0335 \lg Pe_x \qquad Nu_{x*} = cPe_x^n \qquad (4.35)$$

analogous to the relation for the friction coefficient [10]. Simpler power relation (the second formula 4.35) can be obtained by approximating this relation using three straight lines with the following constants:

$$c = 0.282, 0.036, 0.00575, n = 0.62, 0.8, 0.9$$

$$(4.36)$$

$$\text{for } Pe_x = 10^3 - 10^5, 10^5 - 5 \cdot 10^8, 5 \cdot 10^8 - 2.5 \cdot 10^{12}$$

These results are obtained using Pr_{tb} = 1. The calculations for Pr_{tb} = 1.5 and Pr_{tb} = 0.5 shown in Figure 4.7 give results that are respectively lower and larger than experimental data.

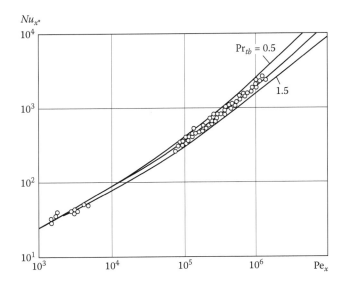

FIGURE 4.7
Comparison between calculation (first Equation 4.35) and experimental data for gradientless flow and Pr < 1, for air [21] and liquid metal [23].

4.3.3 Relations for Heat Transfer from an Isothermal Surface in Gradient Flows

Results obtained for isothermal surfaces in gradient flows and Pr < 1 can also be approximated by formulae of the type in Equation (4.35) by using the Peclet number calculated for average free stream velocity. The latter is a product of variable Φ and the Prandtl number:

$$Pe_\Phi = \Phi Pr = \frac{1}{\nu} \int_0^x U d\xi \cdot \frac{\nu}{\alpha} = \frac{U_{av} x}{\alpha} \qquad (4.37)$$

The variable Φ can be calculated using Equation (4.19) if β and Re_{δ^*} are known. This can be more easily done by using the plotted dependence $\Phi = f(Re_{\delta^*}, \beta)$ in Figure 4.8 computed by Equation (4.19).

Dependence $St_* = f(Pe_\Phi)$ obtained using calculation results for gradient and gradientless flows for Pr < 1 is presented in Figure 4.9. It can be seen that all the points related to different flows form a single curve that can be approximated by the following formula:

$$St_* = cPe_\Phi^{n-1} \qquad (4.38)$$

with the same constants c and n (Equation 4.36) as in Equation (4.35). Calculation shows that for gradientless and favorable gradients for large Reynolds

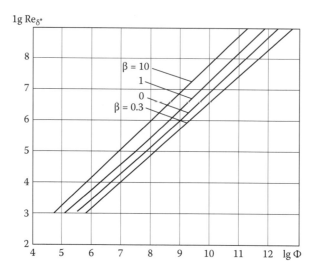

FIGURE 4.8
Dependence $\Phi = f(\text{Re}_{\delta^*}, \beta)$ for equilibrium turbulent boundary layers.

numbers, the error of this formula is less than 5%. For low Reynolds numbers, adverse gradients, and Prandtl numbers close to 1 the error increases
and reaches about 35% for $\text{Pr} = 1, \text{Re}_{\delta^*} = 10^3$, and $\beta = 10$.

As follows from Figures 4.1, 4.2, and 4.3, the effect of the pressure gradient
on heat transfer intensity is relatively small at large Prandtl numbers; therefore, in this case the relations (4.34) for gradientless flows can be used. The
error is less than 10% for the whole range of Reynolds numbers if the Prandtl
number is larger than 10, while for the range $1 < \text{Pr} < 10$, this is true only for
$\text{Re}_{\delta^*} > 10^5$. For the range of small Reynolds numbers and Prandtl numbers
close to 1, where the accuracy of Equations (4.34) and (4.38) is insufficient, the

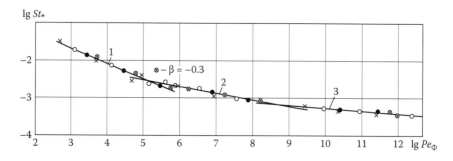

FIGURE 4.9
Dependence between St_* and Pe_Φ for $\text{Pr} < 1$ for gradient and gradientless flows
$\otimes - \beta = -0.3, \bullet - \beta = 0, \circ - \beta = 1, \times - \beta = 10$, lines 1, 2, 3 — according to Equation (4.38).

well-known relations of the type in Equation (4.33) can be used. In our notations, such a formula can be written as

$$St_* = 0.0295\Phi^{-0.2}\,Pr^{-0.6} \tag{4.39}$$

The coefficients in expressions (4.34), (4.38), and (4.39) do not depend on the pressure gradient. Therefore, these relations can not only be used for equilibrium boundary layers, but they can also be applied in the general case of pressure gradient if the flow is not close to separation. To estimate the value of β in this case, one can use the same approach as in the case of laminar flow. Because Φ and Re_{δ^*} are known, Figure 4.8 can be used to estimate β.

4.4 The Effect of the Turbulent Prandtl Number on Heat Transfer on Flat Plates [25]

In this section, the results of numerical investigation of the effect of the turbulent Prandtl number on heat transfer are given. Computations have been carried out for two values of the turbulent Prandtl number $Pr_{tb} = 0.5$ and $Pr_{tb} = 1.5$, four values of Prandtl numbers $Pr_{tb} = 10^{-1}$, 1, 10^2, 10^3, and three Reynolds numbers $Re_{\delta^*} = 10^3 (Re_x = 2.95 \cdot 10^5)$, $10^5 (7.93 \cdot 10^7)$, $10^9 (2.56 \cdot 10^{12})$.

Results of calculation in the form of the ratio $St_*/(St_*)_{Pr=1}$ are given in Figure 4.10 as a function of Prandtl number. It follows from Figure 4.10 that an increase in Pr_{tb} leads to a reduction, while a decrease in Pr_{tb} leads to an increase in heat transfer compared to that at $Pr_{tb} = 1$. An increase

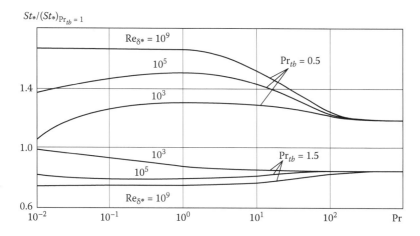

FIGURE 4.10
Dependence between $St_* / (St_*)_{Pr_{tb}=1}$ and Pr for different Re_{δ^*} and Pr_{tb}.

in Pr_{tb} to 1.5 yields less reduction in the Stanton number than the corresponding increase in the Stanton number by the decrease of Pr_{tb} to 0.5. The most significant effect of turbulent Prandtl number is when the physical Prandtl number is close to unity. For $Pr_{tb} = 0.5$, the maximum increase in Stanton number in comparison with that at $Pr_{tb} = 1$ is at $Pr = 1$ and $Re_{g*} = 10^9$, and it reaches 67%. The corresponding decrease for $Pr_{tb} = 1.5$ is less at 25%. These differences decrease with an increase of Pr and for $Pr > 10^2$, the ratio $St_*/(St_*)_{Pr=1}$ becomes practically independent of Pr as well as of Re (Figure 4.10). The comparatively small influence of Pr_{tb} at large Prandtl numbers is a result of a thin thermal boundary layer lying basically in the viscous sublayer. The effect of Pr_{tb} is also low at small Prandtl and Reynolds numbers, which is a result of significant molecular heat conduction in this case.

It follows from the computations presented here that Stanton numbers obtained with a turbulent Prandtl number equal to unity agree with experimental data. Furthermore, computed results at $Pr_{tb} = 0.5$ and $Pr_{tb} = 1.5$ do not agree with experimental data as do these for $Pr_{tb} = 1$. However, it should be indicated that this conclusion is only a recommendation for numerical computations, because as shown by many studies (e.g., [12,26,27]), turbulent Prandtl number varies appreciably across the layer (Section 4.2).

4.5 Coefficients g_k of Heat Flux Series for Nonisothermal Surfaces [9]

The coefficients g_k are calculated for the same broad range (Equation 4.32) of Reynolds and Prandtl numbers and the same four values of pressure gradient parameters: $\beta = -0.3$ (flow at stagnation point), $\beta = 0$ (gradientless flow), $\beta = 1$ and $\beta = 10$ (flows with weak and strong adverse pressure gradients). The results are given in Figures 4.11 and 4.12.

Analyzing the graphs and using Equation (4.26) yields the following conclusions:

1. The values of the coefficients g_k rapidly decrease with increasing k as in the case of laminar flow; hence, one may use only a few of the first terms in series (Equation 4.26).

2. The coefficients decrease with increasing Prandtl number as in the case of laminar flow. However, in contrast to the laminar flow, where at large Prandtl numbers the values of coefficients g_k become independent of Pr, but finite, in the case of turbulent flow they tend to zero with increasing Pr, so that beginning with some value of Prandtl number ($\approx 10^2$), the nonisothermicity effect becomes negligible.

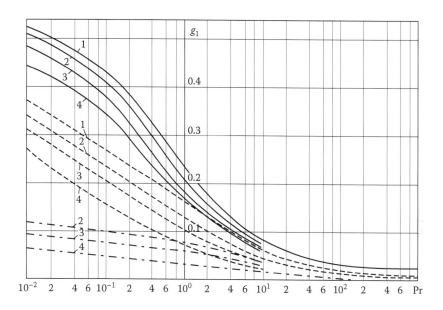

FIGURE 4.11
Dependence of coefficient g_1 on Prandtl and Reynolds numbers for turbulent flow
$(1 - \beta = -0.3, 2 - \beta = 0, 3 - \beta = 1, 4 - \beta = 10, — \text{Re}_{\delta^*} = 10^3, --- 10^5, ---- 10^9)$.

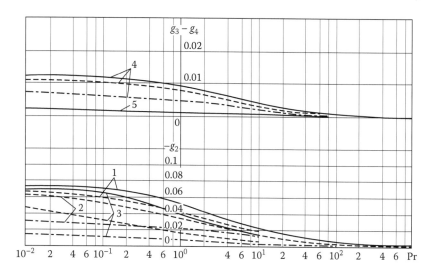

FIGURE 4.12
Dependence of coefficient g_k on Prandtl and Reynolds numbers for turbulent flow
$(1, 2, 3 - g_2, 4 - g_3, 5 - g_5, 1 - \beta = -0.3, 2 - \beta = 0, 3 - \beta = 1, 4 - \beta = 10, — \text{Re}_{\delta^*} = 10^3, ---$
$10^5, ---- 10^9)$.

3. The coefficients g_k are significantly smaller than the corresponding coefficients for laminar flow, and they decrease with increasing Reynolds number, so that the influence of nonisothermicity turns out to be greatest for laminar flow.

4. The coefficients g_1 and g_2 depend weakly on β, while the others are practically independent of β. This allows one to use Equation (4.26) to calculate heat transfer in the case of arbitrary pressure gradient, as for laminar flow. To estimate the value of β in this case, one can use the same approach as in the case of laminar flow. Because Φ and Re_{δ^*} are known, Figure 4.8 can be used to estimate β. As only two first coefficients slightly depend on β and the others are independent of it, such approximate determining of β is usually sufficient. Otherwise, the value of β can be refined by Equations (4.18) and (4.10) that were used to compute the data of Figure 4.8:

$$\beta = \frac{\mathrm{Re}_{\delta^*}[\kappa + \sqrt{C_f/2}(H-1)]}{\Phi(C_f/2)(\kappa + \sqrt{C_f/2})(H+1)} - \frac{H}{H+1} \qquad (4.40)$$

Exponents C_1 and C_2 for influence function $f(\xi/\Phi)$ in integral form (Equation 4.28) can be found in the same way as for laminar boundary layer using Equation (3.49). As before, these exponents should correspond to the values g_1 and g_2. Results of such calculations are depicted in Figures 4.13 and 4.14.

It is seen that the exponents increase with decreasing Prandtl and Reynolds numbers. The exponent C_2 increases with decreasing pressure gradient, while it turns out that the exponent C_1 is practically independent of the pressure gradient. It follows from Figures 4.13 and 4.14 that $C_1 = 1, C_2 = 0.18$ for gradientless flow when $\mathrm{Pr} = 1$ and $\mathrm{Re}_{\delta^*} = 10^3$, but under the same conditions

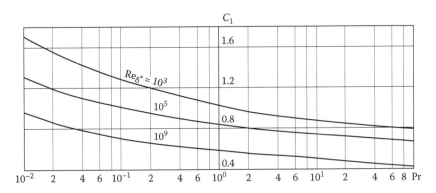

FIGURE 4.13
Dependence of exponent C_1 in the influence function on Prandtl and Reynolds numbers for turbulent flow.

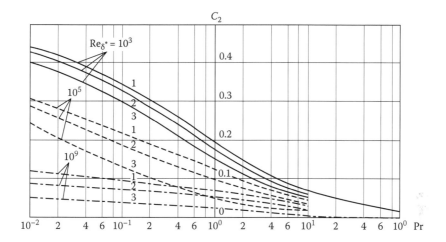

FIGURE 4.14

Dependence of exponent C_2 in the influence function on Prandtl and Reynolds numbers for turbulent flow $1 - \beta = -0.3$, $2 - \beta = 0$, $3 - \beta = 1$.

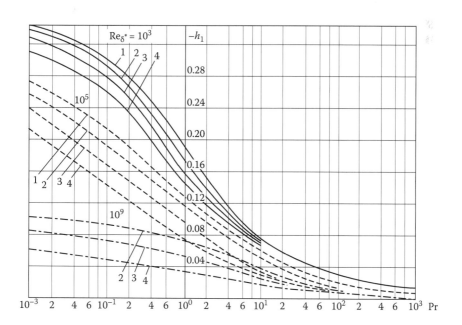

FIGURE 4.15

Dependence of coefficient h_1 on Prandtl and Reynolds numbers for turbulent flow $1 - \beta = -0.3$, $2 - \beta = 0$, $3 - \beta = 1$, $4 - \beta = 10$.

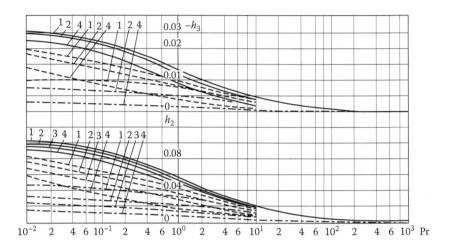

FIGURE 4.16
Dependence of coefficient h_k on Prandtl and Reynolds numbers for turbulent flow $(1 - \beta = -0.3, 2 - \beta = 0, 3 - \beta = 1, 4 - \beta = 10, — \mathrm{Re}_{\delta^*} = 10^3, - - -10^5, — - — -10^9)$.

and $\mathrm{Re}_{\delta^*} = 10^5$ the exponents are $C_1 = 0.84, C_2 = 0.1$. Hence, our calculation based on the integration of the boundary layer differential equation leads to a well-known first expression (1.46) [28,29] for a relatively low Reynolds number ($\mathrm{Re}_x \approx 5 \cdot 10^5$) and to a first expression (1.47) for a greater Reynolds number ($\mathrm{Re}_x \approx 10^8$) as experimentally obtained in the literature [30,31].

In the case of known heat flux distribution, the temperature head is determined by the same Equations (3.59) and (3.64). Coefficients h_k are obtained using Equation (3.62) and coefficients g_k for turbulent flow from Figures 4.11 and 4.12. The results for h_1 and h_2 are plotted in Figures 4.15 and 4.16.

4.6 Approximate Relations for Heat Flux in a Transition Regime [2]

If Φ_1 is the value of Görtler's variable for the point of transition, the integral expression (3.40) can be presented in the following form:

$$q_w = h_* \left[\int_0^{\Phi_1} f_1 \left(\frac{\xi}{\Phi} \right) \frac{\partial \theta_w}{\partial \xi} d\xi + \int_{\Phi_1}^{\Phi} f_2 \left(\frac{\xi}{\Phi} \right) \frac{\partial \theta_w}{\partial \xi} d\xi + \theta_w(0) \right] \qquad (4.41)$$

where indices 1 and 2 denote the first and the second regimes, respectively. It was shown in Section 3.3 that the integral (Equation 3.40) and differential (Equation 3.32) forms are equivalent. However, this is only true when the upper limit of an integral is the same variable Φ as that in the influence function. To find the equivalent differential form for the first integral of sum (Equation 4.41), it is transformed similarly using integration by parts. Corresponding processing leads to the following result:

$$\int_0^{\Phi_1} f\left(\frac{\xi}{\Phi}\right)\frac{\partial \theta_w}{\partial \xi} d\xi = \theta_w - \theta_w(0) + \sum_1^\infty g_k\left(\Phi_1/\Phi\right)\Phi^k\left(\frac{d^k\theta_w}{d\Phi^k}\right)_{\Phi=\Phi_1}$$ (4.42)

Coefficients g_k in this relation are determined by an equation similar to Equation (3.47):

$$g_k(\Phi_1/\Phi) = \frac{(-1)^{k+1}}{k!}\left[k\int_0^{\Phi_1/\Phi} (1-\zeta)^{k-1} f(\zeta)d\zeta + (\Phi_1/\Phi-1)^k - (\Phi_1/\Phi)^k\right]$$ (4.43)

Equation (4.42) gives the differential form that is equivalent to an integral with different variables Φ_1 and Φ in the upper limit and in the influence function. In the case of equality $\Phi_1 = \Phi$, Equation (4.42) transforms into Equation (3.40), and Equation (4.43) transforms into Equation (3.47). Using Equation (4.42), the differential form that is equivalent to integral one (Equation 4.41), is obtained as

$$q_w = h_*\left\{\theta_w + \sum_1^\infty g_{k2}\Phi^k\frac{d^k\theta_w}{d\Phi^k} + [g_{k1}(\Phi_1/\Phi) - g_{k2}(\Phi_1/\Phi)]\Phi^k\left(\frac{d^k\theta_w}{d\Phi^k}\right)_{\Phi=\Phi_1}\right\}$$ (4.44)

If the transition range is negligibly short, and one counts that the regime changes at the point with coordinate Φ_1, indices 1 and 2 of the influence functions in Equation (4.41) and at coefficients g_k in Equation (4.44) relate to laminar and turbulent layers, respectively. Otherwise, formulae (4.41) and (4.44) determine heat fluxes in the laminar and transition ranges. The expression for turbulent range is obtained similarly. Then, both integral and differential forms for the turbulent range ($\Phi > \Phi_2$) become as follows:

$$q_w = h_*\left[\int_0^{\Phi_1} f_1\left(\frac{\xi}{\Phi}\right)\frac{\partial\theta_w}{\partial\xi}d\xi + \int_{\Phi_1}^{\Phi_2} f_2\left(\frac{\xi}{\Phi}\right)\frac{\partial\theta_w}{\partial\xi}d\xi + \int_{\Phi_2}^{\Phi} f_3\left(\frac{\xi}{\Phi}\right)\frac{\partial\theta_w}{\partial\xi}d\xi + \theta_w(0)\right]$$ (4.45)

$$q_w = h_*\left\{\theta_w + \sum_1^\infty g_{k3}\Phi^k\frac{d^k\theta_w}{d\Phi^k} + [g_{k1}(\Phi_1/\Phi) - g_{k2}(\Phi_1/\Phi)]\Phi^k\left(\frac{d^k\theta_w}{d\Phi^k}\right)_{\Phi=\Phi_1}\right.$$

$$\left. + [g_{k2}(\Phi_2/\Phi) - g_{k3}(\Phi_2/\Phi)]\Phi^k\left(\frac{d^k\theta_w}{d\Phi^k}\right)_{\Phi=\Phi_2}\right.$$ (4.46)

Here, indices 2 and 3 relate to transition and turbulent layers, respectively. The influence function and coefficients g_k for the transition regime can be approximately estimated using any interpolation relation, for example, $g_{k,tr} = g_{k,lm}[1 - \gamma] + g_{k,tb}\gamma$, where γ is the intermittency factor [10].

Examples of calculations and analyses using formulae presented in this chapter and comparisons with experimental data are given in Chapters 5 and 6.

References

1. Clauser, F. H., 1956. The turbulent boundary layer, *Advances in Applied Mechanics*, vol. IV, pp. 1–51. Academic Press, New York.
2. Rotta, J. C., 1962. Turbulent boundary layers in incompressible flow, *Progress in Aerospace Sciences*, vol. 2, p.1.
3. Townsend, A. A., 1976. *The Structure of Turbulent Shear Flow*, 2nd ed. Cambridge University Press, Cambridge.
4. Mellor, G. L., 1966. Effects of pressure gradients on turbulent flow near a smooth wall, *Fluid Mach.* 24: 255–274.
5. Mellor, G. L., and Gibson, D. M., 1966. Equilibrium turbulent boundary layers, *J. Fluid Mach.* 24: 225–253.
6. Wilcox, D. C., 1994. *Turbulence Modeling for CFD*. DCW Industries, Inc., La Canada, California.
7. Laufer, J., 1954. The structure of turbulence in fully developed pipe flow. NACA, Rep. No. 1174.
8. Dorfman, A. S., 1971. Solution of heat transfer equation for equilibrium turbulent boundary layer when the temperature distribution on the streamlined surface is arbitrary, *Fluid Dynamics* 6: 778–785.
9. Dorfman, A. S., and Lipovetskaya, O. D., 1976. Heat transfer of arbitrary nonisothermic surface with gradient turbulent flow of an incompressible liquid within a wide range of Prandtl and Reynolds numbers, *High Temperature* 14: 86–92.
10. Schlichting, H., 1979. *Boundary-Layer Theory*, 7th ed. McGraw-Hill, New York.
11. Reynolds, A. J., 1974. *Turbulent Flows in Engineering*. Wiley, New York.
12. Kutateladze, S. S., 1973. *Near-Wall Turbulence* (in Russian). Nauka, Novosibirsk.
13. Kestin, J., and Richardson, P. D., 1963. Heat transfer across turbulent incompressible boundary layers, *Int. J. Heat Mass Transfer* 6: 147–189.
14. Zhukauskas, A. A., and Shlanchyauskas, A. A., 1973. *Heat Transfer in Turbulent Flow of Liquids* (in Russian). Mintas, Vilnyus.
15. Patankar, S. U., and Spalding, D. B., 1970. *Heat and Mass Transfer in Boundary Layers*, 2nd ed. Intertext Books, London.
16. Monin, A. S., and Yaglom, A. M., 1971. *Statistical Fluid Mechanics*, Vol. 1, edited by John Lumley.
17. Loytsyanskiy, L. G., 1960. Heat transfer in turbulent flow, *Prikl. Mat. Mekh.* 24: 950–964.
18. Kutateladze, S. S., Borishansii, V. M., Novikov, I. I., and Fedynsii, O. S., 1958. *Liquid-Metal Heat-Transfer Agents* (in Russian). Atomisdat, Moscow.

19. Dorfman, A. S., and Lipovetskaya, O. D., 1976. Heat transfer to an isothermal flat plate in turbulent flow of a liquid over a wide range of Prandtl and Reynolds numbers, *J. Applied Mechan. Technic. Physics* 17: 530–535.

20. Shi, S. W., and Spalding, D. B., 1966. Influence of temperature ratio on heat transfer to a flat plate through a turbulent boundary. *Chemical Engineering Progress* 7: 80–88.

21. Petukhov, B. S., Detlaf, A. A., and Kirilov, V. V., 1954. Experimental investigation of local heat transfer from a plate to subsonic turbulent air flow, *J. Engn. Phys. Thermod.* 24: 1761–1772.

22. Popov, V. N., 1970. Heat transfer and frictional drag in longitudinal flow of a gas with variable physical roperties past a plate, *High Temperature* 8: 311–318.

23. Fedorovich, E. D., 1959. Heat transfer to a flat plate streamlined by a turbulent boundary layer of incompressible fluid with $Pr \ll 1$, *J. Engn. Phys. Thermod.* 2: 3–11.

24. Reynolds, W. C., Kays, W. M., and Kline, S. T., 1960. A summary of experiments on turbulent heat transfer from nonisothermal flat plate, *Trans. ASME, J. Heat Transfer, Ser. C* 4: 341–348.

25. Dorfman, A. S., 1984. Influence of turbulent Prandtl number on heat transfer of a flat plate, *J. Applied Mechan. Technic. Physics* 25: 572–575.

26. Leontev, A. I., Shishov, E. V., Belov, V. M., and Afanas'ev, V. N., 1977. Mean and fluctuating characteristics of thermal turbulent boundary later and heat transfer in a diffuser. In *Teplomassoobmen–V (Heat and Mass Transfer–V [Proceedings of the Fifth All-Union Conference on Heat and Mass Transfer])*, Pt. 1, pp. 125–132. [Engl. trans. *Heat Transfer–Soviet Research* 9: 48–56]. Minsk.

27. Maksin, A. L., Petukhov, B. S., and Polyakov, A. F., 1977. Computation of turbulent heat transfer in stabilized pipe flow. In Teplomassoobmen–V (*Heat and Mass Transfer–V [Proceedings of the Fifth All-Union Conference on Heat and Mass Transfer]*), Pt. 1, pp. 73–81. [Engl. trans. *Heat Transfer–Soviet Research* 9: 1–10]. Minsk.

28. Eckert, E. R. G., and Drake, R. M., 1959. *Heat and Mass Transfer*. McGraw-Hill, New York.

29. Kays, W. M., 1969, 1980. *Convective Heat and Mass Transfer*. McGraw-Hill, New York.

30. Ambrazovichus, A. B., and Zukauskas, A. A., 1959. Investigation of heat transfer in a flow of liquid, *Trudy Akad. Nauk Lit. SSR, Ser. B* 3: 111–121.

31. Mironov, B. P., Vasechkin, V. N., and Yarugina, N. I., 1977. Effect of an upstream adiabatic zone on heat transfer in a subsonic and supersonic downstream boundary layer at different flow histories. In *Teplomassoobmen–V (Heat and Mass Transfer–V [Proceedings of the Fifth All-Union Conference on Heat* and Mass Transfer])*, Pt. 1, pp. 87–97. [Engl. trans. *Heat Transfer–Soviet Research* 9: 57–65]. Minsk.

5

General Properties of Nonisothermal and Conjugate Heat Transfer

At the beginning of Chapter 3, it is stated that in essence, a theory of conjugate heat transfer is a theory of an arbitrary nonisothermal surface. In this chapter, such a theory of conjugate convective heat transfer is formulated by studying the general properties of heat transfer of arbitrary nonisothermal surfaces using the exact and highly accurate approximate solutions obtained in Chapters 3 and 4.

The effect of different factors on conjugate heat transfer characteristics is investigated, and general relations and conclusions are obtained. In particular, some rules and general statements are summarized in the conclusion of the book in the form of an answer to the question: "Should any convective heat transfer problem be considered as conjugate?"

5.1 The Effect of Temperature Head Distribution on Heat Transfer Intensity

Presentation of the heat flux in the form of series of successive derivatives, like Equation (3.32) and others of this type, makes it possible to investigate the general effect of temperature head distribution on heat transfer intensity. Physically, such series can be considered as a sum of perturbations of the surface temperature distribution. The case when the all derivatives are equal to zero corresponds to the isothermal surface with an undisturbed temperature field. The series containing only the first derivative presents the linear disturbed temperature distribution. The series with two derivatives describes the quadratic temperature distribution, and so on. In the general case, the series consist of an infinite number of derivatives. However, because the coefficients g_k rapidly decrease with an increase in the derivative number, it is possible to consider only the first few terms for practically accurate calculations.

As indicated in Chapter 3, in the simplest case of a gradientless flow, the series contain derivatives with respect to longitudinal coordinate, while for the flows with pressure gradient, the role of longitudinal coordinate plays Görtler's variable Φ. In cases using this variable, coefficients g_k of series practically are independent on pressure gradient, telling us that Görtler's variable naturally takes into account the effect of the pressure gradient. To understand

TABLE 5.1

Relation between Coefficients g_1 and g_2

	g_1	g_2	$\dfrac{g_2}{g_1}$
1	1	1/6	1/6
2	0.6123	0.1345	0.22
3	0.380	0.135	0.36
4	½	3/16	3/8
5	2.4	0.8	1/3
6	≈ 0.5	≈ 0.05	≈ 0.1
7	≈ 0.1	≈ 0.01	≈ 0.1
8	≈ 0.2	≈ 0.04	≈ 0.04
9	≈ 0.8	≈ 0.2	≈ 0.25
10	≈ 0.4	≈ 0.06	≈ 0.15
11	1.25	0.15	0.12

Notes: Laminar layer: arbitrary $\theta_w - 1 - \Pr \to \infty$, $2 - \Pr \to 0$, arbitrary $q_w - 3 - \Pr \to \infty$, $4 - \Pr \to 0$; unsteady laminar layer: $5 - \Pr = 1$; turbulent layer: $6 - \Pr \to 0$, $\mathrm{Re}_{\delta^*} = 10^3$, $7 - \mathrm{Re}_{\delta^*} = 10^9$, $8 - \Pr = 1$, $\mathrm{Re}_{\delta^*} = 10^3$; non-Newtonian fluid: $9 - n = 1.8$, $10 - n = 0.2$, $11 - \Pr \approx 1$, $\varepsilon = 0$ moving surface.

it, one should recall that Φ is determined by the integral (Equation 3.3) of free stream velocity; hence, it takes into account the flow history.

5.1.1 The Effect of the Temperature Head Gradient

The results of calculation show that the first coefficient g_1 is significantly larger than others in all studied cases. The relations between the first and the second coefficients for different regimes and parameters are given in Table 5.1.

It follows from these data that the second coefficient is less than the first from three to ten times. This result indicates that the first derivative (i.e., temperature head gradient) basically determines the effect of nonisothermicity. Because the first coefficient is positive, it means, according to Equation (3.32), that positive temperature head gradients lead to an increase in heat flux, while the negative gradients cause a decrease in heat flux. More precisely, if the temperature head increases in the flow direction or in time, the heat transfer coefficient is greater than the isothermal coefficient, whereas the decrease of the temperature head along the flow direction or in time yields a decrease of the heat transfer coefficient compared with the isothermal one (see Equation 3.33).

However, as will be clear in what follows, an identical change of increasing and decreasing in temperature head leads to significantly different variations in the heat transfer coefficients. The reason for this is that the same absolute difference in temperature head and in corresponding heat transfer coefficient yields much greater change in relative difference in the case of falling than that of growing temperature head and heat transfer coefficient. Moreover, as shown below, the heat transfer coefficient may become zero in

the case of decreasing temperature head if the streamlined surface is sufficiently long. These general properties are shown and discussed further in
different examples. First, the examples of steady laminar flow are considered, and then the effect of unsteady and turbulent regimes is examined.

Example 5.1: Linear Temperature Head along the Plate

The results of calculation of nonisothermicity coefficient for dependence

$$\theta_w/\theta_{wi} = 1-(\theta_{we}/\theta_{wi}-1)(x/L) = 1-K(x/L), \qquad (5.1)$$

Pr > 0.5 and gradientless laminar flow are given in Figure 5.1.

Each curve corresponds to a fixed value of θ_{we}/θ_{wi}, where θ_{wi} and θ_{we}
denote the initial and ending temperature heads determining the coeffi-
cient in a linear dependence. It is seen how much curves for positive and
negative values of θ_{we}/θ_{wi} differ. For instance, if the temperature head
at the end is 1.5 to 2 times less, the heat transfer coefficient is 1.5 to 2.5
times less than that for the isothermal plate, while the same increasing
of temperature head leads only to 20% to 30% growth of an isothermal
heat transfer coefficient. Figure 5.1 also shows that the triple decreased

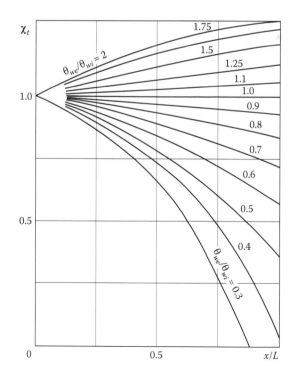

FIGURE 5.1
Variation of nonisothermicity coefficient along a plate for linear temperature head distribution
Pr > 0.5.

temperature head yields almost zero heat flux at the end of the plate. The effect of nonisothermicity for small Prandtl numbers is even more considerable. Thus, for Pr = 0.01, double decreased temperature head leads to six times less heat transfer coefficient at the end of a plate, but in the case of Pr = 0, such a decrease in temperature head turns out to be sufficient to reduce that heat flux to zero.

Example 5.2: A Plate Heated from One End (Qualitative Analysis)

This example helps us clearly understand the role of the temperature head. If the plate is passed from the heated end, the temperature head decreases in the direction of the flow. Otherwise, when the flow runs in the opposite direction, the temperature head grows in the flow direction. Because the heat transfer coefficient on an isothermal surface and the nonisothermicity coefficient in the first case both decrease in flow direction, the total heat transfer coefficient severely falls with increasing distance from the heated end. In the second case, the nonisothermicity coefficient grows in the flow direction, while the isothermal coefficient decreases in the same direction; therefore, the final heat transfer coefficient determined as a sum of these two may increase or decrease on different parts of the surface. The quantitative results can be obtained by solving a conjugate problem (see Example 6.8).

Example 5.3: Power Law Temperature Head along the Plate

There are exact self-similar solutions for this case and the power law free stream velocity distribution [1,2]. In this case, for integer exponent, $\varepsilon = m_1/(m+1)$, where m and m_1 are exponents in the power law free stream velocity and temperature head, Equation (3.33) gives an exact solution in the form of a finite sum:

$$\chi_t = 1 + \sum_1^{k=\varepsilon} g_k \varepsilon(\varepsilon-1)\dots(\varepsilon-k+1) \tag{5.2}$$

Figure 5.2 shows the results of calculation by this formula and presents the numerical solution from Reference [1] and experimental data [3] for comparison. It is seen as agreement between different data. One can see also that the numerical computing results for different Prandtl numbers are practically the same independent of Pr as it should be for Pr > 0.5, according to Figures 3.2 and 3.3.

It is known that the heat flux equals zero if the exponent in the thermal power law of self-similar solutions is $m_1 = -(m+1)/2$ (Section 1.2). This result also follows from Equation (5.2). Because in such a case $\varepsilon = -1/2$, after using for g_k relation (3.25), Equation (5.2) becomes

$$\chi_t = 1 - \sum_1^{\infty} \frac{(2k-1)!}{2^k k!(2k-1)} \tag{5.3}$$

This sum equals 1, and thus one gets the result $\chi_t = 0$.

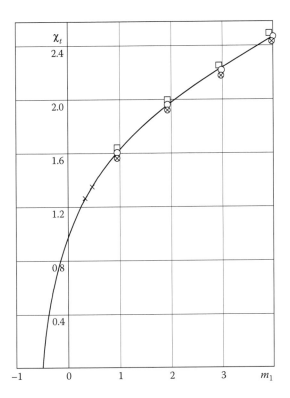

FIGURE 5.2
Variation of nonisothermicity coefficient along a plate for power law temperature head distribution Pr > 0.5. ——— Equation (5.2), numerical integration [1] for Pr: □ — 0.7, ○ — 10, ⊗ — 20, × — experiment [3].

Example 5.4: Sinusoidal Temperature Head Variation along the Plate

Because there are points with zero temperature head in which coefficients of heat transfer and nonisothermicity become infinity and, hence, meaningless, the heat flux should be calculated directly using Equation (3.32) or Equation (3.40). Using Equation (3.32) in the case of gradientless flow and temperature head distribution in the sinusoidal form yields

$$\theta_w = \theta_{w.m} \sin 2\pi(x/L) = \theta_{w.m} \sin \phi \qquad (5.4)$$

$$\tilde{q}_w = \frac{q_w L}{\lambda \theta_{wm} g_0 \sqrt{Re}} = \frac{1.45}{\sqrt{\phi}}[(1 - g_2\phi^2 + g_4\phi^4 - \ldots) \sin \phi$$
$$+ (g_1 - g_3\phi^2 + g_5\phi^4 - \ldots)\phi \cos \phi] \qquad (5.5)$$

The heat flux distribution is asymmetric (Figure 5.3) and is shifted on $3\pi/16$ regarding temperature head distribution. At the points $\phi = 2.55$

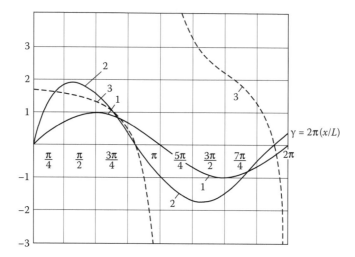

FIGURE 5.3
Variation of basic heat transfer characteristics along a plate for sinusoidal temperature head distribution $1 - \theta_w/\theta_{w.m}$, $2 - \tilde{q}_w$, $3 - \chi_t$.

and $\phi = 5.9$, heat flux equals zero. At the points $\phi = \pi$ and $\phi = 2\pi$, the temperature head is zero, and the heat flux is finite. Hence, here $h = \chi_t \to \infty$ and therefore are meaningless. At $\phi = 0$, both temperature head and heat flux are zero, but θ_w tends to zero as ϕ, while q_w according to Equation (5.5) approaches zero as $\sqrt{\phi}$. Consequently, at the leading edge, the heat transfer coefficient goes to infinity as $\sqrt{\phi}$, while the nonisothermicity coefficient at this point is finite, and according to Equations (3.33) and (5.4), it equals $(1 + g_1)$.

Example 5.5: Linear Temperature Head near Stagnation Point

It is known that velocity distribution is linear in this case $U/U_\infty = C(x/D)$, where D is cylinder diameter and U_∞ is the velocity of running on flow (see Equation 5.8). Because all coefficients g_k except g_1 practically are constant, the formula (3.25) can be used. Then, the expression for the nonisothermicity coefficient according to Equation (3.33) and linear dependence (Equation 5.1) takes the following form:

$$\chi_t = 1 - \frac{K(2\Phi/C\,\mathrm{Re})^{1/2}}{1 - K(2\Phi/C\,\mathrm{Re})^{1/2}}\left[\frac{g_1}{2} + \sum_{k=2}^{\infty}\frac{(2k-3)!}{2^k k!(2k-1)}\right] \qquad \Phi = \mathrm{Re}\int_0^{x/L}\frac{U}{U_\infty}d\xi \qquad (5.6)$$

where the coefficient in linear dependence (5.1) is $K = \theta_{we}/\theta_{wi} - 1$. Performing summation and returning to variable x gives

$$\chi_t = 1 - \frac{g_1 + \pi - 3}{2}\frac{K(x/D)}{1 - K(x/D)} \qquad (5.7)$$

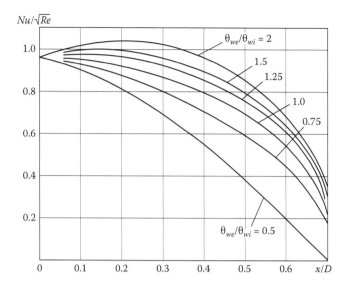

FIGURE 5.4
Local heat transfer for a transversely streamlined cylinder. Newtonian fluid. Linear temperature head distribution Pr = 0.7.

One can see the numerical results as initial parts of curves in Figure 5.4 for a cylinder.

Example 5.6: Transverse Flow Past Nonisothermal Cylinder — Linear Temperature Head

Newtonian Fluid

The free stream velocity distribution around a cylinder experimentally established is approximated by polynomial [2,4]

$$U/U_\infty = 3.631(x/D) - 3.275(x/D)^3 - 0.168(x/D)^5 \tag{5.8}$$

To obtain the function $\theta_w(\Phi)$, which corresponds to linear distribution (5.1), it is necessary to have the function $(x/D) = f(\Phi)$. This relation is found by constructing the function $\Phi(x/D)$ using the second equation in Equation (5.6), and Equation (5.8) and then deriving a function:

$$x/D = 0.74(\Phi/\mathrm{Re})^{1/2} + 0.1(\Phi/\mathrm{Re}) \tag{5.9}$$

which approximates an inverse to the $\Phi(x/D)$ function.

The nonisothermicity coefficient computed by Equation (3.33) is obtained in the form

$$\chi_t = 1 - \frac{K(\Phi/\mathrm{Re})^{1/2}[0.37(g_1 + 0.14) + 0.1g_1(\Phi/\mathrm{Re})^{1/2}]}{1 - K(\Phi/\mathrm{Re})^{1/2}[0.74 + 0.1(\Phi/\mathrm{Re})^{1/2}]} \tag{5.10}$$

The result is given in Figure 5.4 as a function $Nu/\sqrt{Re} = f(x/L)$ obtained by the following equation:

$$\frac{Nu}{\sqrt{Re}} = \frac{St_* \sqrt{Re}}{Pr} \chi_t = \frac{g_0}{Pr^2} \left(\frac{C_f \sqrt{Re}}{2} \right)^{1/2} \left(\frac{Re}{\Phi} \right)^{1/4} \chi_t \qquad (5.11)$$

which is found by using the relation $Nu = St\,Re/Pr$ and Equation (3.12). The value of $g_0(\beta)$ is given in Figure 3.1, and β is found by Equation (3.37).

It follows from Figure 5.4 that nonisothermicity significantly deforms the Nusselt number distribution along an isothermal cylinder. For $K = \theta_{we}/\theta_{wi} - 1 < 1$ when the temperature head decreases in the flow direction, the Nusselt number falls much more intensely than in the case of an isothermal surface. Thus, for $\theta_{we}/\theta_{wi} = 0.5(K = -0.5)$, the heat flux becomes almost zero at the point close to separation. Vice versa, an increasing temperature head slows the fall in the Nusselt number. Therefore, in this case at small positive values of K, the Nusselt number decreases slower than on an isothermal cylinder, while for greater nonisothermicities, the heat transfer intensity increases first (as, for example, for $\theta_{we}/\theta_{wi} = 2$) and then goes down.

Non-Newtonian Fluid

In this case, the heat transfer from a nonisothermal cylinder was computed for a theoretical free stream velocity distribution given by a sinusoidal function [5]:

$$U/U_\infty = 2\sin 2(x/D) \qquad (5.12)$$

Heat transfer from an isothermal cylinder and the nonisothermicity coefficient can be obtained by Equations (3.101) and (3.33) or (3.48), respectively. Results plotted in Figure 5.5 show that the value of heat flux and its variation along the cylinder strongly depend on the type of fluid determined by an exponent n in the law (Equation 3.94). In particular, the heat transfer coefficient is a finite value at the stagnation point for Newtonian fluids, while for non-Newtonian fluids it becomes zero for $n > 1$ or tends to the infinite for $n < 1$.

The temperature head variation also considerably affects the heat transfer intensity. Thus, in the case of decreasing temperature head, heat flux becomes zero at $\phi = 80°$, yet when the temperature head increases or is constant, the heat transfer intensity at the same point is close to the average value over a cylinder. As a result, the curve of the distribution of the heat transfer coefficient varies from parabolic for $n > 1$ and constant or increasing temperature head to s-shaped form for $n < 1$ and decreasing temperature head.

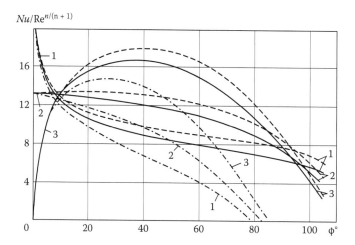

FIGURE 5.5

Local heat transfer for a transversely streamlined cylinder: non-Newtonian fluid. $s = n-1$, Pr = 1000, 1 — $n = 0.6$, 2 — $n = 1$, 3 — $n = 1.8$, ——— $\theta_{we}/\theta_{wi} = 1$, - - - - $\theta_{we}/\theta_{wi} = 1 + x/D$, - . - . -. $\theta_{we}/\theta_{wi} 1 - x/D$.

Example 5.7: Transverse Flow Past Nonisothermal Cylinder with q_w = Constant

In the case of known arbitrary heat flux distribution, the temperature head distribution can be determined by Equations (3.59) or (3.64), which for specified constant heat flux take the forms (Equation 3.66). In general, integrals in these formulae can be computed numerically. Analytical expressions can be obtained in some simple cases. For example, if $1/h_*$ can be approximated by polynomial, the value of $1/h_q$ according to Equation (3.66) is also presented by polynomial with coefficients b_i expressed via beta functions:

$$\frac{1}{h_*} = \sum_0^{i=k} a_i \Phi^i \qquad \frac{1}{h_q} = \sum_0^{i=k} b_i \Phi^k \qquad b_i = \frac{a_i B \left[\dfrac{C_1(1-C_2)+i}{C_1}, C_2 \right]}{B(C_2, 1-C_2)} \qquad (5.13)$$

Now, consider the case of constant heat flux for a cylinder using data presented in Figure 5.4. The function $Nu_*(x/D)$ for this isothermal cylinder ($\theta_{we}/\theta_{wi} = 1$) can be approximated by the following polynomial:

$$\sqrt{Re}/Nu_* = 1.04 + 0.75(\Phi/Re) - 0.83(\Phi/Re)^2 + 3.4(\Phi/Re)^3 \qquad (5.14)$$

The coefficients of the corresponding polynomial determining $1/Nu_q$ are given by Equation (5.13). Because the values of exponents C slightly depend on β, they are estimated from Figure 3.4 approximately: $C_1 = 0.92$ and $C_2 = 0.4$ for the surface part near the stagnation point

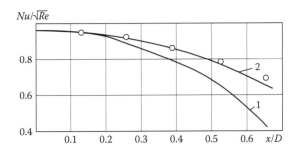

FIGURE 5.6
Local heat transfer from a transversely streamlined cylinder. $1 - \theta_w = const.$, $2 - q_w = const.$
$\circ -$ experiment.

$(\beta \approx 1)$, and $C_1 = 0.9$ and $C_2 = 0.38$ for the rest of the cylinder ($\beta \approx 0$). Then, Equation (3.66) yields

$$\sqrt{Re}/Nu_q = 1.04 + (0.44 \div 0,46)(\Phi/Re) - (0.39 \div 0.42(\Phi/Re)^2$$
$$+ (1.4 \div 1.5)(\Phi/Re)^3 \tag{5.15}$$

where a sign ($\div$) is used to indicate two values of coefficients obtained for different β and hence for different C_1 and C_2.

Calculation shows that both values of these coefficients lead to practically the same result (Figure 5.6) which is in agreement with experimental data obtained by E. P. Diban at the Institute of Technical Thermalphysics of the Ukrainian Academy of Science. For comparison of the same figure, results obtained by Equation (5.14) for an isothermal cylinder are plotted.

Example 5.8: A Jump of Heat Flux on a Transversally Streamlined Cylinder

In this case, the heat transfer coefficient after jump is determined by the influence function (Equation 3.70). This equation for the heat transfer coefficient given by Equation (5.14) becomes

$$\frac{1}{(h_q)_\xi} = \frac{1}{B(C_2, 1 - C_2)} \sum_0^{i=k} a_i \Phi^i B_\sigma \left[\frac{C_1(1 - C_2) + i}{C_1}, C_2 \right] \tag{5.16}$$

where $B_\sigma(i, j)$ is an incomplete beta function (Equation 3.71). The computing results for a cylinder with distribution (Equation 5.14) of the isothermal Nusselt number are presented in Figure 5.7.

It is assumed that the jump occurs at a point with $\phi = 30°$ before which the unheated zone exists. The distribution of the Nusselt number after a heat flux jump on the plate calculated by Equation (3.71) is also plotted on Figure 5.7. It is seen that the pressure gradient significantly affects the variation of the heat transfer intensity.

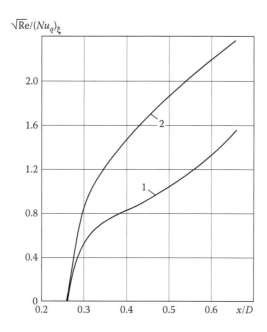

FIGURE 5.7

Variation of Nusselt number after a jump of heat flux Pr = 0.7, 1 — cylinder; 2 — plate.

Example 5.9: Transverse Flow Past Nonisothermal Cylinder — Effect of Dissipation

The results obtained with regard to energy dissipation are given in Figure 5.8. They are computed using Equation (3.12) and coefficients g_d from Figure 3.13. The calculations are made using the theoretical free stream velocity distribution (Equation 5.12) for the case of non-Newtonian fluid flow past a cylinder. This is because the effect of dissipation is significant only in the flows of incompressible fluids with large Prandtl numbers, which are typical for non-Newtonian fluids (Section 3.9.2).

The results are given in the form of the ratio $\Delta q_w / q_w$, where Δq_w is additional energy dissipation when the Eckert number equals one to heat flux q_w calculated without regard to dissipation. The additional dissipation energy is proportional to the Eckert number, and because the velocity of incompressible flows is usually relatively small, the additional heat flux associated with dissipation is small as well. Nevertheless, in some cases, the effect of dissipation is appreciable. In particular, this is true in the case of decreasing temperature head when the self heat flux without regard to dissipation is small. For constant or increasing temperature head, the dissipation energy is usually small. For example, if the Eckert number is 1/500, the additional heat flux is about 10% for a constant temperature head and about 5% for increasing one at the point with $\phi \approx 70°$ where the effect of dissipation reaches maximum (Figure 5.8).

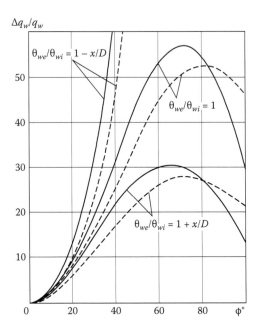

FIGURE 5.8
Additional heat flux caused by energy dissipation on a cylinder transversely streamlined by non-Newtonian fluid. $Ec = (U_\infty^2/c_p\theta_w) = 1$, $s = n - 1$, $Pr = 1000$, ——$n = 0.6$, - - - - $n = 1.8$.

Note that the expressions obtained ignoring dissipation are valid for the flows with significant dissipation if one defines the temperature head using the adiabatic wall temperature $\theta_{ad} = T_{ad} - T_\infty$ instead of the usual temperature head $\theta_w = T_w - T_\infty$. The adiabatic wall temperature is computed by applying the recovery factor given by Equations (3.104) or (3.105).

5.1.2 The Effect of Flow Regime

Table 5.1 shows that the first coefficient g_1 in series, which basically determined the nonisothermicity effect, significantly depends on flow regime. Its value varies from largest $g_1 = 2.4$ at the time derivative in the case of laminar unsteady flow to negligible, small magnitude for turbulent flow of fluids with large Prandtl numbers. Nevertheless, in all cases, the qualitative effect of the temperature head gradient on the heat flux intensity is the same as in the case of laminar flow discussed above. The quantitative results for different flows can be seen in examples given below.

Example 5.10: Comparison between the Effects of Nonisothermicity in Turbulent and Laminar Flows — Linear Temperature Head [6]

Coefficients g_k for turbulent flow are less than those in the case of a laminar regime. The higher are the Reynolds and Prandtl numbers, the less

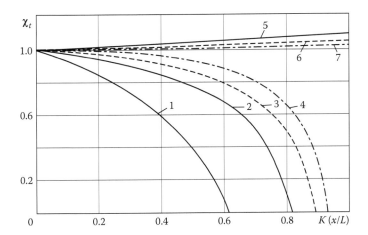

FIGURE 5.9
Variation of nonisothermicity coefficient for different flow regimes. Laminar flow: 1-$K < 0$, turbulent flow: 2, 3, 4-$K < 0$, 5, 6, 7- $K > 0$. ———Re$_{\delta^*} = 10^3$, - - - - 10^5, -. -. -. 10^9.

are coefficients g_k, and, correspondingly, the less the effect of nonisothermicity. The difference between these effects for two regimes is shown in Figure 5.9 where the values of the nonisothermicity coefficient are defined for the linear temperature head (Equation 5.1) and Pr = 1.

It is seen that in spite of small coefficients g_k, nonisothermicity strongly affects the heat transfer intensity in turbulent flows if the temperature head decreases. In that case, the effect of nonisothermicity is not as strong as in laminar flows, but if the surface is long enough, the heat flux also reaches zero. This is true for all cases except the turbulent flows of fluid with large Prandtl numbers, for which the effect of nonisothermicity is negligible (Figures 4.11 and 4.12).

Example 5.11: Different Temperature Head and Heat Flux Variations for Turbulent Flow — Comparison with Experimental Data [6]

The effect of nonisothermicity was studied experimentally in gradientless turbulent flows of air by Leontev et al. [7]. They obtained data for the increasing and decreasing linear temperature heads as well as for exponential variations of temperature head and heat flux. In Figure 5.10, the results of calculation are compared with these data.

Line 1 corresponds to isothermal gradientless flow and is computed by Equation (4.39). The rest of the curves are obtained using series (Equation 4.25) or integral form (4.28). For gradientless flow and linear temperature head (Equation 5.1), expression (4.25) after using Equation (4.26), becomes

$$St = St_* \left[1 - \frac{g_1}{(Re/K\,Re_x) - 1} \right] \tag{5.17}$$

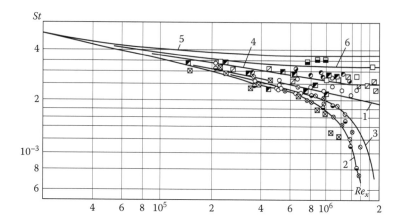

FIGURE 5.10
Comparison between results obtained by calculation for different temperature heads and experimental data [7] for turbulent flow. 1 — 6 calculation, experiment: temperature heads θ_w
7 — ○ — 158.6; 8 — ⊙ — [222 — 150(x/L)]; 9 — ⊗ — [137 — 81(x/L)]; 10 — ● — [204 — 140(x/L)]; 11 — ● — [44 + 170(x/L)]; 12 — □ — 0.19 exp 7(x/L); 13 — ⊠ — [34 + 159(x/L)]; 14 — ⊠ — [159 − 100(x/L)]; 15 — ■ — [3.8 + 110(x/L)]; 16 — ■ — q_{CT} = 3.4 exp 5.8(x/L).

Experiments are carried out under low Reynolds numbers. According to Figure 4.11 for Pr ≈ 1 and low Reynolds number, the estimation gives $g_1 = 0.2$. The coefficient K in the linear law in experiments corresponds to the following inverse values: $1/K = 1.46, 1.49, 1.59, 1.68$ for decreasing and $1/K = 0.215, 0.258, 0.346$ for increasing temperature heads. Curves 2, 3, 4, and 5 are calculated by formula (Equation 5.17) for two limiting values of coefficient K for decreasing (curves 2 and 3) and for increasing (4 two curves coincide) temperature heads. Line 5 represents exponential increasing temperature head $\theta_w = \theta_{wi} \exp[K(x/L)]$ with $K = 7$. Because series diverge slowly for the large values of K, the integral form is used in this case. According to Figures 4.13 and 4.14 for small Reynolds numbers and Pr ≈ 1, the exponents are $C_1 = 1$ and $C_2 = 0.2$. Transforming Equation (4.28) to Stanton number, yields for line 5 the following expression:

$$St = St_* \left\{ \exp[-K(x/L)] + K(x/L)^{0.2} \int_0^{x/L} \xi^{-0.2} \exp(-K\xi)d\xi \right\} \qquad (5.18)$$

To calculate temperature head distribution that corresponds to exponential dependence of the heat flux $q_w = q_{w.i} \exp[K(x/L)]$ with $K = 5.8$, one should apply Equation (3.65). Using the same values of $C_1 = 1$ and $C_2 = 0.2$ as in Equation (5.18), and computing the corresponding Stanton number, one gets a formula to which corresponds curve 6:

$$\frac{1}{St} = 6.35\,Re^{0.2}\,Pr^{0.6}\int_0^{x/L}\xi^{-0.8}\exp(K\xi)d\xi \tag{5.19}$$

Figure 5.10 shows that there is a reasonable agreement between both results. The points representing the experimental data are close to the corresponding theoretical results: curves 2 and 3 to points 8, 9, 14 (linear

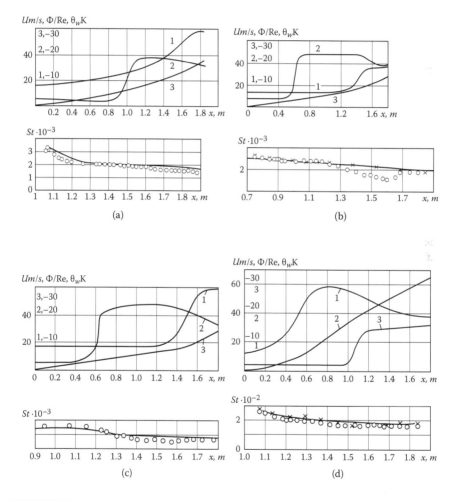

FIGURE 5.11

Comparison between results obtained by calculation for different stepwise temperature heads and various pressure gradients and experimental data [9] for turbulent flow, ×-numerical integration, (b) [10], (e) [11] $1-U$, $2-\theta_w$, $3-\Phi/Re$. Different temperature head variations containing one jump under increasing free stream velocity gradually (a) or stepwise (b) and (c) or under first increasing and then decreasing free stream velocity (d); several temperature head jumps under gradually increasing free stream velocity (e).

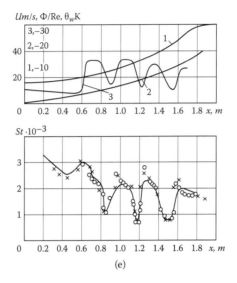

FIGURE 5.11 (*Continued*)

decreasing temperature head); curve 4 to points 11, 13, 15 (linear increasing temperature head); curve 6 to point 12 (exponential growing temperature head); and curve 5 to point 16 (exponential growing heat flux).

Example 5.12: Stepwise Temperature Head — Comparison with Experimental Data [8]

Moretti and Kays [9] experimentally studied heat transfer in the case of stepwise temperature head variations. In Figure 5.11, the comparison between calculations and their experimental results is shown. Experimental conditions differ from each other by temperature head variation or by free stream velocity distribution. The free stream velocity along the surface increases gradually (Figures 5.11a and 5.11e) or stepwise (Figures 5.11b and 5.11c), or first increases and then decreases as in Figure 5.11d. The temperature head variations contained one jump, as in four cases (Figures 5.11a, 5.11b, 5.11c, and 5.11d), or several jumps as shown in Figure 5.11e. Because the temperature head distributions contain jumps, the calculations are performed using integral form (Equation 4.28). For example, in the case presented in Figure 5.11a, this equation transformed to Stanton number becomes

$$St = \frac{St_*}{\theta_w}\left\{1+\left[1-\left(\frac{\Phi_1}{\Phi}\right)^{C_1}\right]^{-C_2}\Delta\theta_w + \int_{\Phi_2}^{\Phi}\left[1-\left(\frac{\xi}{\Phi}\right)^{C_1}\right]^{-C_2}\frac{d\theta_w}{d\xi}d\xi\right\} \qquad (5.20)$$

Here, Φ_1 and Φ_2 relate to the start and end points of the temperature head jump. In Figure 5.11 by sign × are given results obtained numerically in the literature [10,11].

The comparison shows that calculations by formula (Equation 4.28) practically coincide with numerical results and are basically in agreement with experimental data. More importantly, the calculations by Equation (4.28), unlike other analytical methods, show agreement not only for increasing temperature heads but for decreasing ones, as well (Figure 5.11e). More noticeable difference between calculation and experimental results exists in the case of sharp increasing free stream velocity (Figures 5.11b and 5.11c). This is associated with changes in boundary layer structure, which could not be taken into account using only the turbulent flow model. The correction can be made by empirical relations [9].

5.1.3 The Effect of Pressure Gradient

The effect of pressure gradient on the nonisothermicity coefficient is determined basically by the first term of the series that is convenient to use in the following form:

$$\Phi \frac{d\theta_w}{d\Phi} = \frac{U_{av}}{U} x \frac{d\theta_w}{dx} \tag{5.21}$$

It follows from this expression that the value of ratio U_{av}/U can be used as an approximate measure of the pressure gradient effect on heat transfer of a nonisothermal surface. For example, for self-similar flows $U_{av}/U = 1/(m+1)$. This means that other conditions being equal, the first term of the series for the flow near the stagnation point is half as much as in the case of the gradientless boundary layer. In general, under known free stream velocity distribution, the value of ratio U_{av}/U is easy to estimate.

5.2 Gradient Analogy and Reynolds Analogy

Some investigators [7,12,13] noticed that the temperature head gradient has the same qualitative effect on the heat transfer coefficient as the free stream velocity gradient on the friction coefficient. Below it is shown that this analogy holds not only for the first derivatives, which characterize the effect of the corresponding gradients, but also for all other subsequent derivatives [14].

Let us introduce a nonisotachicity coefficient $\chi_f = C_f/C_{f*}$ that is similar to a nonisothermicity one and shows how much the friction coefficient in a flow with a variable free stream velocity is more or less than that in a flow with constant velocity. Because the dynamic boundary layer equation is nonlinear, the exact solution could not be presented in the form of a sum of subsequent derivatives like series (3.32) for heat flux. However, the approximate solution in such a form can be obtained using a method suggested in Reference [15] which is based on a linearized dynamic boundary layer equation.

Linearizing is achieved by substituting the self-similar velocity profile for the actual one in the right part of the Prandtl-Mises-Görtler boundary layer equation. Transforming the Prandtl-Mises equation in the form (Equation 1.3) to Görtler's independent variables (Equation 3.3) leads to the following equation with corresponding boundary conditions:

$$2\Phi\frac{\partial Z}{\partial\Phi}-\varphi\frac{\partial Z}{\partial\varphi}-\frac{u}{U}\frac{\partial^2 Z}{\partial\varphi^2}=0 \quad \varphi=0, Z=U^2(\Phi) \quad \varphi\to\infty Z\to 0 \quad Z=U^2-u^2 \quad (5.22)$$

Linearization yields a system containing linear differential and integral equations:

$$2\Phi\frac{\partial Z}{\partial\Phi}-\varphi\frac{\partial Z}{\partial\varphi}-\omega(\varphi,\beta)\frac{\partial^2 Z}{\partial\varphi^2}=0 \qquad \int_0^\infty\left[\frac{u}{U}-\omega(\varphi,\beta)\right]\frac{\partial^2 Z}{\partial\varphi^2}=0 \qquad (5.23)$$

Although this integral equation is similar in form to integral equation (3.36), this equation cannot be easily integrated like the latter, because $Z=U^2-u^2$ depends on the free stream velocity and, hence, parameter β depends on it as well. This is a significant difference from the case of a thermal problem where β does not depend on the temperature head if physical properties are considered as independent of temperature. This difference arises from the fact that the dynamic boundary layer equation unlike the thermal one is nonlinear. Despite the difficulties, examples of approximate solutions to system (Equation 5.23) can be found in References [15] and [16].

For the present purpose of comparing coefficients $\chi_t=h/h_*$ and $\chi_f=C_f/C_{f*}$, there is no need to solve the system (Equation 5.23). Because differential Equation (5.23) is analogous to Equation (3.1), its solution can be presented in a similar form of series of subsequent derivatives of free stream velocity square:

$$Z=\sum_{k=0}^\infty B_k(\varphi)\Phi^k\frac{d^k U^2}{d\Phi^k} \qquad (5.24)$$

Substituting Equation (5.24) into Equation (5.23) and proceeding as in Section 3.1, one finally gets

$$C_f=2\Phi^{-1/2}\sum_{k=0}^\infty b_k\frac{\Phi^k}{U^2}\frac{d^k U^2}{d\Phi^k} \qquad \chi_f=1+\sum_{k=1}^\infty\tilde{b}_k\frac{\Phi^k}{U^2}\frac{d^k U^2}{d\Phi^k} \qquad (5.25)$$

Equations (5.25) and (3.33) for χ_t are analogous in form, and for each dynamic term with derivative of velocity square U^2 in the first, there is a corresponding and similar thermal term with derivative of temperature head θ_w in the second. Equation (5.25) yields a familiar result: when gradientless $(dU/dx=0)$ flow is compared with favored $(dU/dx>0)$ and unfavored $(dU/dx<0)$ flows under otherwise equal conditions, the friction coefficients are greater in the first case and lesser in the second. Similar, as it follows from Equation (3.33), the temperature head affects the heat transfer coefficient. Because the first coefficient g_1 is positive, increasing $(d\theta_w/dx>0)$ or decreasing $(d\theta_w/dx<0)$ the temperature head in the flow direction leads to growing or lessening the heat transfer coefficient in comparison with that for an isothermal surface. The same is valid for unsteady heat transfer, because in Equation (3.83), both coefficients g_{10} and g_{01} are positive. Increasing or decreasing the temperature head in the flow direction or in time leads to growing or lessening the heat transfer coefficient in comparison with that for an isothermal surface.

It is known that the different influence of the favored and unfavored velocity gradients on the friction coefficient is explained by deformation of velocity profiles in a dynamic boundary layer. Similarly, the different effect of increases and decreases in temperature head on the heat transfer coefficient is explained by deformation of the temperature profiles in a thermal boundary layer.

Let the surface temperature be higher than the temperature of flowing fluid. If the wall temperature increases in the flow direction, the descended layers of fluid of the adjoining wall come into contact with the increasingly hotter wall. Because of the fluid inertness, these layers warm up gradually. As a result, the cross-sectional temperature gradients near a wall turn out to be greater than in the case of constant wall temperature, which leads to higher heat transfer coefficients than those obtained for an isothermal surface. Analogously, in the case of decreasing surface temperature in the flow direction, the cross-sectional temperature gradients near a wall and the heat transfer coefficients as well become less than those for an isothermal surface. The same situation exists in the case of cooler than fluid surface temperature. The difference is that in this case, the absolute values of the falling temperature head and lesser heat transfer coefficients correspond to an increase in the flow direction wall temperature; inversely, the growing absolute values of the temperature head and higher heat transfer coefficients correspond to a decrease in the flow direction wall temperature.

Considering the second terms of the series for χ_t and for χ_f, one concludes that because coefficients g_2 and b_2 are negative, the effect of the second terms that depend on the curvature of the $\theta_w(\Phi)$ and $U^2(\Phi)$ curves is opposite: a positive curvature leads to a reduction of the friction and heat transfer coefficients, and a negative curvature yields in increasing these coefficients under otherwise equal conditions. However, this case is more complicated, and the result of comparing depends on a concrete situation. For example, if we compare nearly linear convex and concave $\theta_w(\Phi)$ and $U^2(\Phi)$ curves with linear dependence of Φ, we arrive at the opposite result: the friction and heat transfer coefficients turn out to be smaller in the second case (negative curvature) and larger in the first (positive curvature). The reason for this is that the gradient, which has a more significant role, also changes with a change in the curvature.

It is seen from Equations (3.33) and (5.25) that the effect of the third, fifth, and other odd derivatives is of the same nature as that of the first, while the effect of even derivatives is of the same nature as that of the second. This follows from the fact that all odd coefficients of both series are negative, and all even coefficients are positive.

In the case of unsteady heat transfer, the effect of the second derivatives is the same as in the case of steady heat transfer, because contrary to the first coefficients, (as well as in steady heat transfer), the coefficients g_{20} and g_{02} are negative. The other coefficients in Equation (3.83), g_{k0} and g_{0i} are positive for odd and negative for even numbers, and coefficients g_{ki} are positive if $(k+i)$ is odd and negative if $(k+i)$ is even. According to the sign of coefficients g_{ki}, the derivatives of higher order influence the intensity of heat transfer like the first or the second derivative.

Although Equations (3.33) and (5.25) are similar in form, there is significant difference between them: the coefficients g_k in Equation (3.33) are either essentially constant or depend weakly on Pr and β, but they do not depend on the temperature head, while the coefficients b_k in Equation (5.25) are functions of the free stream velocity. In this dependence of the coefficients on the velocity in series (5.25), determining the effect of the same velocity is a difference between a nonlinear dynamic and a linear thermal problem.

The difference between nonlinear dependence of χ_f on the first term f_1 of series (5.25) and the linear dependence of χ_t on the first term f_{t1} of series (3.33) is clear from Figure 5.12. It follows from Figure 5.12 that negative temperature head gradients, like negative free stream velocity gradients, lead to a reduction of the corresponding coefficients, χ_t and χ_{f1}, and positive gradients lead to an increase of these coefficients. However, the effect of the velocity gradient on the friction coefficient is always greater than the effect of the temperature head on the heat transfer coefficient. It follows also from this graph that the effect of nonisothermicity increases with a decrease in Prandtl number, so the highest effect of nonisothermicity in steady flows is in the case of liquid metals.

Applying nonisothermicity and nonisotachicity coefficients makes it possible to study how much the pressure and temperature head gradients violate the

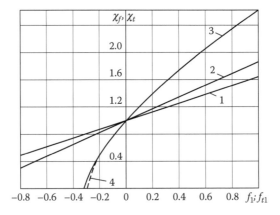

FIGURE 5.12
Dependence of nonisotachicity and nonisothermicity coefficients on first terms of series (5.25) and (3.33). $\chi_t(f_{t1})$: $1 - \text{Pr} > 0.5$, $2 - \text{Pr} = 0.01$, $\chi_f(f_1)$ 3 — Equation (5.25), 4 — numerical integration [2].

Reynolds analogy according to which ratio $2St/C_f$ equals one for gradientless laminar flow and isothermal surface for Pr = 1. Using Equation (3.12) yields

$$\frac{2St}{C_f} = \frac{St_* \chi_t}{C_f/2} = \frac{g_0}{\Pr(C_f/2)^{1/2} \Phi^{1/4}} = \frac{g_0}{0.576\,\Pr} \frac{\chi_t}{\sqrt{\chi_f}} \qquad (5.26)$$

It follows from this equation that the Reynolds analogy coefficient varies as the nonisothermicity coefficient and is inversely proportional to the square root of the nonisotachicity coefficient. In particular, this means that the Reynolds analogy coefficient is greater for flows with positive pressure gradients and lesser for flows with negative pressure gradients than for gradientless flow. Analogously, it follows from Equation (5.26) that an increasing temperature head leads to a greater Reynolds analogy coefficient and a decreasing temperature head leads to a lesser Reynolds analogy coefficient in comparison with that for an isothermal surface. Because the effects of both factors are opposite, there should be a condition that yields the same Reynolds analogy coefficient as for gradientless flow past an isothermal surface. Corresponding values of χ_t and χ_f can be easily obtained from Equation (5.26).

5.3 Heat Flux Inversion [8,22]

It can be shown that for a certain relationship between terms of series, coefficients χ_f and χ_t vanish. In this case, the local friction or the local heat flux vanishes as well. Ordinarily, when the primary role is played by the first series term, this may occur at negative gradients, in which case the free stream velocity or temperature head decreases along a surface. Vanishing of the friction accompanied by separation of the boundary layer is known to be associated with a deformation of the velocity profiles in the boundary layer of divergent flows [5]. Analogously, the vanishing of the heat flux in a flow with decreasing temperature head is associated with the deformation of the temperature profiles in a thermal boundary layer. Some examples of such flows are given in the literature [17,18].

Figure 5.13 shows profiles of the relative excess temperature in the thermal boundary layer for gradientless flow for linear temperature head distribution (Equation 5.1) and Pr ≈ 1. They are calculated by the first term of series (3.4) in the following form:

$$\frac{T_w - T_\infty}{T_{wi} - T_\infty} = [1 - G_0(\varphi)]\left\{1 - K\frac{x}{L}\right\} + G_1(\varphi)K\frac{\dot{x}}{l} \qquad (5.27)$$

Functions $G_0(\varphi)$ and $G_1(\varphi)$ are determined by Equation (3.7) with boundary conditions (Equation 3.8). The nonisothermicity coefficient and dimensionless

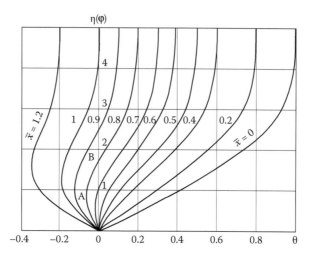

FIGURE 5.13
Deformation of the excess temperature profile in laminar thermal boundary layer for the linear decreasing temperature head, $Pr = 0.7$, $\bar{x} = K(x/L)$.

heat flux for the linear temperature head distribution (Equation 5.1) are defined according to Equations (3.33) and (3.32) as follows:

$$\chi_t = 1 - \frac{g_1 K(x/L)}{1 - K(x/L)} \qquad \bar{q}_w = \frac{q_w L}{\lambda \theta_{wi} g_0 \sqrt{K \, Re}} = \frac{0.576}{\sqrt{K(x/L)}} [1 - (1 + g_1) K(x/L)]$$

(5.28)

These relations are plotted in Figure 5.14.

Figure 5.13 shows how the initial profile deforms into a profile with an inflection point and then converts into a profile with a vertical tangent at a wall. Although the temperature head is finite at this point, the local heat flux vanishes and changes its direction (Figure 5.14). The coordinate of this point at which the inversion of heat flux occurs is obtained from the first part of Equation (5.28) by equating it to zero: $K(x/L)_{\text{inv}} = 1/(1 + g_1)$.

The deformation-profile pattern for the temperature shown in Figure 5.13 is analogous to a familial deformation-profile pattern of the velocity which leads to a separation of the boundary layer. Nevertheless, these phenomena are radically different. Separation leads to restructuring of the flow, to the appearance of a reverse flow, and to the actual destruction of the boundary layer, so that boundary layer equations are no longer valid beyond the separation point. In contrast with this situation, the thermal boundary layer equations remain valid beyond the point of zero heat flux, because only the direction of the heat flux changes at this point, and the hydrodynamics

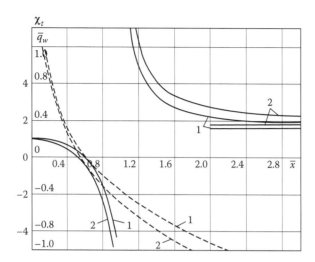

FIGURE 5.14
Variation of the heat flux and nonisothermicity coefficient along the plate for linear decreasing temperature head. —— χ_t, - - - - $\overline{q}_w$, 1 — Pr > 0.5, 2 — Pr = 0.01, $\overline{x} = K(x/L)$.

remains the same. Beyond the inverse point of the heat flux in the region before the point $K(x/L) = 1$, at which the temperature head vanishes, the heat flux is directed from the liquid to the wall, although the wall temperature in this region is higher than that of the fluid outside of the boundary layer.

Physically, this is explained as follows [17]: Because the profiles in Figure 5.13 are plotted in excess temperatures, it means that the temperature of the surface is higher than that of the fluid. In such a case of wall temperature decreasing, the descended layers of fluid of the adjoining wall come into contact with a cooler wall. As a result, the temperature difference between the wall and the layers of fluid near the wall decreases and finally becomes zero at the inverse point with coordinate $K(x/L)_{inv} = 1/(1 + g_1)$. After this point, the temperature of the fluid near the wall turns out to be above the wall temperature, because the wall temperature continues to decrease. Thus, before the inverse point, the heat flux is directed from the wall to the fluid, and after this point, close to the surface the heat flux direction changes, so that the heat flux near the wall up to the point B in Figure 5.13 is directed from the fluid to the wall. Because of this, the boundary layer near the wall is divided vertically by point A, where the heat flux vanishes, into two regions. In the region adjacent to the wall, the heat flux is directed toward the wall; in the other region, it is directed away from the wall. At the end of this surface region, at the point $K(x/L) = 1$, the flow temperature outside the boundary layer and the surface temperature become equal, and the temperature head vanishes. Nevertheless, the heat flux at this point does not vanish, so the concept of a heat transfer coefficient and of a nonisothermicity coefficient becomes, as

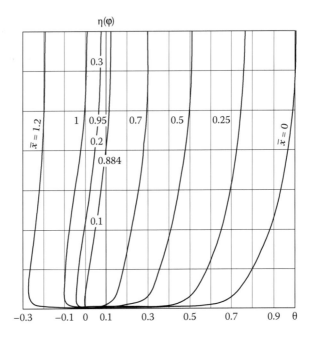

FIGURE 5.15
Deformation of the excess temperature profile in turbulent thermal boundary layer for the linear decreasing temperature head. Pr = 0.7.

first indicated by Chapman and Rubesin [18], meaningless. The functions $h(x)$ and $\chi_t(x)$ at this point become discontinuous (Figure 5.14). At point $K(x/L) = 1$, the temperature head changes its direction, and the heat flux, which changed direction before at point $K(x/L)_{inv}$, continues to increase to infinity as $x \to \infty$.

An analogous temperature deformation-profile pattern in the turbulent boundary layer is shown in Figure 5.15.

Example 5.13: Heat Transfer Inversion in Unsteady Flow — Linear Temperature Head [19]

As mentioned above, the effect of nonisothermicity is the highest for the case of a time-dependent temperature head. According to Figure 3.6, the coefficient in Equation (3.83) at derivative with respect to time is $g_{01} \approx 2.4$ when the dimensionless time is $z > 2.4$, and coefficients g_{ki} do not depend on time. The coefficient in this equation at the derivative with respect to the coordinate for the same case of gradientless flow and Pr ≈ 1 is $g_{10} \approx 0.6$ (Figure 3.2). Thus, if both derivatives are of the same order, the effect induced by the time variation temperature head is $g_{01}/g_{10} \approx 2.4/06 = 4$ times greater than that caused by coordinate variation.

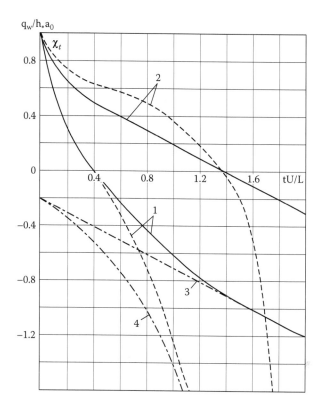

FIGURE 5.16
Variation of surface heat flux and nonisothermicity coefficient for temperature head linearly decreasing in time $\theta_w = a_0 - a_1 t$. ------q_w/h_*a_0, - - - - χ_t, $1 - x/L = 1$, $2 - x/L = 0.25$, - . - . - . solution without initial conditions [20] $3 - q_w/h_*a_0$, $4 - \chi_t$.

Figure 5.16 shows the time variation of heat flux and nonisothermicity coefficient obtained by Equations (3.83) and (3.87) for a linear decreasing temperature head with time $a_0 - a_1 t$. In this case, these equations take the following form:

$$\frac{q_w}{h_*a_0} = 1 - C_t \left[Z + g_{01}(z)\frac{x}{L} \right], \qquad \chi_t = 1 - \frac{C_t g_{01}(z)(x/L)}{1 - C_t z},$$

$$C_t = \frac{a_1 L}{a_0 U}, \qquad Z = \frac{tU}{L} \tag{5.29}$$

where coefficient $g_{01}(z)$ is given by Figure 3.6. It is seen that the heat flux depends not only on time but also on the coordinate x, despite the surface temperature depending only on time. It follows from Equation (3.83) that this is always true for unsteady heat transfer because the terms containing derivatives with respect to time depend on coordinate x as well.

As follows from Equation (5.29) and from a linear function $\theta_w = a_0 - a_1 t$, the heat flux and temperature head become zero at different dimensionless times determined by the first and second equations, respectively:

$$C_t \left[Z + g_{01}(z) \frac{x}{L} \right] = 1 \qquad \frac{C_t g_{01}(z)(x/L)}{1 - C_t z} = 1 \qquad (5.30)$$

In the case of data plotted in Figure 5.16 which are calculated for $x/L = 1 \, x/L = 0.25$, and $C_t = 0.5$, one finds $z_{inv} = 0.4 \, (x/L = 1), z_{inv} = 1.4 \, (x/L = 0.25)$, and $z_{h \to \infty} = 2$. It follows from Figure 5.16 that in this case as well as in the case of steady flow, the temperature head is not zero at the time of inversion z_{inv} when the heat flux becomes zero, and the heat transfer and nonisothermicity coefficients become zero. Correspondingly, the heat flux is not zero at the time $z_{h \to \infty}$ when the temperature head becomes zero and the heat transfer coefficient becomes infinite. Then, for the time $z > z_{inv}$, the heat flux becomes negative as well as in the case of steady heat transfer after inversion (Figure 5.14).

Example 5.14: Heat Flux Inversion in Gradientless Flow for a Parabolic and Some Other Temperature Heads [2]

For a parabolic temperature head, according to Equation (3.32), one obtains

$$\theta_w / \theta_{wi} = 1 + K_1(x/L) + K_2(x/L)^2,$$

$$\chi_t = 1 + (1 + g_1)K_1(x/L) + [1 + 2(g_1 + g_2)]K_2(x/L)^2 \qquad (5.31)$$

Setting $\chi_t = 0$ leads to a quadratic equation. Knowing that coordinates of inversion x_{inv1} and x_{inv2} are positive, one reaches the following conclusions:

1. There are two inversion points if three inequalities are satisfied:

$$\frac{x_{inv1}}{L} + \frac{x_{inv2}}{L} = -\frac{K_1(1 + g_1)}{K_2 g_\Sigma} > 0,$$

$$\frac{x_{inv1}}{L} \cdot \frac{x_{inv2}}{L} = \frac{1}{K_2 g_\Sigma} > 0. \quad \left[\frac{K_1(1 + g_1)}{K_2 g_\Sigma} \right]^2 - \frac{1}{K_2 g_\Sigma} > 0, \qquad (5.32)$$

where $g_\Sigma = 1 + 2g_1 + 2g_2$. Solution of inequalities (5.32) and calculation of the ordinate of vortex of the parabola (5.31) for $\mathrm{Pr} \approx 1$ yields

$$K_1 < 0, \ K_2 > 0, \ K_1^2 / K_2 > g_\Sigma / (1 + g_1)^2 \approx 3,$$

$$\theta_{w.\min} / \theta_{w.i} = \left[1 - K_1^2 / 4K_2 \right] < 1/4 \qquad (5.33)$$

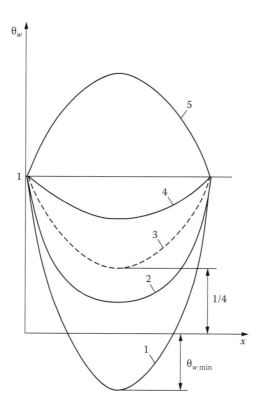

FIGURE 5.17
Different parabolic temperature head distributions.

It follows from these results that two inversion points exist if the parabola is located below line $\theta_w = 1$ so that the temperature head first decreases at least four times and then increases (Figure 5.17, curves 1 and 2). Thus, the inversion may occur even in the case when the temperature head does not change the sign, but the last inequality (Equation 5.33) is satisfied, and the parabola is located under the limiting curve (curve 3 in Figure 5.17). This is explained by the fluid inertness as well as the other phenomena associated with deformation of temperature profiles.

2. There are no inversion points. Any parabola (Equation 5.31) with $K_1 > 0$ conforms to this case. Such curves (curve 4) are located under line $\theta_w = 1$ but above the limiting curve 3 (Figure 5.17). This follows from Equation (5.32) because in this case, if $K_2 > 0$, the first inequality (Equation 5.32) is not valid, and if $K_2 < 0$, the second one is not satisfied.

3. There is one inverse point. In this case, the first and the third inequalities (5.32) and the second inverse inequality should be satisfied. Thus, it should be $K_2 < 0$, and hence, such parabola is located above line $\theta_w = 1$ so that the temperature head first increases and

then decreases (curve 5). In this case, the inverse point is located on the falling down part of the parabola. To show this, consider a solution of quadratic equation $\chi_t = 0$ and take into account that $g_1 > |g_2|$. Then, the following inequality is obtained:

$$\frac{x_{inv}}{L} = \frac{1+g_1}{g_\Sigma}\frac{x_v}{L}\left(1+\sqrt{1-\frac{4K_2 g_\Sigma}{K_1^2\left(1+g_1\right)^2}}\right) \tag{5.34}$$

$$\frac{x_{inv}}{L} > 2\frac{1+g_1}{g_\Sigma}\frac{x_v}{L} \approx 1.5\frac{x_v}{L} \tag{5.35}$$

where x_v is the vertex coordinate. Thus, the coordinate of inverse point is greater than that of parabola vertex, and hence, it is located on the falling down part of the curve.

For a polynomial temperature head, the number of inverse points can be equal n or less depending on polynomial coefficients. Similarly, the inverse for other temperature head distributions can be studied. For example, for sinusoidal distribution (Equation 5.4), one finds

$$\frac{tg\phi}{\phi} = -\frac{g_1 - g_3\phi^2 + g_5\phi^4 - \cdots}{1 - g_2\phi^2 + g_4\phi^4 - \cdots} \tag{5.36}$$

This equation has two roots: 2.55 and 5.9 if $Pr > 0.5$ and 2.33 and 5.5 if $Pr = 0.01$. Consequently, there are two inverse points for sinusoidal temperature head located where the temperature head decreases: the second and fourth quarters of the sinusoid. As another example, consider the exponential distribution of the temperature head $\theta_w / \theta_{w.i} = \exp[-K(x/L)]$ for which exists one inverse point at $1.19/K$ for $Pr > 0.5$ or $0.87/K$ for $Pr = 0.01$.

5.4 Zero Heat Transfer Surfaces

It is known that the friction coefficient lessens in flows with decreasing free stream velocities and becomes zero at the separation point. This nature is used in preseparated diffusers featuring small energy losses. The shape of such a diffuser is designed so that in each section the flow is close to separation. This provides small friction coefficients and, as a result, low total losses [21].

Because a heat transfer coefficient has a similar feature decreasing in the case of a negative temperature head gradient and becoming zero at the inverse point, it makes it possible to create a surface with theoretically zero but practically very small heat losses. The temperature head along such surface should vary so that the conditions of heat flux inversion are performed in each section of the thermal boundary layer. In such a case, the fluid layer adjoining the wall becomes nonconducting and serves as an insulating layer. The insulating effect is achieved due to specific distribution of temperature in the boundary layer which is constant across the adjoining wall fluid layer. As a result, the heat flux becomes zero in the normal to the wall direction.

In self-similar flows, the heat flux becomes zero in the case of the power law temperature head distribution with exponent $m_1 = (1-m)/2$ (Example 5.3). Assuming that in general, zero heat flux is achieved by power law temperature head distribution and using Equations (3.40) and (3.50) leads to the following results:

$$\theta_w = K\left(\frac{\Phi}{Re}\right)^s, \quad q_w = h_* \left[\frac{Ks(\Phi/Re)^s}{C_1}\int_0^1 [1-\sigma]^{-C_2}\sigma^{(s/C_1)-1}d\sigma + \lim_{\Phi\to 0} K\left(\frac{\Phi}{Re}\right)^s\right]$$

(5.37)

where $\sigma = (\xi/\Phi)^{C_1}$. Because $C_1 > 0$ and $0 < C_2 < 1$, integral (5.37) in the case of $s > 0$ converges and can be expressed by beta function. For $s < 0$, integral (5.37) diverges. Considering this integral as an improper integral, one gets

$$q_w = h_* \left[\frac{Ks(\Phi/Re)^s}{C_1}\lim_{\zeta\to 0}\int_\zeta^1 [1-\sigma]^{-C_2}\sigma^{(s/C_1-1)}d\sigma + \lim_{\Phi\to 0} K\left(\frac{\Phi}{Re}\right)^s\right]$$

(5.38)

As $\sigma \to 0$, the integrant tends to $\sigma^{(s/C_1)-1}$. Hence, because $C_1 > 0, 0 < C_2 < 1$, this expression is also finite as $s < 0$. Thus, using gamma functions, Equation (5.37) can be presented in the following form:

$$q_w = \frac{q_{w*}s}{C_1}\frac{\Gamma(1-C_2)\Gamma(s/C_1)}{\Gamma(1-C_2+s/C_1)}$$

(5.39)

There is only one possibility to obtain $q_w = 0$ — namely, because $\Gamma(0) = \pm\infty$, one should set $1 - C_2 + s/C_1 = 0$. So, the zero heat transfer surface exists in general if the temperature head decreases according to the power law:

$$\theta_w = K\left(\frac{\Phi}{Re}\right)^{-C_1(1-C_2)}$$

(5.40)

This result is valid for laminar and turbulent flows as well as for other cases if the influence function has the form (3.50). Using data from Figure 3.4, one estimates that for laminar flow, the exponent in Equation (5.40) is practically independent of β and Pr and equals $-1/2$.

However, temperature distribution (Equation 5.40) is difficult to implement in reality because according to Equation (5.40), the surface temperature should be infinite at the starting point where $\Phi \to 0$ [22]. Therefore, only laws close to Equation (5.40) can be realized practically. For instance, a law that more exactly approximates the relation (Equation 5.40) gains greater distance from the starting point:

$$\theta_w = K\left(K_1 + \frac{\Phi}{\mathrm{Re}}\right)^s \quad q_w = q_{w*}\left\{(1-z)^s + s\int_0^z\left[1-\left(\frac{\zeta}{z}\right)^{C_1}\right]^{-C_2}\frac{d\zeta}{(1-z+\zeta)^{1-s}}\right\}$$

$$\zeta = \frac{\xi}{K_1 + \Phi/\mathrm{Re}}$$

(5.41)

where z is the value of ζ when $\xi = \Phi$. As $\Phi \to \infty$, the last expression becomes Equation (5.40), and heat flux becomes zero.

Although there are other distributions close to that of Equation (5.40), strictly speaking, to find the temperature distribution providing the zero heat transfer, it is necessary to consider the corresponding variation problem.

5.5 Examples of Optimizing Heat Transfer in Flow over Bodies

The heat transfer rate depends not only on the heat transfer coefficient, but also on the distribution of the temperature head because the heat flux is defined by the integral of the product of the heat transfer coefficient and temperature head. Therefore, by appropriate selection of the temperature head distribution, one can ensure that a given heat transfer system satisfies the desired conditions, for example, the maximum or minimum of heat transfer rate [23].

In the general case of gradient flow over a flat, arbitrary nonisothermal surface, the heat flux and temperature head are related by Equations (3.40) and (3.64). Mathematically, the problems of optimum heat transfer modes reduce to finding a function $\theta_w(x)$ or $q_w(x)$ corresponding to the extreme of one of the integrals (3.40) or (3.64) or quantities that are their functions (e.g., the total heat flux). The solution of such a problem in general is difficult and becomes more complicated if the conjugate problems should be considered. It is much simpler to use Equations (3.40) and (3.64) for comparing different outcomes and selecting the best. Although this approach does not utilize all capabilities of optimization, the results are interesting and useful for practical applications.

Example 5.15: Distributing Heat Sources for Optimal Temperature

There exist several heat sources or sinks, for instance, electronic components with linear varying strengths. How should these be arranged on a plate so that the maximum temperature of the plate would be minimal?

Gradientless Flow

For linear varying heat flux $q_w = K + K_1(x/L)$ and gradientless flow, Equation (3.64) becomes

$$\theta_w = \frac{1}{h_*}(KI_1 + K_1I_2), \quad I_i = \frac{C_1}{\Gamma(C_2)\Gamma(1-C_2)}\int_0^1 (1-\zeta^{C_1})^{C_2-1}\zeta^{C_1(1-C_2)+n+i-2}d\zeta, \quad (5.42)$$

where $\zeta = \xi/x$ and n is the exponent in the expression $\mathrm{Nu}_{x*} = C\mathrm{Re}_x^{1-n}$. This integral can be expressed using gamma functions like other similar integrals above:

$$I_i = \frac{\Gamma[1-C_2+(n+i-1)/C_1]}{\Gamma(1-C_2)\Gamma[1+(n+i-1)/C_1]} \quad (5.43)$$

Expressing coefficients K and K_1 in the relation for heat flux by the total heat flux Q_w and difference $\Delta q = q_{max} - q_{min}$ gives, instead of Equation (5.42),

$$\theta_w = \frac{1}{h_*}\left\{I_1 q_{av} \pm \Delta q\left[\frac{1}{2}I_1 - I_2(x/L)\right]\right\} \quad (5.44)$$

where $q_{av} = Q/BL$ is the average heat flux, B is the width of the plate, and the plus sign (+) is assigned to decreasing while the minus sign (−) to increasing heat flux along the plate.

For gradientless flow $h_* \to \infty$ as $x \to 0$. Hence, the temperature head at the beginning becomes zero. It then follows from Equation (5.44) that the maximum temperature head for increasing heat flux is located at the end of the plate, at $x \to L$. This is not obvious in the case of decreasing heat transfer. Differentiating Equation (5.44) with consideration of the fact that $h_*(x) = h_*(L)(x/L)^{-n}$ and equating the result to zero yields the point coordinate at which the temperature head is maximum if heat flux decreases.

$$x_m/L = \frac{I_1 n[1+0.5(\Delta q/q_{av})]}{I_2(1+n)(\Delta q/q_{av})} \quad (5.45)$$

Setting in this expression $x_m = L$, one finds the limiting value:

$$\left(\frac{\Delta q}{q_{av}}\right)_{lim} = \frac{I_1 n}{I_2(1+n)-0.5I_1 n} \quad (5.46)$$

For all ratios $\Delta q/q_{av}$ lower than limiting, the maximum temperature head is located at the end of the plate, and for all those higher than the limiting, at $x < L$. In the case of laminar flow at $\mathrm{Pr} \geq 1 : C_1 = 3/4, C_2 = 1/3$ (Figure 3.4) and $n = 1/2$; in the case of turbulent flow at $\mathrm{Pr} \approx 1 : C_1 = 1, C_2 = 0.18$ at $\mathrm{Re}_{\delta*} = 10^3$ ($\mathrm{Re} = 5 \cdot 10^5$), $C_1 = 0.84$, $C_2 = 0.1$ at $\mathrm{Re}_{\delta*} = 10^5$ ($\mathrm{Re} = 10^8$) (Figures 4.13 and 4.14) and $n = 1/5$; for laminar wall

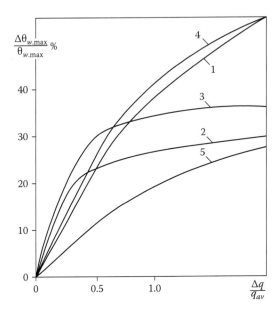

FIGURE 5.18
Absolute value of difference between maximum temperature heads under linear decreasing and increasing heat fluxes. 1 — laminar flow, 2 and 3 — turbulent flow $Re = 5 \cdot 10^5$ and $Re = 10^8$, 4 — laminar wall jet, 5 — stagnation point.

jet at $Pr \approx 1 C_1 = 0.42, C_2 = 0.35$, and $n = 3/4$. At these values of the exponents, Equation (5.43) yields are as follows: for laminar flow, $I_1 = 0.73$ and $I_2 = 5/9$; for turbulent flow, $I_1 = 0.93$ and $I_2 = 0.8$ at $Re_{\delta^*} = 10^3$; $I_1 = 0.96$ and $I_2 = 0.88$ at $Re_{\delta^*} = 10^5$; and for laminar wall jet $I_1 = 0.55$ and $I_2 = 0.43$ [24].

Using these values of integrals, Equation (5.44) gives the results plotted in Figure 5.18. In Figure 5.18 is given the absolute value of the difference $\Delta\theta_{w.max} = |\theta_{w.max.in} - \theta_{w.max.de}|$ of the maximum temperature heads in the cases of increasing $\theta_{w.max.in}$ and decreasing $\theta_{w.max.de}$ heat fluxes, referred to the higher of these two, vs $\Delta q / q_a$. If the strength of sources or sinks is given, the value of ratio $\Delta q / q_{av}$ is known, and by using Figure 5.18, one can estimate the difference of the maximum temperature heads in the cases of increasing and decreasing heat fluxes. According to these data, the maximum of the temperature head in the first case is greater than in the second, and it follows from this fact that the strength of sources encountered by the cold coolant should decrease in the direction of flow, while the strength of the sinks encountered by a hot flow should increase in that direction. It follows from Figure 5.18 that by an appropriate positioning of the sources or sinks one can significantly decrease the maximum surface temperature without decreasing the heat flux.

The Flow near the Stagnation Point

The free stream velocity in this case is proportional to coordinate $U = Cx$. The heat transfer coefficient from the isothermal surface is independent of x ($n = 0$), and the Görtler's variable is $\Phi = Cx^2/2$. Taking this into account, transforming the linear dependence for heat flux to variable Φ, and substituting the result into Equation (5.42) for I_i gives again Equation (5.44). For integrals I_j, the same formula (5.43) is valid if $(n+i-1)$ is replaced by $(i-1)/2$. Using this formula and neglecting the slight dependence of C_1 and C_2 on the pressure gradient, one finds $I_1 = 1$ and $I_2 = 0.73$. Equation (5.44) then becomes

$$\theta_w = \frac{1}{h_*}\left\{q_{av} \pm \Delta q\left[\frac{1}{2} - 0.73(x/L)\right]\right\} \tag{5.47}$$

In this case, h_* is independent of x, so the temperature head under decreasing heat flux is maximum at $x = 0$, whereas in the case of increasing heat flux, the temperature head is the highest at $x = L$. Unlike the previous case of a gradientless flow, the maximum temperature head at increasing heat flux is smaller than that at decreasing heat flux. The corresponding relationship for the relative absolute value of difference in maximum temperature heads is plotted in Figure 5.18, from which it is seen that the difference in the temperature heads is smaller than in the case of a plate. It is obvious that in this case, in order to reduce the maximum surface temperature, the strength of the sources should increase in the direction of flow in the case of a cold coolant, and the strength of sinks should decrease in the direction of a hot flow.

Example 5.16: Mode of Change of Temperature Head

Given the maximum allowable surface temperature, find the mode of change of the temperature head at which the quantity of the heat removed (or supplied) from the surface is maximum.

Gradientless Flow

The desired distribution of the temperature head is approximated by a quadratic polynomial. To determine the heat flux, it is convenient to use the differential form (Equation 3.32), because in this case only two terms are retained in a series:

$$\theta_w = a_1 + a_2(x/L) + a_3(x/L)^2$$
$$q_w = h_*[a_0 + a_1(1+g_1)(x/L) + a_2(1+2g_1+2g_2)(x/L)^2] \tag{5.48}$$

Integrating this equation after using $h_* = h_*(L)(x/L)^{-n}$ yields the total heat flux:

$$Q_w = h_*LB\left[\frac{a_0}{1-n} + \frac{a_1(1+g_1)}{2-n} + \frac{a_2(1+2g_1+2g_2)}{3-n}\right] \tag{5.49}$$

Two cases are considered:

1. The maximum temperature head is located at the beginning or at the end of the plate. In this case, the first part of Equation (5.48) has one of two forms:

$$\theta_w = \theta_{w.max} + a_1(x/L) + a_2(x/L)^2 \text{ or}$$
$$\theta_w = \theta_{w.max} + a_1[1-(x/L)] + a_2[1-(x/L)]^2 \tag{5.50}$$

The sum of the last two terms should be negative at all the $0 \le x \le L$:

$$a_1(x/L) + a_2(x/L)^2 \le 0, \quad a_1[1-(x/L)] + a_2[1-(x/L)]^2 \le 0 \tag{5.51}$$

If $a_2 \le 0$, then in order to satisfy the first of these inequalities at $x \to 0$ and the second at $x \to 1$, it is necessary that $a_1 \le 0$. It then follows from Equation (5.49) that the maximum heat flux is attained at $a_1 = a_2 = 0$ and $a_0 = \theta_{w.max}$ (i.e., when heating the plate uniformly to the temperature equal to the specified maximum temperature). If $a_2 > 0$, then $a_1 < 0$; satisfaction of the first of the inequalities at $x = L$ and of the second at $x = 0$ requires $|a_1| > a_2$. It is easy to check that under these conditions, the sum of the two last terms in Equation (5.49) in the case of laminar (e.g., for $\text{Pr} \approx 1$, $g_1 = 0.62$, $g_2 = -0.135$ and $n = 1/2$) or turbulent (e.g., for $\text{Pr} \approx 1$, $\text{Re}_{g^*} = 10^3$, $g_1 = 0.2$, $g_2 = -0.05$ and $n = 1/5$) gradientless flow is negative, and hence, the total heat flux is again maximum at $a_1 = a_2 = 0$ and $a_1 = \theta_{w.max}$.

2. The maximum temperature head occurs at $0 < x < L$. In this case, $a_2 < 0$ because parabola (Equation 5.48) is convex toward the positive ordinate. Therefore, the plus sign in the last term in Equation (5.48) should be changed to a minus sign, and then only for the case of positive a_2 when the parabola vortex coordinates are $x_m/L = a_1/2a_2$, $\theta_{w.max} = a_0 + a_1^2/4a_2$ should be considered. Deriving from these equations a_0 gives then instead of Equation (5.49),

$$Q_w = h_*(1)LB\left[\frac{\theta_{w.max}}{1-n} + a_2 F\left(\frac{a_1}{a_2}\right)\right],$$

$$F\left(\frac{a_1}{a_2}\right) = -\frac{1}{4(1-n)}\left(\frac{a_1}{a_2}\right)^2 + \frac{1+g_1}{2-n}\frac{a_1}{a_2} - \frac{1+2g_1+2g_2}{3-n} \tag{5.52}$$

The discriminant of the last quadratic trinomial is

$$\Delta = \frac{1+2g_1+2g_2}{(1-n)(3-n)} - \left(\frac{1+g_1}{2-n}\right)^2 \tag{5.53}$$

This discriminant turns out to be positive for laminar, turbulent, and laminar wall jet over the plate, which means that trinomial (5.52) has no roots, and because this trinomial is negative at $a_1/a_2 = 0$, it follows that $F(a_1/a_2)$ is negative at all ratios a_1/a_2.

According to the first part of Equation (5.52), this means that heat flux is maximum at $a_1 = a_2 = 0$.

Thus, in the above cases, the largest heat flux is removed if the plate temperature is uniform and equal to the specified maximum temperature. At first sight, this conclusion appears obvious. In fact, this is not true, because the result of the analysis depends on the relationship between coefficients $\mathcal{S}_1$ and $\mathcal{S}_2$, which govern the effect of nonisothermicity, and exponent n in the expression for the heat transfer coefficient on an isothermal wall. Two examples in which a nonuniform distribution of temperature head is optimal are given below.

The Flow Near the Stagnation Point

Using integral formula (Equation 3.40) for heat flux, integrals (Equation 5.43) are calculated to find $I_1 = 2.9$ and $I_2 = 1.62$. For the stagnation point, $n = 0$. The trinomial corresponding to Equation (5.52) is

$$F\left(\frac{a_1}{a_2}\right) = -\frac{1}{4}\left(\frac{a_1}{a_2}\right)^2 + 0.73\frac{a_1}{a_2} - 0.533 \qquad (5.54)$$

The discriminant of this trinomial is $\Delta \approx 0$, and the root is 1.46. It follows from the first part of Equation (5.52) that $Q_{w.\max} = h_* \theta_{w.\max}$ is attained at any a_2 as long as $a_1/a_2 = 1.46$ and $F(a_1/a_2) = 0$. The corresponding distributions of the temperature head and heat flux are

$$\theta_w = \theta_{w.\max} - a_2[0.53 - 1.46(x/L) + (x/L)^2],$$
$$q_w = h_*\{\theta_{w.\max} - a_2[0.53 - 2.12(x/L) + 1.62(x/L)^2]\} \qquad (5.55)$$

Thus, the same maximum heat flux that is transformed in the case of uniform heating of the plate to a maximally permissible temperature can be transferred in the case of nonuniform heating, providing that the temperature at all points, except one, of the plate is lower than maximally permitted. The physical explanation for this is that heat transfer coefficients in the case of increasing temperature head are higher than those in the case of an isothermal plate, and this compensates for the decrease in the temperature heads. Figure 5.19 shows distributions (5.55) for several values of $a_2/\theta_{w.\max}$. The minimum temperature head is zero at the beginning of the plate in the case of the highest value of this ratio.

The Jet Wall Flow at Low Prandtl Numbers

In this case, $C_1 = 4$ and $C_2 = 0.03$ [24]. With these values, the sum of the two last terms in Equation (5.49) becomes zero at any value of $a = -a_1 = a_2$. Hence, if $a_0 = \theta_{w.\max}$, then the transferred heat flux is maximum for any distribution of temperature head in the following form:

$$\theta_w = \theta_{w.\max} - a(x/L)[1 - (x/L)] \qquad (5.56)$$

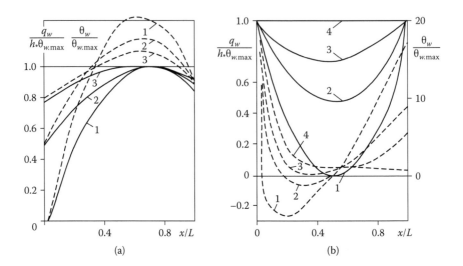

FIGURE 5.19
Different distributions of the temperature head (solid curves) and the corresponding heat flux distributions (dashed curves) providing the same total heat removed from the surface. (a) Stagnation point $a_2/\theta_{w.max}$: 1–1.88, 2–1.0, 3–0.4, (b) jet wall: 1–4, 2–3, 3–1.0, 4–0.

The corresponding distribution of heat flux follows from Equation (3.32):

$$q_w = h_*\{\theta_{w.max} - a(x/L)[5 - 9(x/L)]\} \tag{5.57}$$

Figure 5.19 shows these distributions for several values of $a/\theta_{w.max}$. For the highest of these, the minimum temperature head at $x = L/2$ is equal to zero.

Thus, these examples show that for a given total heat flux, there exists a heat flux distribution at which the maximum surface temperature is lower than in the case of uniform heating.

Example 5.17: Heat Flux Pattern

It is required to remove (supply) from the plate a heat flux Q_w. Find a heat flux pattern that minimizes the maximum plate temperature.

The desired heat flux distribution is approximated by a quadratic polynomial. Integrating this expression gives the value of heat flux for uniform distribution:

$$q_w = a_0 + a_1(x/L) + a_2(x/L)^2 \qquad q_{w.un} = \frac{Q_w}{BL} = a_0 + \frac{a_1}{2} + \frac{a_2}{3} \tag{5.58}$$

The corresponding expression for the temperature head is found by Equation (3.64):

$$\theta_w = \frac{1}{h_*}[a_0 I_1 + a_1 I_2(x/L) + a_2 I_3(x/L)^2] \tag{5.59}$$

Integrals I_i are given by Equation (5.43). Equating the derivative of θ_w to zero and taking into account that $h_* = h_*(L)(x/L)^{-n}$, one finds an equation for the coordinate where the temperature has maximum

$$\left(\frac{x_m}{L}\right)^2 + \frac{I_2 a_1(n+1)}{I_3 a_2(n+2)}\left(\frac{x_m}{L}\right) + \frac{I_1 n}{I_3(n+2)}\left(\frac{q_{w.un}}{a_2} - \frac{a_1}{2a_2} - \frac{1}{3}\right) = 0 \tag{5.60}$$

Two cases are considered:

1. Equation (5.60) has no roots. In this case, the following inequality is valid:

$$\left(\frac{I_2 a_1(n+1)}{2I_3 a_2(n+2)}\right)^2 - \frac{I_1 n}{I_3(n+2)}\left(\frac{q_{w.un}}{a_2} - \frac{a_1}{2a_2} - \frac{1}{3}\right) < 0 \tag{5.61}$$

The maximum temperature head is in this case attained at the end of the plate at $x = L$ and, according to Equations (5.58) and (5.59), can be represented as

$$\frac{\theta_{w.max}}{\theta_{w.max.un}} = 1 - \frac{a_2}{q_{w.un}}\left[\frac{a_1}{a_2}\left(\frac{I_2}{I_1} - \frac{1}{2}\right) + \frac{I_3}{I_1} - \frac{1}{3}\right] \tag{5.62}$$

Differentiating this equation with respect to a_1/a_2, substituting $q_{w.un}/a_2$ derived from condition (5.61), and equating the derivative to zero, leads to the equation

$$\frac{n+2}{n}\left(\frac{I_2}{I_1} - \frac{1}{2}\right)\frac{I_3}{I_1}\left(\frac{I_2(n+1)}{2I_3(n+2)}\right)^2\left(\frac{a_1}{a_2}\right)^2 + 2\left(\frac{I_3}{I_1} - \frac{1}{3}\right)$$

$$\frac{(n+2)}{n}\frac{I_3}{I_1}\left(\frac{I_2(n+1)}{2I_3(n+2)}\right)^2\frac{a_1}{a_2} + \frac{1}{2}\frac{I_3}{I_1} - \frac{1}{3}\frac{I_2}{I_1} = 0 \tag{5.63}$$

from which the optimum a_1/a_2 is determined. The results of calculations are listed in Table 5.2. It is seen that an appropriate selection of the distribution of the removed heat flux allows one to significantly reduce the maximum temperature, particularly in the case of a wall jet flow, in which this decrease may exceed 30%. The corresponding heat flux distribution in this case, according to Table 5.2, is

$$q_w = q_{w.un}[1.95 - 2.85(x/L) + 1.45(x/L)^2] \tag{5.64}$$

2. Equation (5.60) has roots. It follows from Equation (5.58) that

$$q_w = a_2\left[\frac{q_{w.un}}{a_2} - \frac{1}{2}\frac{a_1}{a_2} - \frac{1}{3} + \frac{a_1}{a_2}(x/L) + (x/L)^2\right] \tag{5.65}$$

TABLE 5.2

Comparison between Maximum Temperatures Obtained by Different Distributions of Removed Heat Flux

Kind of Flow	I_1	I_2	I_3	n	a_1/a_2	$q_{w.un}/a_2$	$\theta_{w.max}/\theta_{w.max.un}$
Laminar	0.73	0.56	0.48	½	−2.14	1.18	0.79
Turbulent $Re_{\delta^*} = 10^3$	0.93	0.80	0.74	1/5	−2.31	3.23	0.88
Turbulent $Re_{\delta^*} = 10^5$	0.96	0.88	0.84	1/5	−2.38	3.33	0.86
Laminar wall jet	0.55	0.43	0.37	¾	−1.97	0.69	0.67

To make sure that the heat flux does not change sign, the expression

$$\frac{1}{2}\left(\frac{a_1}{a_2}\right)^2 - \left(\frac{q_{w.un}}{a_2} - \frac{1}{2}\frac{a_1}{a_2} - \frac{1}{3}\right) < 0 \tag{5.66}$$

should be satisfied. Using Equation (5.61), one obtains the inequality

$$\frac{1}{4}\left(\frac{a_1}{a_2}\right)^2 + \frac{I_3}{I_1}\frac{n+2}{n}(x_m/L)^2 + \frac{I_2}{I_1}\frac{a_1}{a_2}\frac{n+1}{n}(x_m/L) < 0 \tag{5.67}$$

that should be satisfied if both conditions (5.60) and (5.61) are to be satisfied. Obviously, at $a_1/a_2 > 0$, inequality (5.67) is not satisfied at $x_m > 0$. This is not obvious in the case $a_1/a_2 < 0$. However, one can easily check by direct calculation that at $0 < x_m/L < 1$, inequality (5.67) is not satisfied for the condition under study even at $a_1/a_2 < 0$.

References

1. Evance, H. L., 1968. *Laminar Boundary Layer Theory*, Addison-Wesley, Reading, MA.
2. Loytsyanskiy, L. G., 1962. *The Laminar Boundary Layer* (in Russian). Fizmatgiz Press, Moscow [Engl. trans., 1966].
3. Zhukauskas, A. A., and Zhuzhgda, I. I., 1969. *Heat Transfer from Laminar Stream of Fluids* (in Russian). Mintas, Vilnyus.
4. Spalding, D. B., and Pun, W. M., 1962. A review of methods for predicting heat transfer coefficients for laminar uniform-property boundary layer flows, *Int. J. Heat Mass Transfer* 5: 239–244.
5. Schlichting, H., 1979. *Boundary Layer Theory*, McGraw-Hill, New York.
6. Dorfman, A. S., and Lipovetskaya, O. D., 1976. Heat transfer of arbitrary nonisothermic surface with gradient turbulent flow of an incompressible liquid within a wide range of Prandtl and Reynolds numbers, *High Temperature* 14: 86–92.

7. Leontev. A. I., Mikhin, V. A., Mironov, B. P., and Ivakin, V. P., 1968. Effect of boundary condition on development of turbulent thermal boundary layer (in Russian). In *Teplo i Massoperenos*, 1: 125–132, Energiya, Moscow.

8. Dorfman, A. S., 1982. *Heat Transfer in Flow around Nonisothermal Bodies* (in Russian). Mashinostroenie, Moscow.

9. Moretti, P. M., and Kays, W. M., 1965. Heat transfer to a turbulent boundary layer with varying free stream velocity and varying surface temperature — An experimental study, *Int. J. Heat and Mass Transfer* 8: 1187–1202.

10. Patankar, S. V., and Spalding, D. B., 1970. *Heat and Mass Transferin Boundary Layers*, 2nd ed. Intertext, London.

11. Solopov, V. A., 1972. Heat transfer in turbulent boundary layer incompressible fluid with varying pressure gradient and varying surface temperature, *Izv. AN SSSR, Mech. Zhidkosti i gasa* 5: 166–167.

12. Ambrok, G. S., 1957. The effect of surface temperature variability on heat transfer exchange in laminar flow in a boundary layer, *Soviet Phys.-Tech. Phy.* 2: 738–748.

13. Zhugzhda, I. I., and Zukauskas, A. A., 1962. Experimental investigation of heat transfer of a plate in a laminar flow of liquid, *Trudy Akad. Nauk Lit. SSR, Ser. B* 4: 117–126.

14. Dorfman, A. S., 1971. Exact solution of the equation for a thermal boundary layer for an arbitrary surface temperature distribution, *High Temperature* 9: 870–878.

15. Dorfman, A. S., 1967. Application of Prandtl-Mises transformation in boundary layer theory, *High Temperature* 5: 761–768.

16. Dorfman, A. S., and Vishnevskii, V. K., 1970. Boundary layer of non-Newtonian power-law fluids at arbitrary pressure gradients, *J. Engn. Phys.* 20: 398–405.

17. Grober, H., Erk, S., and Grigull, U., 1955. *Die Grengesetze der Wärmeubetragung*, 3rd ed. Springer, Berlin.

18. Chapman, D., and Rubesin, M., 1949. Temperature and velocity profiles in the compressible laminar boundary layer with arbitrary distribution of surface temperature, *J. Aeronaut Sci.* 16: 547–565.

19. Dorfman, A. S., 1995. Exact solution of nonsteady thermal boundary equation, *ASME J. Heat Transfer* 117: 770–772.

20. Sparrow, E. M., 1958. Combined effects of unsteady flight velocity and surface temperature on heat transfer, *Jet Propulsion* 28: 403–405.

21. Ginevskii, A. S., 1969. *Theory of Turbulent Jets and Wakes* (in Russian). Mashinostroenie, Moscow.

22. Eckert, E. R. G., and Drake, R. M., 1959. *Heat and Mass Transfer*, McGraw-Hill, New York.

23. Dorfman, A. S., 1988. Solution of certain problems of optimizing the heat transfer in flow over bodies, *Applied Thermal Sci.* 1: 25–34.

24. Grechannyy, O. A., Nagolkina, Z. I., and Senatos, V. A., 1984. Heat transferin jet flow over an arbitrary nonisothermal wall, *Heat Transfer-Soviet Res.* 16: 12–22.

6

Analytical Methods for Solving Conjugate Convective Heat Transfer Problems

In this chapter, the theory presented in Chapters 3, 4, and 5 is used for developing the general methods of solution of conjugate convective heat transfer problems. Solutions of different conjugate problems are given to illustrate the application of these methods. The other analytical methods used for solving heat transfer problems and examples of applying these methods are discussed in this chapter.

6.1 A Biot Number as a Criterion of the Conjugate Heat Transfer Rate

Examples presented here and in Chapter 5 as well as the numerical and analytical solutions of conjugate problems considered in what follows show that the character of heat transfer behavior is basically determined by variation of temperature head. When the temperature head increases in the flow direction or in time, the heat transfer coefficients are higher, and when it decreases in the flow direction or in time, the heat transfer coefficients become smaller than these in the case of an isothermal surface. In the first case, the increasing of heat transfer coefficients is relatively mean, while in the second the heat transfer coefficients decrease vigorously so that in some situations the heat flux becomes zero and even changes direction (Section 5.3).

The other parameter defining the intensity of conjugate heat transfer is the ratio of thermal resistances of the body and flowing fluid. In the case of the given temperature head variation, this ratio largely specifies the absolute value of heat transfer change caused by the nonisothermicity of the surface. This can be shown by using the conjugate conditions and Equation (3.32) for the heat flux. Taking into account that the nonisothermicity effect is basically determined by the second term of a series, one gets the following relation after applying the conjugate condition for heat fluxes:

$$g_1 \Phi \frac{d\theta_w}{d\Phi} = \frac{1}{\mathrm{Bi}_*} \frac{\partial T}{\partial (y/\Delta_{av})}\bigg|_{y=0}, \qquad \mathrm{Bi}_* = \frac{h_* \Delta_{av}}{\lambda_w} \qquad (6.1)$$

where Δ_{av} is an average thickness of a body.

Equation (6.1) shows that the value of temperature head gradient is proportional to inverse Biot number (Equation 6.1) for an isothermal surface which is a ratio of the thermal resistances of a fluid $(1/h_*)$ and a body (Δ_{av}/λ_w). Therefore, the Biot number under a given temperature head variation determines the rate of heat transfer in the situation described by a particular conjugate problem. It follows from Equation (6.1) that in both limiting cases $\mathrm{Bi}_* \to \infty$ and $\mathrm{Bi}_* \to 0$, the conjugate problem degenerates, because in these cases, only one resistance is finite, while another is either infinite as in the first case or becomes zero as in the second.

In the first case, a conjugate problem transforms into a problem with isothermal surface, because according to Equation (6.1), the temperature gradient in this case is zero. This case corresponds to a situation when the body of infinite thickness (or negligible conductivity) is streamlined by the fluid with finite heat transfer coefficient or when a body of finite thickness and conductivity is streamlined by the fluid with infinite heat transfer coefficient.

In the other case, a conjugate problem transforms into a problem of a body streamlined by a fluid that changes temperature in a stepwise manner, because according to Equation (6.1), the temperature head gradient is infinite in this case. This case corresponds to a situation when the body of finite thickness and conductivity is streamlined by the fluid with zero heat transfer coefficient, or the body with infinite conductivity (or negligible thickness) is streamlined by a fluid with finite heat transfer coefficient.

In both limiting cases, one of the thermal resistances is negligible, and as a result, the conjugate problem degenerates, so one concludes that usually the greatest effect of nonisothermicity should be when both resistances are of the same order and the Biot number (Equation 6.1) is close to unity. The Biot number (Equation 6.1) can be presented in various forms suitable for one or another conjugate problem. In such a case, the Biot number characterizes the relation between the resistances of the body and fluid as well, but the quantitative results can be different (see, for instance, Example 6.7). This is also the reason why there are several similar other criteria to characterize the relation of body–fluid thermal resistances. For example, Luikov in an early work [1] suggested the Brun number $\mathrm{Br} = (\Delta/x)(\lambda/\lambda_w)\,(\mathrm{Pr}\,\mathrm{Re}_x)^{1/3}$ or later Cole [2] proposed another criterion $(\lambda/\lambda_w)(\mathrm{Pe})^{1/3}$. However, it is obvious that all those relations are, in fact, Biot numbers.

6.2 General Boundary Condition for Convective Heat Transfer Problems [3,4]: Errors Caused by Boundary Condition of the Third Kind [5]

As mentioned previously, the presentation of the heat flux in the form of a series of successive derivatives, like Equations (3.32), is a sum of perturbations of the surface temperature distribution or, in other words, a sum of perturbation of surface boundary conditions. The case when all derivatives are equal to zero corresponds to the isothermal surface where boundary conditions are undisturbed. The series containing only the first derivative presents the linear boundary condition. The series with two derivatives describes the quadratic boundary condition, and so on. In a general case, the series consists of an infinite number of derivatives and describes arbitrary boundary conditions.

From such considerations, one concludes that expressions (3.32) can be considered as a general boundary condition that describes different types of surface temperature distribution. In the case of an isothermal surface, in Equations (3.32) it retains only the first term, and it becomes the boundary condition of the third kind. This well-known boundary condition is still often employed despite existing numerical methods due to its simplicity and the fact that there are some problems in which the nonisothermicity slightly affects the final results.

If the solution obtained with the boundary condition of the third kind is known, the error of this approximate result can be estimated by computing the second term in the general boundary condition (3.32) or in the others of this type obtained in Chapter 3. Comparing the value of the second term with the known approximate solution gives an understanding of the necessity of a conjugate solution.

Example 6.1: Heat Transfer from Fluid to Fluid in a Flow Past Two Sides of a Plate

Using the boundary condition of the third kind, one finds the temperature head:

$$\theta_{w1} = \frac{T_{\infty 1} - T_{w1}}{T_{\infty 1} - T_{\infty 2}} = \frac{q_w}{h_{*1}(T_{\infty 1} - T_{\infty 2})} = \frac{1}{1 + h_{*1}/h_{*2} + h_{*1}\Delta/\lambda_w} \tag{6.2}$$

at one side of the wall, where the isothermal heat transfer coefficients are determined as

$$\mathrm{Nu}_{x*} = 0.332\,\mathrm{Pr}^{1/3}\,\mathrm{Re}_x^{1/2} \quad \mathrm{Nu}_{x*} = 0.0295\,\mathrm{Pr}^{0.4}\,\mathrm{Re}_x^{0.8} \tag{6.3}$$

for laminar and turbulent flows, respectively. If the flow regimes on both sides are the same, the ratio h_{*1}/h_{*2} does not depend on x. Because the last term in the denominator of Equation (6.2) is the Biot number $\mathrm{Bi}_* = h_*\Delta/\lambda_w$, which, according to Equation (6.3) is a power law function, $\mathrm{Bi}_* \sim x^{-n}$, Equation (6.2) may be written in the form

$$\theta_{w1} = \frac{1}{D_1 + D_2 x^{-n}} \tag{6.4}$$

where D_1 and D_2 are constants. Using this expression and taking into account that $\Phi = \mathrm{Re}_x$ for gradientless flow, it is easy to calculate the second term of the series and, by comparing it with Equation (6.4), to estimate the relative error arising when a boundary condition of the third kind is used:

$$\sigma = g_1 \frac{x}{\theta_w} \frac{d\theta_w}{dx} = \frac{g_1 n D_1}{D_1 x^n + D_2} \qquad (6.5)$$

This expression takes the maximum value $\sigma_{max} = g_1 n$ at $x = 0$. Thus, for laminar flow, the greatest error is $g_1/2$, and for turbulent flow it is $g_1/5$. For laminar gradientless flow and $Pr > 0.5$, the coefficient g_1 is practically independent of Pr, and it is 0.62 (Figure 3.2), so the maximum error in this case is $\approx 30\%$. For turbulent gradientless flow and $Pr = 0.5$, $g_1 = 0.22$ (Figure 4.11) and decreases when Pr increases. Therefore, in this range of Prandtl numbers, the maximum error is $\sigma_{max} \approx 4\%$. However, for $Pr = 0.01$ $g_1 = 0.52$ and, hence, $\sigma_{max} \approx 10\%$.

Thus, for laminar flow, the error may be moderate, while for turbulent flow, when $Pr > 0.5$, the use of the boundary condition of the third kind does not lead to significant errors. The estimates obtained are in agreement with corresponding conjugate problems solutions. For laminar flow, such a solution gives the maximum error of 20% to 25% (Figure 6.3); for turbulent flow, the maximum error is 7% [6].

The errors in this problem are relatively small because the temperature head increases in flow direction on both sides of the plate. If the temperature head decreases, the problem should be considered as a conjugate one even in the case of turbulent flow, because in such a situation, the errors may be large and even qualitatively different.

Example 6.2: Heat Transfer from Thermally Thin Plate Heated from One End

A steel plate of length 0.25 m and of thickness 0.01 m in a symmetric airflow is considered. The air temperature is 300 K, velocity is 3 m/s. The left-hand end is insulated, and the temperature of the right-hand end is maintained at $T_w(L)$.

The differential equation (Equation 2.7) for a thermally thin plate and its solution with the boundary condition of the third kind for the problem in question are

$$\frac{d^2\theta_w}{dx^2} + \mathrm{Bi}^2\theta_w = 0 \qquad \theta_w = \frac{T_w - T_\infty}{T_w(L) - T_\infty} = \frac{1}{ch(\mathrm{Bi})} ch\left(\mathrm{Bi}\frac{x}{L}\right) \qquad \mathrm{Bi} = \frac{\mathrm{Nu}_{*av} L\lambda}{\Delta\lambda_w} \qquad (6.6)$$

Differentiating the expression for θ_w, an error and its maximum are obtained:

$$\sigma = g_1 \frac{x}{\theta_w} \frac{d\theta_w}{dx} = g_1 \mathrm{Bi}\frac{x}{L} th\left(\mathrm{Bi}\frac{x}{L}\right) \qquad \sigma_{max} = g_1 \mathrm{Bi} = g_1 \frac{\mathrm{Nu}_{*av} L\lambda}{\Delta\lambda_w} \qquad (6.7)$$

For laminar gradientless flow, $g_1 = 0.62$. Calculating Nu_{*av} for $Re = 5 \cdot 10^4$ yields $\mathrm{Bi} = 1.64$, and $\sigma_{max} \approx 1$. Hence, the solution of this problem with the boundary condition of the third kind is inadequate because the temperature head decreases along the plate (Example 5.2). The conjugate solution of this problem is given in Example 6.8.

Example 6.3: Heat Transfer from Elliptic Cylinder with Semiaxis Ratio $a/b = 4$ and Uniformly Distributed Source of Power q_v

The problem of laminar flow past a nonisothermal elliptical cylinder with the boundary condition of the third kind is considered in the next section. To estimate the error, it is necessary to calculate by numerically differentiating the derivative of the temperature head $d\theta_w/dS$, where S is a coordinate measured along the cylinder surface. The error for the flow with pressure gradient is estimated using variable Φ:

$$\sigma = g_1 \frac{\Phi}{\theta_w} \frac{d\theta_w}{d\Phi} = g_1 \frac{\Phi}{\theta_w} \frac{d\theta_w}{dS} \frac{dS}{dv} \bigg/ \frac{d\Phi}{dv}$$

$$dS = \sqrt{a^2 \sin^2 v + b^2 \cos^2 v}\, dv, \qquad \frac{v\Phi}{U_\infty a} = \left(1 + \frac{b}{a}\right)(1 - \cos v) \qquad (6.8)$$

where v is the elliptical coordinate. The calculation is made for the case of the equal thermal resistances of cylinder and flowing fluid $\mathrm{Bi} = \lambda\sqrt{\mathrm{Re}}/\lambda_w = 1$. The coefficient g_1 varies along the elliptical cylinder and is the largest at the stagnation point: $g_1 = 0.68$. The estimation gives a maximum error of 25% at the point $S/L \approx 0.3$, which is in agreement with the results of solving the conjugate problem (Figure 6.9). The error is moderate because the temperature head increases along the cylinder.

Example 6.4: A Continuous Plate (Strip) of Polymer at Temperature T_0 Extruded from a Die and Passed at Velocity U_w through a Bath with Water (Pr = 6.1) at Temperature T_∞

The problem of such a continuous plate with the boundary condition of the third kind is solved in Section 8.2. The result is presented in the form $\theta_w(\tilde{x})$, where $\tilde{x} = x\alpha_w/\Delta^2 U_w$, $\theta_w = (T_w - T_\infty)/(T_0 - T_\infty)$. Calculations of error are made for the ratio of thermal characteristics of the plate and coolant $(c\rho\lambda)_w/c_p\rho\lambda = 8.51$ and $\phi = U_\infty/U_w = 0$. Numerical differentiation of the curve $\theta_w(\tilde{x})$ (Figure 8.2) gives the derivative $d\theta_w/d\tilde{x}$. Then, using Equation (6.5), the error is estimated taking into account that the effect of nonisothermicity in the case of the moving plate is greater ($g_1 = 1.3$; Figure 3.9) than for the fixed plate. The error grows as the distance from the die increases and finally reaches $\sigma_{max} \approx 2.6$. It is evident that this problem must be solved as conjugate. The reason for this is the decreasing temperature head.

It follows from examples that in the case of the temperature head decreasing, any problem should be considered as a conjugate. In another case when the temperature head increases, the estimation of error may help to answer this question.

6.3 Reduction of a Conjugate Convective Heat Transfer Problem to an Equivalent Heat Conduction Problem [3,4]

Because the basic relation (Equation 3.32) or (Equation 3.40) and others of this type determines the connection between the temperature head and heat flux on the body–fluid interface in general, there is no need to solve the boundary layer equations when one of these relations is used. Thus, in such a case, only the heat conduction equation remains to be solved using conjugate conditions. In other words, the solution of the conjugate problem is reduced to the integration of the heat conduction equation with the general boundary condition in differential (Equation 3.32) or integral (Equation 3.40) form.

As shown in the previous section, the relation (Equation 3.32) is a general boundary condition. If the heat conduction equation is solved using this condition with the first term only, an approximate solution of the conjugate problem as with the boundary condition of the third kind is obtained. By retaining the first two terms in Equation (3.32) and solving the heat conduction equation, a more accurate solution of the conjugate problem is obtained. This process of refining can be continued by retaining a larger number of terms in Equation (3.32). However, this entails difficulties posed by the calculation of higher-order derivatives, and therefore, the integral form of general boundary condition (Equation 3.40) should be used for further approximations.

In practical calculations, it is convenient to retain the first few terms of the series and to calculate the error term from the results of previous approximations. When, in this case, the first three terms of the series are retained, the conjugate problem is reduced to a heat conduction equation for a solid with the following boundary condition:

$$q_w = h_* \left[\theta_w + g_1 \Phi \frac{d\theta_w}{d\Phi} + g_2 \Phi^2 \frac{d^2\theta_w}{d\Phi^2} + \varepsilon(\Phi) \right], \qquad \varepsilon(\Phi) = \frac{1}{h_*}(q_w^{int} - q_w^{diff}) \qquad (6.9)$$

$$\frac{d\theta_w}{d\Phi} = \frac{v}{U}\frac{d\theta_w}{dx}, \qquad \frac{d^2\theta_w}{d\Phi^2} = \frac{v^2}{U^2}\frac{d^2\theta_w}{dx^2} - \frac{v^2}{U^2}\frac{dU}{dx}\frac{d\theta_w}{dx} \qquad (6.10)$$

Quantities q_w^{int} and q_w^{diff} are defined by integral relation (3.40) and by differential relation (3.32) in the form of Equation (6.9), respectively. The first approximation is found by assuming that $\varepsilon(\Phi) = 0$. Calculating the error term using the results of the first approximation makes it possible to introduce the error into Equation (6.9) and to find the second approximation. By continuing this process, the solution with the desired accuracy can be obtained.

Retaining in Equation (6.9) terms with derivatives not higher than second, leads to differential equations of the second order in any approximation. As a result, in this case, the conjugate problem is reduced to the ordinary differential equation in the case of thin body and to Laplace or Poinsot equation in the general case. This makes it possible to use well-known effective methods of solving such equations.

In using this method, named the method of successive differential-integral approximation, one should start from the most simple form of Equations (6.9) and, after each approximation, assess the error either by comparing the results of the successive approximations or by evaluating the error term $\varepsilon(\Phi)$. The suggested method of reducing the conjugate problem to the equivalent conduction problem by successive differential-integral approximation can be used to solve any linear conjugate convective heat transfer problem. Examples are given below in this and following chapters. However, this method could not be used for nonlinear problems, because in that case, the method of superposition is not applicable.

Example 6.5: Heat Transfer from Liquid to Liquid in a Flow Past a Thin Plate [7]

If the plate is thin ($\Delta/L \ll 1$) and its thermal resistance is comparable with that of liquids, the longitudinal conductivity of the plate is negligible. In this case, the temperature distribution across the plate thickness can be considered as linear, and hence, the heat fluxes on both sides of the plate are taken to be equal (Figure 6.1):

$$-q_{w1} = q_{w2} = \frac{\lambda_w}{\Delta}(T_{w1} - T_{w2})$$

(6.11)

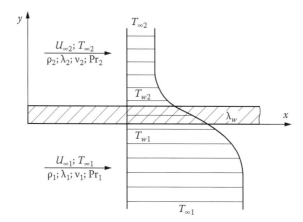

FIGURE 6.1
Scheme of heat transfer between concurrent flows.

Substituting q_{w1} and q_{w2} defined by Equation (3.32) into Equation (6.11) yields two equations:

$$-h_{*1}\left[T_{w1}-T_{\infty1}+\sum_{k=1}^{\infty}g_{k1}x^k\frac{d^kT_{w1}}{dx^k}\right]=h_{*2}\left[T_{w2}-T_{\infty2}+\sum_{k=1}^{\infty}g_{k2}x^k\frac{d^kT_{w2}}{dx^k}\right] \tag{6.12}$$

$$T_{w2}-T_{w1}=\frac{h_{*1}\Delta}{\lambda_w}\left[T_{w1}-T_{\infty1}+\sum_{k=1}^{\infty}g_{k1}x^k\frac{d^kT_{w1}}{dx^k}\right] \tag{6.13}$$

As the distance from the origin of the plate grows, the boundary layer thickness increases, and the heat flux decreases so that at $x\to\infty$, the heat flux approaches zero. In this case, the plate temperature tends to limiting value $T_{w\infty}$, which is determined from Equation (6.12) if all derivatives of temperature are taken to be zero.

This temperature and Biot number are used to form the dimensionless variables

$$\mathrm{Bi}_*=\frac{h_*\Delta}{\lambda_w}, \qquad \theta=\frac{T_w-T_\infty}{T_{w\infty}-T_\infty}, \qquad T_{w\infty}=\frac{\mathrm{Bi}_{*1}T_{\infty1}+\mathrm{Bi}_{*2}T_{\infty2}}{\mathrm{Bi}_{*1}+\mathrm{Bi}_{*2}} \tag{6.14}$$

Then, Equations (6.12), (6.13), and (6.11) defining temperature heads and heat flux become

$$\theta_1+\sum_{k=1}^{\infty}\overline{g}_{k1}\mathrm{Bi}_*^k\frac{d^k\theta_1}{d\mathrm{Bi}_*^k}=\theta_2+\sum_{k=1}^{\infty}\overline{g}_{k2}\mathrm{Bi}_*^k\frac{d^k\theta_2}{d\mathrm{Bi}_*^k} \tag{6.15}$$

$$\mathrm{Bi}_{*2}(1-\theta_1)+\mathrm{Bi}_{*1}(1-\theta_2)=\mathrm{Bi}_{*1}\mathrm{Bi}_{*2}\left(\theta_1+\sum_{k=1}^{\infty}\overline{g}_{k1}\mathrm{Bi}_*^k\frac{d^k\theta_1}{d\mathrm{Bi}_*^k}\right) \tag{6.16}$$

$$\mathrm{Bi}_K=\frac{\mathrm{Bi}_{*1}(1-\theta_1)+\mathrm{Bi}_{*2}(1-\theta_2)}{\mathrm{Bi}_{*1}+\mathrm{Bi}_{*2}},$$

$$\overline{g}_k=\sum_{n=k}^{\infty}g_n\sum_{j=1}^{k}\frac{(-1)^{k-j}}{j!(k-j)!}\left(-\frac{1}{2}j\right)\left(-\frac{1}{2}j-1\right)\cdots\left(-\frac{1}{2}j-n+1\right) \tag{6.17}$$

where Bi_K is the Biot number determining the overall heat transfer coefficient. The boundary conditions follow: one from the fact that at the origin the temperature of each side of the plate and of the corresponding fluid should be equal, and the second from the asymptotic behavior of temperature at $x\to\infty$:

$$\mathrm{Bi}_*=\infty \quad \theta_1=\theta_2=0, \qquad \mathrm{Bi}_*\to0 \quad \theta_1=\theta_2=1, \quad \theta_1'=\theta_2'=\theta_1''=\theta_2''=\cdots\to0 \tag{6.18}$$

For the fluids with $\mathrm{Pr}>0.5$, all coefficients g_k including the first two are practically independent of Pr (Figures 3.2 and 3.3). Therefore, if both fluids have $\mathrm{Pr}>0.5$, all coefficients of the left-hand and right-hand sides

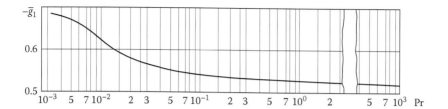

FIGURE 6.2
Dependence $\bar{g}_1(\mathrm{Pr})$ for laminar flow.

of Equation (6.15) are equal. For fluids with $\mathrm{Pr} < 0.5$, the first two coefficients slightly depend on Pr: with the change of Pr from 0.5 to 0, the first
coefficient changes from 0.63 to 1, and the second changes from 0.14 to
0.17. Consequently, in this case also, the coefficients g_k may be assumed
to be approximately equal if the mean value for both fluids is adopted.
In this case, both dimensionless temperatures are equal, and Equations
(6.15) and (6.17) for temperature heads and heat flux can be presented in
the form using the overall Biot number Bi_Σ:

$$\theta(1+\mathrm{Bi}_\Sigma)+\bar{g}_1\mathrm{Bi}_\Sigma^2\frac{d\theta}{d\mathrm{Bi}_\Sigma}+\bar{g}_2\mathrm{Bi}_\Sigma^3\frac{d^2\theta}{d\mathrm{Bi}_\Sigma^2}+\cdots=1,$$

$$\mathrm{Bi}_K=1-\theta,\qquad \mathrm{Bi}_\Sigma=\frac{1}{1/\mathrm{Bi}_{*1}+1/\mathrm{Bi}_{*2}}\qquad(6.19)$$

The first part of Equation (6.19) is solved using in sequence one, two, and
three terms [4,8]. The first two approximations are obtained as follows:

$$\theta=\frac{1}{1+\mathrm{Bi}_\Sigma},$$

$$\theta=\frac{(-1/\bar{g}_1\mathrm{Bi}_\Sigma)\exp(-1/\bar{g}_1\mathrm{Bi}_\Sigma)}{2-(1/\bar{g}_1)}F[1-(1/\bar{g}_1),\,2-(1/\bar{g}_1),\,(-1/\bar{g}_1\mathrm{Bi}_\Sigma)]\qquad(6.20)$$

where $F(a,b,c)$ is the confluent hypergeometric function. Dependence
$\bar{g}_1(\mathrm{Pr})$ is given in Figure 6.2. For $\bar{g}_1=-1\,(\mathrm{Pr}\to0)$ and $\bar{g}_1=-1/2\,(\mathrm{Pr}\to\infty)$,
Equation (6.20) may be presented using the exponential function:

$$\theta=1-\mathrm{Bi}_\Sigma[1-\exp(-1/\mathrm{Bi}_\Sigma)]\qquad \theta=1-\mathrm{Bi}_\Sigma-0.5\mathrm{Bi}_\Sigma^2[1-\exp(-2/\mathrm{Bi}_\Sigma)]\qquad(6.21)$$

The results of the calculation are plotted in Figure 6.3 in the form
$(\Delta q_w/q_{w*})=f(\mathrm{Bi}_\Sigma)$, where Δq_w and q_{w*} are the error and heat flux obtained
by a nonconjugated solution. The difference between the third and sec-
ond approximations is practically negligible (Figure 6.3). The maximum
error (20%) is for the Biot number (Equation 6.19) defined for both fluids

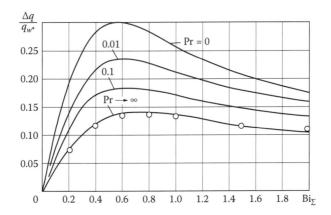

FIGURE 6.3
Relative error in a heat flux arising due to neglecting the effect of conjugation (obtained with a boundary condition of the third kind). Laminar flow, o— third approximation.

$\mathrm{Bi}_\Sigma = 0.5-0.7$. The error in this problem is moderate, because the temperature head increases along both sides of the plate.

For the case of turbulent flow, this conjugate problem is solved in Reference [6] where the calculation gives the maximum error of about 7%.

Example 6.6: Heat Transfer between Two Fluids Separated by a Plate — An Exact Solution

In the case when the thermal resistances of the plate in both directions are comparable, it is necessary to consider the full form of the heat conduction equation. In particular, this situation takes place at the initial segment of any plate where, in fact, it is impossible to neglect the transverse thermal resistance of the plate, or its longitudinal conductivity. These approximations are not valid at the initial segment, because in that area, the thermal boundary layer is thin and the longitudinal temperature gradients are high due to vigorous variations of the heat transfer coefficient.

The solution of the heat conduction equation may be presented in the following form [9]:

$$\vartheta_1 + \vartheta_2 = \frac{1}{\pi} \int_0^\infty (\mathrm{Bi}_1 + \mathrm{Bi}_2) \ln[4sh\,|\,z-\zeta\,|\,\pi\,sh(z+\zeta)\pi]dz$$

$$\vartheta_1 - \vartheta_2 = \frac{1}{\pi} \int_0^\infty (\mathrm{Bi}_2 - \mathrm{Bi}_1) \ln\left[cth\frac{|z-\zeta|}{2}\pi\,cth\frac{z+\zeta}{2}\pi \right]dz$$

(6.22)

$$\vartheta = \frac{T_w - T_{w\infty}}{T_{\infty 1} - T_{\infty 2}}, \qquad \mathrm{Bi} = \frac{q_w \Delta}{\lambda_w (T_{\infty 1} - T_{\infty 2})}, \qquad \zeta = x/\Delta$$

(6.23)

The solution of thermal boundary layer equations for fluids can be obtained by using Equations (3.32) or (3.40):

$$Bi_1 = Bi_{\Delta*1}\zeta^{-1/2}\left(\vartheta_1 - \frac{1}{\sigma+1} + g_{11}\zeta\vartheta_1' + g_{21}\zeta^2\vartheta_1'' + \cdots\right)$$

$$= Bi_{\Delta*1}\zeta^{-1/2}\int_0^\zeta f\left(\frac{\xi}{\zeta}\right)\frac{d\vartheta_1}{d\xi}d\xi + \vartheta_1(0) + \frac{1}{\sigma+1}$$

$$Bi_1 = Bi_{\Delta*2}\zeta^{-1/2}\left(\vartheta_2 - \frac{\sigma}{\sigma+1} + g_{12}\zeta\vartheta_2' + g_{22}\zeta^2\vartheta_2'' + \cdots\right) \qquad (6.24)$$

$$= Bi_{\Delta*1}\zeta^{-1/2}\int_0^\zeta f\left(\frac{\xi}{\zeta}\right)\frac{d\vartheta_2}{d\xi}d\xi + \vartheta_2(0) + \frac{\sigma}{\sigma+1}$$

Here, $Bi_{\Delta*} = h_{\Delta*}\Delta/\lambda_w$, $\sigma = Bi_{\Delta*1}/Bi_{\Delta*2} = h_{\Delta*1}/h_{\Delta*2}$, and $h_{\Delta*}$ is the heat transfer coefficient of an isothermal surface based on the plate thickness.

The set of Equations (6.22) and (6.24) yields an exact solution of the conjugate problem of heat transfer between two fluids separated by a plate. These general equations contain different particular, simple cases. For example, the cases when the temperature or heat flux distribution is contained to only one side of a plate and the cases when the plate can be assumed to be thin or thermally thin.

The system of Equations (6.22) and (6.24) can be solved by successive approximations. As a first approximation, the solution for a thin plate from the previous example is used. From Equations (6.11) and (6.19), one finds

$$-Bi_1^{(1)} = Bi_2^{(1)} = Bi = 1 - \theta^{(1)} \qquad (6.25)$$

Substituting Equation (6.25) into Equation (6.22) gives the temperatures in the second approximation:

$$\vartheta_1^{(2)} = -\vartheta_2^{(2)} = \frac{1}{\pi}\int_0^\infty Bi^{(1)}\ln\left[cth\frac{|z-\zeta|}{2}\pi\, cth\frac{z+\zeta}{2}\pi\right]dz \qquad (6.26)$$

The corresponding results for the heat fluxes $Bi_1^{(2)}$ and $Bi_2^{(2)}$ are found by substitution of $\vartheta_1^{(2)}$ and $\vartheta_2^{(2)}$ into Equations (6.24). Then, one returns to Equation (6.22) to obtain the next approximation for the temperature, and so on.

In Figure 6.4, the results of calculations are presented for the temperatures in the first approximation according to Equation (6.21) and for the second approximation according to Equation (6.26) at $s=1$ (equal thermal resistances for both fluids) and $Pr \to 0$ where the conjugation effect is at maximum.

It follows from Figure 6.4 that, starting from a certain length ζ_0, the second approximation coincides virtually with the first one. This means

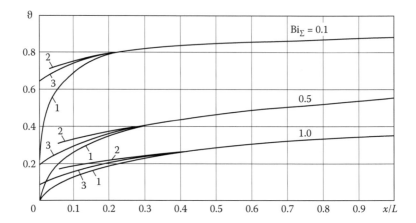

FIGURE 6.4
Temperature distribution along a plate at different ratios of thermal resistance of the plate and fluids: 1 — thin plate, 2 — second approximation for thick plate according to Equation (6.26), 3 — according to Equation (6.30) for the initial segment of a plate.

that for $\zeta > \zeta_0$, a semi-infinite plate can be considered as a thin plate and the heat transfer for this part can be calculated using results obtained in a previous example for a thin plate. Over the initial segment of the plate, the heat transfer should be calculated by solving the full system of equations.

Analogous calculations are performed for the turbulent boundary layer. In this case, the calculations are made for Pr = 0.01 and $Re_{\delta^*} = 10^3$ under which the influence of conjugation also is the maximum. The corresponding value of the first coefficient is $\bar{g}_1 = 0.25$ (Figure 4.11), and the initial segment is smaller than in laminar flow (Figure 6.5).

To study the behavior of the temperature and heat flux near the beginning of the plate, consider again the case of equal thermal resistances of the flow on both sides ($\sigma = 1$). In this case, $\theta_1 = \theta_2, -Bi_1 = Bi_2 = Bi$, and,

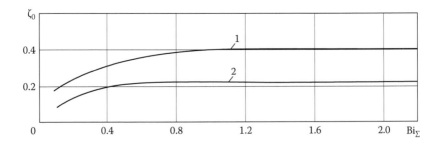

FIGURE 6.5
Dependence of initial length of the plate on the ratio of thermal resistances of the plate and fluids: 1 — laminar flow, 2 — turbulent flow.

therefore, according to Equations (6.14) and (6.23), under $\sigma = 1$, one gets $\vartheta_1 = -\vartheta_2$. Then, taking into account Equations (6.26) and (6.24) leads to the conclusion that at $\sigma = 1$ the problem is governed by two equations:

$$\theta = 1 - \frac{2}{\pi} \int_0^\infty Bi \ln\left[cth \frac{z-\zeta}{2} \pi \, cth \frac{z+\zeta}{2} \pi \right] dz,$$

$$Bi = \frac{1}{2} Bi_{\Delta*} \zeta^{-1/2} (\theta + g_1 \zeta \theta' + g_2 \zeta^2 \theta'' + \cdots)$$

(6.27)

It is shown [4,9] that the solution of this system in the vicinity of $\zeta = 0$ can be presented by the series in integer exponents of variable $\zeta^{1/2}$:

$$\theta = \sum_{n=0}^\infty a_n \zeta^{n/2} \qquad Bi = \sum_{n=-1}^\infty a_{n+1} d_n \zeta^{n/2}$$

$$d_n = \frac{Bi_{\Delta*}}{2}\left[1 + \sum_{k=1}^\infty g_k \frac{(n+1)(n-1)\cdots(n-2k+3)}{2^k} \right], \qquad a_0 = 1 - \frac{1}{\pi} \int_0^\infty Bi \ln cth \frac{z\pi}{2} dz$$

$$a_{4n} = -2 \int_0^\infty \left\{ Bi \, \pi^{2n-1} \sum_{k=1}^n \frac{b_{(n-k),k}}{4^k k}\left[\frac{1}{sh^{2k}(z\pi/2)} + \frac{(-1)^{k+1}}{ch^{2k}(z\pi/2)} \right] \right.$$

(6.28)

$$\left. - \frac{1}{\pi} \left[\sum_0^{k=n-1} a_{4k} d_{4k-1} \frac{z^{2k-2n-1/2}}{n} + a_{4k+1} d_{4k} \frac{z^{2k-2n}}{n} \right] \right\} dz$$

$$b_{ik} = \frac{1}{i} \sum_{j=1}^i [(k+1)j - i] \frac{2b_{(i-j),k}}{[2(j+1)]!}, \qquad b_0 = 1, \qquad a_{4n+1} = \frac{4}{4n+1} a_{4n} d_{4n-1},$$

$$a_{4n+2} = a_{4n+3} = 0$$

One sees that coefficients a_n of series for the temperature at the vicinity of $\zeta = 0$ depend on the integrals with semi-infinite limits. To calculate these integrals, the semi-infinite interval $(0, \infty)$ is divided into two parts: the first part for limits $(0, \varepsilon)$ in which the series are valid, and the second for limits (ε, ∞) in which the results for the thin plate can be used. Then, taking into account the first four coefficients, a_n yields

$$a_0 = \frac{\pi - 4 I_\infty}{\pi + 4 d_{(-1)}[I_{(-1)} + 4 d_0 I_0]}, \qquad I_\infty = \int_\varepsilon^\infty Bi \ln cth \frac{z\pi}{2} dz \quad I_n = \int_0^\varepsilon z^{n/2} \ln cth \frac{z\pi}{2} dz$$

$$a_1 = 4 a_0 d_{(-1)}, \qquad a_2 = a_3 = 0.$$

(6.29)

Considering the first eight terms in series (6.28) leads to more complicated final relations, but the results are practically the same [9]. Therefore, in the case of equal thermal resistances of both flows, the heat transfer on

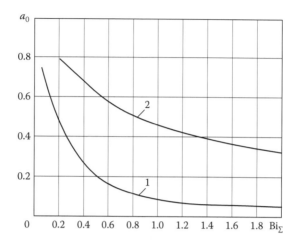

FIGURE 6.6
Dependence of the coefficient a_0 in Equations (6.30) and (6.31) on the ratio of thermal resistances of the plate and fluids: 1 — laminar flow; 2 — turbulent flow.

the initial segment can be computed by the simple formulae obtained by the series with four terms:

$$\theta = \frac{T_w - T_\infty}{T_{w\infty} - T_\infty} = a_0(1 + 4\mathrm{Bi}_{\Sigma\Delta}\zeta^{1/2}), \quad \mathrm{Bi} = a_0\mathrm{Bi}_{\Sigma\Delta}(\zeta^{-1/2} + 4d_0),$$

$$\mathrm{Bi}_{\Sigma\Delta} = \frac{1}{1/\mathrm{Bi}_{\Delta*1} + \mathrm{Bi}_{\Delta*2}} \tag{6.30}$$

The values of a_0 are given in Figure 6.6. The temperatures calculated by Equation (6.30) are plotted in Figure 6.4 (curve 3). These differ in principle on the initial segment from the results obtained for a thin plate. The temperature head at the beginning is finite, so the plate temperature differs substantially from that of corresponding fluid, whereas under the assumption of a thin plate, the temperature head is zero, and consequently, the temperature of the plate and corresponding fluid are equal at the beginning.

Due to the small length of the initial segment in a turbulent flow, it is possible to use only the first terms of the formulae (6.30):

$$\theta = \frac{T_w - T_\infty}{T_{w\infty} - T_\infty} = a_0 \qquad \mathrm{Bi} = a_0\mathrm{Bi}_{\Sigma\Delta}\zeta^{-1/5} \tag{6.31}$$

It should be noted that strictly speaking, in the immediate vicinity of $\zeta = 0$, due to the smallness of the Reynolds and Peclet numbers, the boundary layer theory assumptions are not valid, and in this region, also with the full heat conduction equation, it is necessary to consider the full conservation equations for fluids. However, estimates show that at large values of the Reynolds number for the whole plate, the region in which the boundary layer equations are not valid is many times smaller than the initial segment over which the plate should be considered as a thick plate.

According to Van Dyke [10], the error in the determination of the friction coefficient by the boundary layer theory does not exceed $1.7/\sqrt{Re_x}$. Assuming the maximum admissible error to be 10%, the ultimately admissible Reynolds number is about 300. The segment length over which the boundary layer theory is not valid according to this estimation is $x_{\lim}/L \approx 300/Re$. On the other hand, as calculation shows, the initial segment length in laminar flow is $x_0/L \approx 0.4(\Delta/L)$ (Figure 6.5). Thus, for example, at $Re = 10^5$ and $\Delta/L = 1/100$, $x_{\lim}/L \approx 1/300$, $x_0/L \approx 1/250$. However, at $\Delta/L = 1/10$ and the same Re, the initial segment length turns out to be 12 times greater than $x_{\lim}$. With increasing the plate thickness and Reynolds number, the difference between x_0 and $x_{\lim}$ increases rapidly. Thus, only when a plate is very thin are both dimensions close. In such a case, the initial segment is very small, and only the value of the temperature at $x = 0$ is important.

Example 6.7: Heat Transfer between an Elliptical Cylinder and Flowing Laminar Flow [11]

This problem is considered as an example of using the method of reducing a conjugate problem to the heat conduction equation in the case of gradient flow. The heat conduction equation and the symmetry and surface boundary conditions are

$$\frac{\partial^2 T}{\partial x^2} + \frac{\partial^2 T}{\partial y^2} - \frac{q_v}{\lambda_w} = 0, \qquad -\lambda_w \frac{\partial T}{\partial n}\bigg|_w = q_w, \qquad \frac{\partial T}{\partial y}\bigg|_{y=0} = 0 \qquad (6.32)$$

Here, n is a normal to the surface, and q_v is a heat source. In the elliptical coordinates (u, v):

$$x = cchu \cos v \qquad y = cshu \sin v \qquad c = \sqrt{a^2 - b^2} \qquad (6.33)$$

where a and b are major and minor semiaxes, a half ellipse is mapped into a rectangle, one side of which corresponds to the surface of the semi-ellipse and three others to the axes of symmetry. Then, relation (6.32) becomes

$$\frac{\partial^2 T}{\partial u^2} + \frac{\partial^2 T}{\partial v^2} = Q_v(sn^2 u + \sin^2 v), \qquad \frac{\partial \theta}{\partial v}\bigg|_{u=l} = \sqrt{1 - \frac{c^2}{a^2}\cos^2 v}\,\frac{q_w a}{\lambda_w T_\infty}$$

$$\frac{\partial \theta}{\partial u}\bigg|_{u=0} = \frac{\partial \theta}{\partial v}\bigg|_{v=0} = \frac{\partial \theta}{\partial v}\bigg|_{u=\pi} = 0 \qquad (6.34)$$

Here, $\theta = (T - T_\infty)/T_\infty$, $Q_v = -q_v c^2/\lambda_w T_\infty$, $l = (1/2)\ln[(a+b)/(a-b)]$ are dimensionless temperature difference, power of the internal sources, and the value of the coordinate u corresponding to the surface of the ellipse.

The solution of Equation (6.34) subjected to the last three boundary conditions is obtained by separation of variables [11]:

$$\theta = \frac{Q_v}{8}(ch2u + \cos 2v) + N_0 + \sum_{k=1}^{\infty} N_k chku \cos kv \tag{6.35}$$

The constants $N_0,...N_k$ must be determined from the first boundary condition (Equation 6.34). Because the velocity of a potential flow around a cylinder and the differential of the arc length are as follows:

$$U = \frac{U_\infty(a+b)\sin v}{\sqrt{a^2 \sin^2 u + b^2 \cos^2 v}} \qquad dS = \sqrt{a^2 \sin^2 v + b^2 \cos^2 v}\, vdv, \tag{6.36}$$

the heat flux (Equation 3.32) is defined as

$$q_w = \frac{\lambda Nu_* T_\infty}{a}\left\{ \int_0^v f\left[\frac{\xi(\varepsilon)}{\Phi(v)}\right]\frac{d\theta_w}{d\varepsilon}d\varepsilon + \theta_w(0) \right\},$$

$$\Phi = \left(1+\frac{b}{a}\right)(1-\cos v), \qquad \xi = \left(1+\frac{b}{a}\right)(1-\cos\varepsilon) \tag{6.37}$$

where Φ and ε are Görtler's variables in Equation (6.37) scaled by $U_\infty a$. The temperature head containing the relation for q_w is determined by setting $u = 1$ in Equation (6.35).

$$\frac{d\theta_w}{d\varepsilon} = -\frac{Q_v}{4}\sin 2\varepsilon - \sum_{k=1}^{\infty} kN_k chkl \sin k\varepsilon$$

$$\theta_w(0) = \frac{Q_v}{8}(1+ch2l) + N_0 + \sum_{k=1}^{\infty} N_k chkl \tag{6.38}$$

Substituting these results into Equation (6.37) for heat flux yields

$$q_w = \frac{\lambda Nu_* T_\infty}{a}\left[\frac{Q_v}{4}(ch^2l - J_2) + N_0 + \sum_{k=1}^{\infty} N_k chkl \ (1-kJ_k)\right],$$

$$J_k = \int_0^v \left[1 - \left(\frac{1-\cos\varepsilon}{1-\cos v}\right)^{C_1}\right]^{-C_2} \sin k\varepsilon \, d\varepsilon \tag{6.39}$$

Because the exponent C_2 does not exceed 1, the integrals J_k having a singularity at $\varepsilon = v$ converge. These equations determine the right-hand part of the first boundary condition (6.34). The left-hand part of this condition is as follows:

$$\left.\frac{\partial\theta}{\partial v}\right|_{u=l} = \frac{Q_v}{4}sh2l + \sum_{k=1}^{\infty} kN_k shkl \cos kv \tag{6.40}$$

Substituting the last two equalities into the first boundary condition
(6.34) and combining the terms containing the unknown coefficients N_k
leads to the following expression:

$$\frac{sh2l}{4}+\frac{\text{Nu}_{**}\text{Bi}}{4\sqrt{\text{Re}}}(ch^2 l - J_2) = \int\limits_{w} we - \frac{\text{Nu}_{**}\text{Bi}}{\sqrt{\text{Re}}} n_0$$

$$-\sum_{k} n_k \left[\frac{\text{Nu}_{**}\text{Bi}}{\sqrt{\text{Re}}} chkl (1 - kJ_2) + k\, shkl\, \cos kv \right] \qquad (6.41)$$

$$\text{Bi} = \frac{\lambda\sqrt{\text{Re}}}{\lambda_w} \qquad n_k = \frac{N_k}{Q_v} \qquad \text{Nu}_{**} = \text{Nu}_* \sqrt{1 - \frac{c^2}{a^2}\cos^2 v}$$

$$\text{Nu}_* = \frac{g_0 \tau_w^{1/2}}{[(1+b/a)(1-\cos v)]^{1/4}}$$

The frictional stress τ_w contained in the last relation is determined by
an integral method [12], assuming that for an elliptical cylinder these
are close to that for a plate. It is clear that the thinner the cylinder is, the
better this assumption is satisfied [12].

Knowing the dependence of Nu_* and elliptic coordinate v, it is possible
to use Equation (6.41) to find the coefficients $n_0, \ldots, n_k$. Writing series (6.41)
with the first $(k+1)$ terms for $(k+1)$ points in the interval $(0, \pi)$, in which
Equation (6.41) must be held, one gets a system of linear algebraic equations determining $n_0, \ldots, n_k$. Calculations are performed with $k = 20$.

The effect of the Biot number, ratio a/b, and Prandtl number on the temperature head and heat flux distribution is studied. Figures 6.7, 6.8, and 6.9

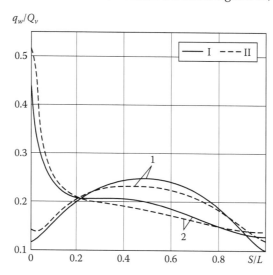

FIGURE 6.7
Distribution of the heat flux over the surface of an elliptical cylinder. $a/b = 4$, $\text{Pr} = 1$, 1 — $\text{Bi} = 10$,
2 — $\text{Bi} = 1$. I — conjugate problem; II — nonconjugate problem.

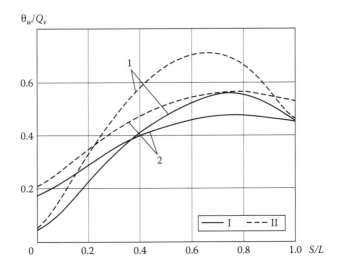

FIGURE 6.8
Distribution of the temperature head over the surface of elliptic cylinder. $a/b = 4$, $Pr = 1$, 1 — $Bi = 10$, 2 — $Bi = 1$. I — conjugate problem; II — nonconjugate problem.

show some of the results. Analysis of the data revealed that the Biot number has the largest effect on the distribution of the heat flux along the cylinder. Thus, for $Bi = 1$, the heat flux has a maximum in the region of the stagnation point, whereas for $Bi = 10$, the maximum is shifted close to the central cylinder section (Figure 6.7). This is explained by

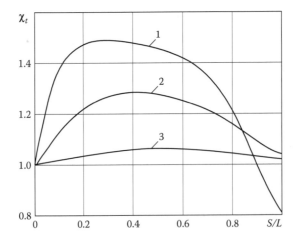

FIGURE 6.9
Distribution of nonisothemicity coefficient over the surface of an elliptical cylinder. $a/b = 4$, $Pr = 1$, 1 — $Bi = 10$, 2 — $Bi = 1$, 3 — $Bi = 0.1$.

the fact that as the Biot number changes, so does the direction in which the total thermal resistance is minimal. It is seen from Figure 6.8 that for large Biot numbers, there are large differences between the temperature distributions obtained in conjugate and nonconjugate solutions which in some points reaches 50%. At the same time, in this case, the heat flux distribution is less affected by the effect of conjugation. However, this effect becomes greater with increasing Biot number.

As seen from Figure 6.9 for Bi < 0.1, the nonisothermicity coefficient did not differ much from unity. For bigger Biot numbers, the distribution of the nonisothermicity coefficient along the cylinder becomes more complicated. Near the stagnation point, the values of χ_t are close to unity. Then, when the distance from the stagnation point increases, the nonisothermicity coefficient grows, reaches the maximum, and then falls steadily.

Over the large part of the surface, with the exception of the end region, bigger Biot numbers correspond to the larger nonisothermicity coefficients. In the end region, the situation is reversed: the value of χ_t for bigger Biot numbers decreases more rapidly than for small Bi and becomes less than unity. As follows from Figure 6.8, the temperature head increases over most of the surface, and this is why the effect of the nonisothermicity is not very big. Nevertheless, the maximum of the heat flux is 40% to 45% greater than that for an isothermal surface, and this maximum value takes place for the special form of Biot number used here, not for Bi close to unity, but for Bi = 10 (Figure 6.9).

6.4 Temperature Singularities on the Solid–Fluid Interface

In early works, Sokolova [13] and later Luikov et al. [14] showed that in the case of laminar flow, the wall temperature of the thermally thin plate near $x = 0$ is not an analytic function of the coordinate but is presented as a series in integer powers of the variable $x^{1/2}$. Here it is shown that in general, the temperature distribution near the origin of the thermally thin plate is an analytical function of variable $x^{1/s}$, where s is determined by the exponent in the relation for an isothermal heat transfer coefficient in the following form [15]:

$$h_* = h_{*L}(x/L)^{-r/s} \qquad (6.42)$$

6.4.1 Basic Equations

Consider a thermally thin plate streamlined past both sides by laminar or turbulent, gradient or gradientless flows of Newtonian or power law non-Newtonian fluids. If the exponent r/s in Equation (6.42) is the same for both flows and temperatures are different, substitution of Equation (3.32) for heat fluxes into Equation (2.7) for the thermally thin plate yields the following equation for the plate temperature:

$$\sum_{k=0}^{\infty} D_k \zeta^k \frac{d^k \vartheta}{d\zeta^k} - \zeta^{r/s} \frac{d^2 \vartheta}{d\zeta^2} - \text{Bi}_{*L2} - \zeta^{r/s} \overline{q}_v = 0,$$

$$\vartheta = \frac{T_w - T_{\infty 1}}{T_{\infty 2} - T_{\infty 1}}, \quad \text{Bi}_{*L} = \frac{h_{*L} L^2}{\lambda_w \Delta} \tag{6.43}$$

$$\overline{q}_v = \frac{q_{v \cdot av} L^2}{\lambda_w (T_{\infty 2} - T_{\infty 1})} \quad D_0 = \text{Bi}_{*L1} + \text{Bi}_{*L1},$$

$$D_k = g_{k1} \text{Bi}_{*L1} + g_{k2} \text{Bi}_{*L1}, \quad \zeta = x/L \tag{6.44}$$

These equations remain valid for the case of equal fluid temperatures if the scale in the definition of ϑ is changed to T_∞, and Bi_{*L2} is omitted. For the symmetric streamlined plate in this equation, one should set, in addition, $\text{Bi}_{*L1} = \text{Bi}_{*L2}$ and $g_{k1} = g_{k2}$.

The exponent r/s usually is not an integer. Therefore, Equation (6.43) has a singularity at the point $x = 0$. If the source distribution near the origin is an analytic function of x, the solution of Equation (6.43) at $x = 0$ can be presented as a series of the variable $x^{1/s}$:

$$\overline{q}_v = \sum_{i=0}^{\infty} b_i \zeta^i, \qquad \vartheta = \sum_{i=0}^{\infty} a_i \zeta^{i/s} \tag{6.45}$$

This follows from the fact that substituting series (6.45) into Equation (6.43) gives the equation for the series coefficients a_i:

$$\sum_{i=0}^{\infty} [D_0 + D_1(i/s) + D_2(i/s)(i/s-1) + \cdots + D_k(i/s)(i/s-1)\cdots(i/s-k+1) + \cdots] a_i \zeta^{\frac{i}{s}}$$

$$- \sum_{i=0}^{\infty} (i/s)(i/s-1) a_i \zeta^{\frac{i+r}{s}-2} - \text{Bi}_{*L2} - \sum_{i=0}^{\infty} b_i \zeta^{i+\frac{r}{s}} = 0 \tag{6.46}$$

To collect the terms with the equal powers of ζ, the indices in the first and third sums are changed so that they and those in the second sum become equal. This transforms indices to $i - (2s - r)$ in the first and to $i/s - 2$ in the third sums. Because these indices must be positive integers, from the first relation it follows that $i > 2s - r$, and hence, the first $2s - r$ coefficients a_i are zero except a_0 and a_s, which are free because the terms in the second sum at $i = 0$ and $i = s$ vanish. The relation for indices in the third sum shows that there should be $i/s = k$, where k is an integer, which means that coefficients b_i should be taken into account only when i is divisible by s and $i/s \geq 2s - r$, because the first $2s - r$ coefficients a_i are zero. The other coefficients are determined by the following equations:

$$D_0 a_0 - (r/s - 1)(r/s - 2)a_{2s-r} - \text{Bi}_{*L2} = 0, \quad i = 2s - r,$$

$$[D_0 + D_1(j-1) + D_2(j-1)(j-2) + \cdots + D_k(j-1)(j-2)\cdots(j-k) + \cdots]a_{s(j-1)}$$

$$-(i/s)(i/s - 1)a_i - b_{(i/s)-2} = 0, \quad j = (i + r - s)/s, \quad i > 2s - r \tag{6.47}$$

All coefficients a_i of series (6.45) can be calculated successively in terms of a_0 and a_s, while the last two are determined from the boundary conditions at the ends of the plate.

Thus, Equation (6.45) gives the actual temperature distribution near the origin of the plate and shows that this temperature is an analytic function in integer powers of the variable $x^{1/s}$, where s is determined by Equation (6.42).

In the case of asymmetric flow, the exponents s/r can be different on two sides. This happens, for instance, when the flow on one side is laminar and on the other side turbulent, or when the pressure gradients or the exponents in the power law for non-Newtonian fluids are different on two sides. For this case, Equation (6.43) becomes

$$\frac{d^2 \vartheta}{d\zeta^2} - \text{Bi}_{*L1}\zeta^{-\frac{r_1}{s_1}}\sum_{k=0}^{\infty} g_{1k}\zeta^k \frac{d^k \vartheta}{d\zeta^k} - \text{Bi}_{*L2}\zeta^{\frac{r_2}{s_2}}\left(\sum_{k=0}^{\infty} g_{2k}\zeta^k \frac{d^k \vartheta}{d\zeta^k} - 1\right) + \bar{q}_v = 0 \tag{6.48}$$

In this case, the temperature head near $\zeta = 0$ is presented by the series (6.45) in power of variable $\zeta^{1/s_1 s_2}$. If the liquids are numbered so that $r_1 s_2 < r_2 s_1$, then the first $(2s_1 s_2 - r_2 s_1)$ coefficients a_i are zero, except a_0 and $a_{s_1 s_2}$ as determined from boundary conditions. The rest of the coefficients are determined by equations similar to Equation (6.47):

$$(r_2/s_2 - 1)(r_2/s_2 - 2)a_{2s_1 s_2 - r_2 s_1} + (1 - a_0)\text{Bi}_{*L2} = 0, \quad i = 2s_1 s_2 - r_2 s_1,$$

$$\text{Bi}_{*L1}[1 + g_{11}(j_1 - 1) + g_{21}(j_1 - 1)(j_1 - 2) + \dots + g_{k1}(j_1 - 1)\dots(j_1 - k) + \dots]a_{s_1 s_2 (j_1 - 1)}$$

$$+\text{Bi}_{*L2}[1 + g_{12}(j_2 - 1) + g_{22}(j_2 - 1)(j_2 - 2) + \dots + g_{k2}(j_2 - 1)\dots(j_2 - k) + \dots]a_{s_1 s_2 (j_2 - 1)}$$

$$-(i/s_1 s_2)[(i/s_1 s_2) - 1]a_i - b_{[(i/s_1 s_2) - 2]} = 0$$

$$j_1 = (i + r_1 s_2 - s_1 s_2)/s_1 s_2, \qquad j_2 = (i + r_2 s_1 - s_1 s_2)/s_1 s_2 \qquad i > 2s_1 s_2 - r_2 s_1 \tag{6.49}$$

6.4.2 Examples of Singularities for Different Flow Regimes and Conditions [15]

It follows from Equation (6.42) that the wall temperature at $x = 0$ is an analytic function of the longitudinal coordinate only when the exponent in Equation (6.42) is an integer. This occurs, for example, for laminar flow at the stagnation point for which the free stream velocity is proportional to x and the isothermal heat transfer coefficient is independent of coordinate ($r/s = 0$).

In this case, there is no singularity, and the wall temperature is presented as a series in integer powers of x:

1. *Laminar gradientless flow.* In this case, $s = 2$, and the wall temperature distribution is presented as a series (Equation 6.45) in integer powers in variable $x^{1/2}$. The first three $(2s - r = 3)$ coefficients except a_0 and a_s should be zero. Therefore, $a_1 = 0$, a_0, and a_2 are determined from the boundary conditions; the coefficient a_3 is found from the first part of Equation (6.47); the rest of the coefficients are found from the second part of Equation (6.47); and the coefficients b are taken into account starting from index 4 and including only those with indices that are divisible by 2 and are greater than or equal to 4 (i.e., 4, 6, 8).

    The expansion (Equation 6.45) for a thermally thin plate differs from series (6.28) obtained in an exact solution of the same problem, but without simplified assumption of the plate thickness. The similar expansion, which can be formed from Equation (6.19) for a thin plate considered in Example 6.5 differs from series (6.28) as well. Although in all three cases the temperature head near the start of the plate is represented by a series in the variable $\zeta^{1/2}$, these series differ substantially. In the most accurate statement of the problem, the series (6.28) contains only terms with $(\zeta^{1/2})^{4k}$ and $(\zeta^{1/2})^{4k+1}$, while the series for the thermally thin plate involves only terms with $(\zeta^{1/2})^{3k}$, and the series for the thin plate, as it is shown in Reference [8], involves all the powers of $\zeta^{1/2}$. Thus, the series for thin and thermally thin plates reflect the character of singularity near the start of the plate only qualitatively.

2. *Turbulent gradientless flow.* Because in this case $s/r = 1/5$, the temperature head is presented as a series in power of $\zeta^{1/5}$. The first nine coefficients, except a_0 and a_s are zero. The coefficients b are taken into account starting from index 10 and considering only those with indices divisible by 5 and those greater than or equal to 10 (i.e., 10, 15, 20, etc.).

3. *Laminar gradient flow with power law free stream velocity.* $U - cx^m$. The heat transfer coefficient for an isothermal surface is defined as $h_* = h_{*L}(x/L)^{(m-1)/2}$. The values of s and r are determined after simplifying the fraction $(1 - m)/2$. For example, for $m = 1/5$ or $m = 1/3$, one gets $s = 5, r = 2$ or $s = 3, r = 1$. In the latter case, the coefficients a_1, a_2 and a_4 are zero, a_0 and a_3 are found from the boundary condition, and b are used when indices are divisible by 3 and are greater than or equal to 6.

4. *Gradientless flow of power law non-Newtonian fluid.* From Equation (3.101), one gets $h_* = h_{*L}(x/L)^{n/(n+1)}$. If n is an integer, then $s = n + 1, r = n$. If n is a fraction such that $n = n_1/n_2$, then $s = n_1 + n_2, r = n_1$. So, for $n = 2, s = 3, r = 2$ and for $n = 3/5, . s = 8, r = 3$.

As an example of an asymmetric flow, consider the case when the flow on one side is laminar $(r_2/s_2 = 1/2)$ and on the other side is turbulent $(r_1/s_1 = 1/5)$. Then, $1/s_1s_2 = 1/10$ and the temperature head distribution is described by series in power of $\zeta^{1/10}$. Because $r_1s_2 = 2 < r_2s_1 = 5$, the first $20 - 5 = 15$ coefficients are zero, but a_0 and a_{10} are determined from the boundary conditions.

6.4.3 Estimation of Accuracy of the Assumption of a Thermally Thin Body [6]

To estimate the error arising when the body is assumed to be thermally thin, the temperature distribution across the body thickness is approximated using the polynomial of the second order. The coefficients of the polynomial are found using the heat fluxes:

$$T/T_{av} = 1 - (\omega_1/3) + (\omega_2/6) + \omega_1(y/\Delta) - (\omega_1 + \omega_2)(y/\Delta)^2/2,$$

$$\omega = q_w \Delta/\lambda_w T_{av} = \text{Bi}(\theta_w/T_{av}) \tag{6.50}$$

Setting $y/\Delta = 0$ and $y/\Delta = 1$, one finds the surface temperature. Then, the condition $T/T_\infty = 1$ yields the inequalities determining a thermally thin body:

$$\frac{1}{T_{av}}\left|\frac{\text{Bi}_1\theta_{w1}}{3} - \frac{\text{Bi}_2\theta_{w2}}{6}\right| \ll 1, \qquad \frac{1}{T_{av}}\left|\frac{\text{Bi}_2\theta_{w2}}{3} - \frac{\text{Bi}_1\theta_{w1}}{6}\right| \ll 1, \qquad \frac{\text{Bi}\theta_w}{6T_{av}} \ll 1 \tag{6.51}$$

The last inequality is attributed to a symmetrical streamlined object. Calculating the left-hand part of these inequalities, one can estimate the error arising when the actual temperature distribution across the body thickness is substituted by an average constant temperature. For the thin metallic plates, inequalities (6.51) are usually satisfied.

6.5 Universal Functions for Solving Conjugate Heat Transfer Problems — Solution Examples [4]

Using series (6.45), one calculates the temperature head to some value of $\zeta > 0$. Then, the numerical integrating of Equation (6.43) or Equation (6.48) yields the solution for the rest of the body. If these equations are used with derivatives not higher than second, the numerical integration can be performed by standard methods. In some cases, these equations can be reduced to well-investigated equations. In particular, Equation (6.43) with the two first derivatives is reduced to a hypergeometric equation. Because $\gamma = (3 - r/s)/(2 - r/s)$ is not usually an integer, the general solution of such an equation is presented as follows:

$$\vartheta = C_1 xF(\alpha, \beta, \gamma, D_2 x^{2-r/s}) + C_2 F(\alpha - \gamma + 1, \beta - \gamma + 1, 2 - \gamma, D_2 x^{2-r/s}) + \sigma_{\text{Bi}} + \vartheta_q$$

$$\tag{6.52}$$

$$\alpha + \beta = \frac{D_1 + D_2}{D_2(2 - r/s)} \qquad \alpha\beta = \frac{D_1 + D_0}{D_2(2 - r/s)} \qquad \sigma_{Bi} = \frac{Bi_{*L2}}{Bi_{*L1} + Bi_{*L2}} \qquad (6.53)$$

Here, α and β are the roots of a quadratic equation, and ϑ_q is a particular solution of inhomogeneous equation (Equation 6.43) with a heat source.

Function (6.52) is independent of boundary conditions of a particular problem, and therefore, in this respect, is universal. Using tabulated functions (Equation 6.52), a simple method is developed for solving the conjugate problems for the thermally thin plates. In the case when the coefficients g_k for both sides are equal, Equation (6.43) reduces to

$$\sum_{k=0}^{\infty} g_k z^k \frac{d^k \vartheta}{dz^k} - z^{r/s} \frac{d^2 \vartheta}{dz^2} - \sigma_{Bi} - z^{r/s} \frac{\overline{q}_v}{z_L^2} = 0,$$

$$z = (Bi_{*L1} + Bi_{*L2})^{1/(2-r/s)}(x/L) = z_L(x/L) \qquad (6.54)$$

According to this equation, the function ϑ depends only on single variable z, and therefore, this function can be easily tabulated. For gradientless laminar and turbulent flows, the two following functions and their derivatives are tabulated (Figures 6.10 and 6.11):

$$\vartheta_1 = F(\alpha - \gamma + 1, \beta - \gamma + 1, 2 - \gamma, g_2 z^{2-r/s})$$

$$\vartheta_2 = zF(\alpha, \beta, \gamma, g_2 z^{2-r/s}) \qquad (6.55)$$

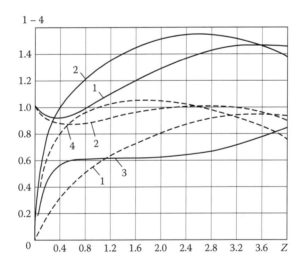

FIGURE 6.10
Universal functions ϑ_1 and ϑ_2 for laminar flow. $Pr > 0.5$, —— ϑ_1, ---- ϑ_2, 1 — $\vartheta/\exp z$, 2 — $\vartheta'/\exp z$, 3 — $\exp z/\vartheta_1''$, 4 — $\vartheta_2''/\exp z$.

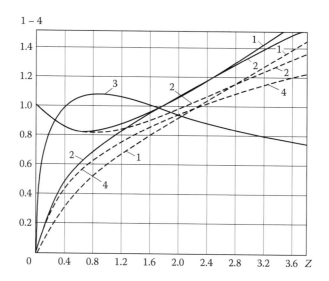

FIGURE 6.11
Universal functions ϑ_1 and ϑ_2 for turbulent flow. $Pr = 0.7$, $Re = 10^6...10^7$, —— ϑ_1, – – – ϑ_2, $1 - \vartheta/\exp(3z/4)$, $2 - \vartheta'/\exp(3z/4)$, $3 - \exp(3z/4)/\vartheta_1''$, $4 - \vartheta_2''/\exp(3z/4)$.

Two other functions ϑ_3 and ϑ_4 determine the functions

$$\vartheta_q = \bar{A}\vartheta_3 + \bar{B}\vartheta_4, \qquad \bar{A} = \frac{AL^2}{\lambda_w (T_{\infty 2} - T_{\infty 1}) z_L^2}, \qquad \bar{B} = \frac{BL^3}{\lambda_w (T_{\infty 2} - T_{\infty 1}) z_L^3} \qquad (6.56)$$

for linear source $\bar{q}_{v.av} = A + B(x/L)$ which are also tabulated (Figures 6.12 and 6.13). For laminar flow, the tabulation is performed for $Pr > 0.5$, for which coefficients g_k are independent of the Prandtl number. For turbulent flow, tabulation is performed for $Pr = 0.7$ and $Re = 10^6 - 10^7$.

For the case of different exponents r/s, the analogous universal functions can be obtained by numerical integration of Equation (6.48) using the same boundary conditions:

$$\vartheta_1(0) = \vartheta_2'(0) = 1, \qquad \vartheta_2(0) = \vartheta_1'(0) = 0 \qquad (6.57)$$

To refine the result obtained by universal functions or to estimate the accuracy of it, one can use Equation (6.9) considering these results as a first approximation. Estimates show that in many cases the universal functions assured practically reasonable accuracy. For example, this occurs when $z < 3$.

To distinguish the temperature head in a particular problem from the universal functions, the following notations are used for the asymmetrical and symmetrical flows, respectively.

$$\theta = \frac{T_w - T_{\infty 1}}{T_{\infty 2} - T_{\infty 1}}, \qquad \theta = \frac{T_w - T_\infty}{T_\infty}$$

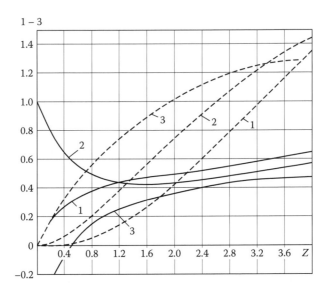

FIGURE 6.12
Universal functions ϑ_3 and ϑ_4 for laminar flow. Pr > 0.5, —— ϑ_3, - - - - ϑ_4, $1 — \vartheta/\exp(3z/4)$, $2 — \vartheta'/\exp(3z/4)$, $3 — \vartheta''/\exp(3z/4)$.

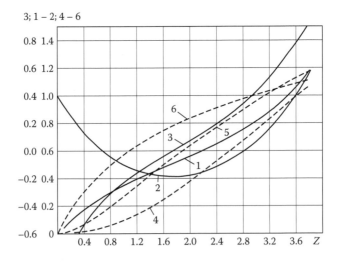

FIGURE 6.13
Universal functions ϑ_3 and ϑ_4 for turbulent flow. Pr = 0.7, Re = 10^6 ... 10^7, ——ϑ_3, - - - - ϑ_4, $1 — \vartheta_3/2$, $2 — \vartheta_3'$, $3 — \vartheta_3''$, $4 — [-\vartheta/\exp(3z/4)]$, $5 — [-\vartheta_4'/\exp(3z/4)]$, $6 — [-\vartheta_4''/\exp(3z/4)]$.

where

$$\theta = C_1 \vartheta_1 + C_2 \vartheta_2 + \sigma_{Bi} + \vartheta_q \qquad (6.58)$$

Provided below are several examples of using universal functions for studying the conjugate heat transfer and solutions of some typical conjugate problems [4].

Example 6.8: A Plate Heated from One End in a Symmetrical Flow

The qualitative analysis of this problem is given in Example 5.2. If the plate is passed from the heated end, the temperature head decreases in the flow direction. Otherwise, when the flow runs in the opposite direction, the temperature head grows in the flow direction. Because the Biot number for an isothermal surface is the same in both cases, this example clearly demonstrates the role of the temperature head variation.

Let the temperature head of heated end be θ_h and the other end be isolated. Determining the constants in Equation (6.58) from these conditions, one finds the temperature head in the first and the second cases, and local heat fluxes from the plate and along the plate:

$$\frac{\theta}{\theta_h} = \vartheta_1 - \frac{\vartheta_1'(z_L)}{\vartheta_2'(z_L)} \vartheta_2, \qquad \frac{\theta}{\theta_h} = \frac{\vartheta_1}{\vartheta_1(z_L)},$$

$$\bar{q}_w = \frac{q_w L^2}{\lambda_w (T_h - T_\infty) z_L^2 \Delta} = \frac{\theta''}{2\theta_h}, \qquad \bar{q}_x \frac{q_x L}{\lambda_w (T_h - T_\infty) z_L} = \frac{\theta'}{\theta_h} \qquad (6.59)$$

Total heat flux from the plate is found by integration. Then, because in the first case $\theta'(z_L) = 0$ and in the second case $\theta'(0) = 0$, the ratio of total heat fluxes is obtained as follows:

$$Q_w = \frac{2L}{\lambda_w (T_h - T_\infty) z_L \Delta} \int_0^L \bar{q}_w dx = \frac{\theta''(z_L) - \theta''(0)}{\theta_h} \qquad \frac{Q_{w1}}{Q_{w2}} = \frac{\vartheta_1(z_L)}{\vartheta_2'(z_L)} \qquad (6.60)$$

In Figure 6.14a are plotted the results for laminar flow. Heat transfer characteristics in both cases differ substantially. In the first case when the temperature head decreases, the heat transfer coefficients are significantly less than the isothermal coefficients and the heat flux sharply decreases along the plate, so that the situation is close to inversion at the end. Here, the heat transfer coefficient is 4.5 times less than an isothermal coefficient. In the second case, the temperature head increases, and according to this, the heat transfer coefficients are greater than the isothermal coefficients but not more than 1.8 times. Nevertheless, the total heat flux in this case is less than that in the other case. This happened because in the first case there are large temperature heads and heat transfer coefficients at the start of flowing, while in the second case at the beginning when the heat transfer coefficients are large, the temperature heads are small, and vice versa. As a result, the heat flux distribution curve has the minimum in this case. The value of the ratio of total heat fluxes Q_{w1}/Q_{w2} depends on Biot number and in the case of

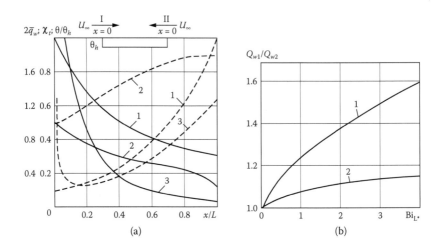

FIGURE 6.14
Heat transfer characteristics for the plate heated from one end. (a) Local characteristics $Bi_{*L} = 1.4$, I —— first case, II - - - - second case, $1 - \theta/\theta_h$, $2 - \chi_t$, $3 - 2\bar{q}_w$. (b) Ratio of total heat fluxes removed from plate: 1 — laminar flow; 2 — turbulent flow.

laminar flow reaches significant values (Figure 6.14b). For instance, for a steel plate with $\Delta/L = 1/10$ past air ($Bi_{*L} = 0.8$) and water ($Bi_{*L} = 4.5$), this ratio is 1.2 or 1.65. In the case of turbulent flow, the difference between total fluxes is smaller (Figure 6.14b), but the distributions of the local heat fluxes along the plate in two opposite directions differ in essence as well.

Note that in a traditional nonconjugate solution with an average isothermal heat transfer coefficient, the results are the same independent of flow direction.

Problem 6.1

A steel plate of length $L = 0.25$ m and thickness $\Delta = 0.01$ m is in the airflow of velocity 3 m/s. The left end of the plate is isolated, and the temperature of the other end is maintained at T_{wL}. The air temperature is 300 K. Calculate the heat transfer characteristics.

It is convenient to use dimensionless temperature head based on its final value $\theta = (T_w - T_\infty)/(T_{wL} - T_\infty)$. Then, applying boundary conditions $q_x(0) = 0$, $\theta(L) = 1$ and relations (Equations 6.58 and 6.59), one finds $\theta = \vartheta_1(z)/\vartheta_1(z_L)$. Because the flow is laminar ($Re = 5 \cdot 10^4$), the Nusselt number for an isothermal surface is $Nu_{*L} = 0.295\sqrt{5 \cdot 10^4}$, and according to Equation (6.43), the Biot number is $Bi_{*L} = \dfrac{Nu_{*L}\lambda L}{\lambda_w \Delta} = 0.656$, $z_L = 1.2$. Heat fluxes are determined by Equation (6.59), $\bar{q}_x = -\vartheta_1'(z)/\vartheta_1(z_L)$ and $\bar{q}_w = \vartheta_1''/2\vartheta_1(z_L)$. The numerical results are given in Table 6.1.

TABLE 6.1

Heat Transfer Characteristics of a Plate Heated from One End (Problem 6.1)

z	x/L	$\vartheta_1(z)$	$\vartheta_1'(z)$	$\vartheta_1''(z)$	θ	$-\bar{q}_x$	$\bar{q}_w$
0	0	1	0	∞	0.278	0	∞
0.2	0.167	1.12	0.949	2.75	0.311	0.264	0.382
0.4	0.334	1.37	1.48	2.66	0.388	0.411	0.369
0.6	0.501	1.72	2.04	3.02	0.478	0.567	0.419
0.8	0.668	2.19	2.70	3.62	0.608	0.750	0.503
1.0	0.835	2.81	3.50	4.42	0.780	0.969	0.614
1.2	1	3.60	4.48	5.12	1	1.244	0.753

Example 6.9: A Plate Streamlined on One Side and Is Isolated from Another and a Plate in a Flow Past Two Sides

Let the temperature T_0 and the heat flux q_0 be given at the starting end of the plate. Using these boundary conditions and the last Equations (6.59), one finds the solution of the problem in question. The same solution can be used in the cases with other boundary conditions at the ends if the corresponding value of q_0 at the starting end is determined. For example, if the temperature T_L or heat flux q_L is given at the other end instead of q_0, the solution of all three problems and relations for values of q_0 are

$$\frac{\theta'}{\theta_0} = \vartheta_1 + \bar{q}_0 \vartheta_2, \qquad \bar{q}_0 = \frac{\theta_L/\theta_0 - \vartheta_1(z_L)}{\vartheta_2(z_L)}, \qquad \bar{q}_0 = \frac{\bar{q}_L - \vartheta_1'(z_L)}{\vartheta_2'(z_L)} \tag{6.61}$$

Figure 6.15 shows the variation of the nonisothermicity coefficient and the temperature head for laminar (a) and turbulent (b) flows in the three cases $\bar{q}_0 = 10, 0$, and (−2). In the first two cases, the temperature head increases along the plate, while in the third one the temperature head first decreases, and after reaching zero its absolute value starts to increase. The same character of heat transfer variation as in other examples is seen. For an increasing temperature head, the heat transfer coefficients are greater than those for an isothermal surface but not more than 75% to 80% in the case of laminar flow and not more than 20% to 25% for turbulent flow. In the third case in which the temperature head decreases, these coefficients are so much smaller that in some points where the temperature head turns to zero, the heat transfer coefficient becomes meaningless, and the corresponding curve $\chi_t(z)$ undergoes discontinuity.

In the case of a symmetrically streamlined plate heated from the end, one gets, after determining the constants in relation (6.58),

$$\bar{q}_x = \theta', \qquad \bar{q}_{w1} = \theta'' + \sigma_{Bi} z^{-r/s}, \qquad \bar{q}_{w2} = q_{w1} - z^{-r/s} \tag{6.62}$$

where the last two relations follow from Equations (2.7) and (3.32).

It follows from Figure 6.16 that in this case the nonisothermicity coefficient variation shows the same pattern. On the side on which the temperature head increases, χ_t is a little more than unity, while on the other side

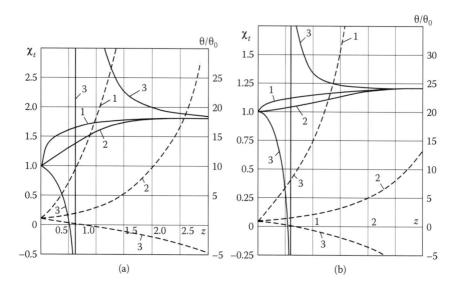

FIGURE 6.15
Heat transfer characteristics for the plate streamlined on one side: (a) laminar flow, (b) turbulent flow —— χ_t, - - - - θ/θ_0, $1 — \bar{q}_0 = 10$, $2 — \bar{q}_0 = 0$, $3 — \bar{q}_0 = -2$.

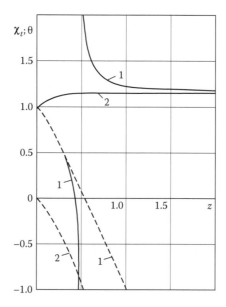

FIGURE 6.16
Variation of temperature head and nonisothermicity coefficient along a plate streamlined on both sides by turbulent flow. $\sigma_{Bi} = 0.5$, $\theta_0 = 1$, $\bar{q}_0 = -2$, —— χ_t, - - - - θ, $1, 2$ — different side of a plate.

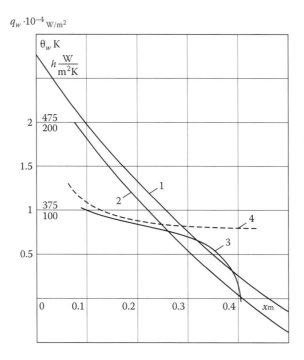

FIGURE 6.17
Variation of heat transfer characteristics along a copper plate streamlined on one side by turbulent flow. $1 — \theta_w(x)$, $2 — q_w(x)$, $3 — h(x)$, $4 — h_*(x)$.

where there is a section with decreasing temperature head, the value of χ_t sharply falls and first becomes zero and then reaches $\pm\infty$.

Problem 6.2

Air at a temperature 313 K flows with a velocity 30 m/s over one side of a copper plate 0.5 m long and 0.02 m in thickness. Another side of the plate is isolated. The temperatures of ends are maintained at $T_{w0} = 593\,\mathrm{K}$ and $T_{wL} = 293\,\mathrm{K}$. Find the distribution of the temperature and heat flux along the plate.

It follows from the boundary conditions $\theta(0) = 1$ and $\theta(L) = \theta_L$ that the dimensionless temperature in the form $\theta = (T_w - T_\infty)/(T_{w0} - T_\infty)$ is determined as $\theta = \vartheta_1 + [\theta_L - \vartheta_1(z_L)][\vartheta_2(z)/\vartheta_2(z_L)]$. For turbulent flow at $Re = 0.88 \cdot 10^6$, one gets $Nu_{*L} = 0.0255\,Re^{4/5}$, $Bi_{*L} = 2.53$, and $z_L = 1.67$, and then the expression for temperature head becomes $\theta = \vartheta_1 - 1.22\vartheta_2(z)$. The results are plotted in Figure 6.17.

Example 6.10: A Plate with Inner Heat Sources [16]

In this case, the solution is presented as a sum of general and particular solutions. The former is found similar to other solutions considered

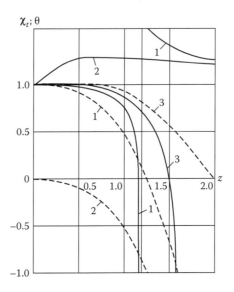

FIGURE 6.18
Heat transfer characteristics for the plate with inner heat sources streamlined by turbulent flow. ——— χ_t, - - - - θ 1, 2 — different sides of a plate, $\sigma_{Bi} = 0.5$, 3 — one side streamlined plate. ($\sigma_{Bi} = 0$)

before using function ϑ_1 and ϑ_2. For the case of uniform or linear distributed sources, the particular solution can be obtained using functions ϑ_3 and ϑ_4 (Figures 6.11 and 6.13). Calculations are performed for the uniform and linear distributed sources in the cases of laminar and turbulent flows. Figure 6.18 shows the results for turbulent flow and the following conditions $\theta_0 = 1$,: $q_0 = 0$, $\bar{A} = 1$, $\bar{B} = 2$ (see Equation 6.56). Although for turbulent flows the heat transfer coefficients do not differ much from those for an isothermal surface in cases of increasing temperature head, for decreasing temperature head the effect of nonisothermicity is as significant as in other similar conditions.

Example 6.11: A Plate in a High-Speed Compressible Flow [4]

In this case, in Equations (3.32) and (3.40) one should substitute an enthalpy difference for temperature head and multiply these equations by the factor $C_x/\sqrt{C}$ (see Equation 3.92). Then, the basic Equation (6.48) takes the following form:

$$\frac{d^2\vartheta}{d\zeta^2} - \frac{Bi_{*L1}C_x}{\sqrt{C}}\zeta^{-r_1/s_1}\left(\vartheta + \sum_{k=1}^{\infty} g_{1k}\zeta^k \frac{d^k\vartheta}{d\zeta^k}\right)$$

$$- \frac{Bi_{*L2}C_x}{\sqrt{C}}\zeta^{r_2/s_2}\left(\vartheta - 1 + \sum_{k=1}^{\infty} g_{2k}\zeta^k \frac{d^k\vartheta}{d\zeta^k}\right) - \bar{q}_v = 0 \qquad (6.63)$$

$$\vartheta = \frac{J_w - J_{ad.1}}{J_{ad.2} - J_{wd.1}}, \qquad Bi_{*L} = \frac{h_{*L} c_p L^2}{\lambda_w \Delta} \qquad \bar{q}_v = \frac{q_{v.av} c_p L^2}{\lambda_w (J_{ad.2} - J_{ad1})} \qquad (6.64)$$

In the case of subsonic and minor supersonic velocities, one may take with reasonable accuracy $C = C_x = 1$ [17]. Then, Equations (6.63) and (6.48) coincide; hence the results obtained for incompressible fluids remain valid for compressible high-speed flows. In such cases, the universal functions can be used for practical calculations. The universal functions can also be used in the more accurate calculations in which for simplicity one assumes $C_x = C$. The value of $\sqrt{C}$ in this case can be included in the relation (6.43) for the Biot number. In order to calculate this value using Equation (2.63), one assumes $C = 1$ and obtains the first approximation that is then used to determine the average surface temperature and then get the second approximation. In the case of calculating without simplifying, the universal functions can be used for first approximation that should be refined by numerical integration of the nonlinear Equation (6.63).

Problem 6.3

Consider the same problem as in Problem 6.2 for an aluminum plate of length 0.3 m and thickness of 0.002 m streamlined by a flow of air of a velocity 250 m/s at an altitude of 20 km. Air temperature is $T_\infty = 223K$; kinetic viscosity is $v = 1.65 \cdot 10^4 \, m^2/s$. The temperature of the front end is a stagnation temperature $T_{w0} = T_{\infty 0} = 254K$, and the other end is maintained at $T_{wL} = 323K$.

Because the Mach number is $M = 250/20.1 \cdot \sqrt{223} = 0.833$, it is necessary to take into account the effect of compressibility. In such a case, instead of dimensionless temperature head, the enthalpy difference should be used. Because the Mach number is not very high, the stagnation temperature difference $\theta = (T_w - T_{wL})/(T_{ad} - T_{wL})$ may be used ignoring the dependence $c_p(T)$. It follows from boundary conditions $\theta(0) = 1$ and $\theta(L) = 0$ that $\theta = \vartheta_1 - [\vartheta_1(z_L)/\vartheta_2(z_L)]$. Estimating the constant in the Chapman-Rubesin formula (Equation 2.63), we get $\sqrt{C} \approx 0.975$. The flow is laminar because $Re = 4.55 \cdot 10^5$. Then one obtains $Bi_{*L} = 2.91, z_L = 2.04$, and $\theta = \vartheta_1(z) - 1.60 \vartheta_2(z)$. The relations for heat fluxes can be obtained from Equation (6.59) substituting $T_{ad} - T_{wL}$ for $T_w - T_\infty$. The results are presented in Table 6.2.

TABLE 6.2

Heat Transfer Characteristics of the Plate in a Compressible Flow (Problem 6.3)

z	x/L	$\vartheta_1(z)$	$\vartheta_1'(z)$	$\vartheta_1''(z)$	$\vartheta_2(z)$	$\vartheta_2'(z)$	$\vartheta_2''(z)$	θ	$\bar{q}_x$	$\bar{q}_w$
0	0`	1	0	∞	0	1	0	1	1.6	∞
0.4	0.196	1.37	1.48	2.66.	0.446	1.29	1.18	0.656	0.584	0.772
0.8	0.392	2.19	2.70	3.62	1.08	1.94	2.10	0.462	0.404	0.260
1.2	0.588	3.60	4.48	5.42	2.06	2.02	3.37	0.304	0.352	0.028
1.6	0.783	5.80	7.16	8.14	3.58	4.71	5.17	0.152	0.378	−0.132
2.04	1	110	11.7	12.6	6.26	7.61	8.04	0	0.476	−0.264

Example 6.12: A Plate in the Flow with a Pressure Gradient [4]

In the general case, when a plate is streamlined by gradient flows with velocities $U_1(x)$ and $U_2(x)$, the equation with two first derivatives similar to Equation (6.48) is

$$(D_{22} + D_{21} - 1)\frac{d^2\vartheta}{d\zeta^2} + (D_{11} + D_{12})\frac{d\vartheta}{d\zeta} + (D_{01} + D_{02})\vartheta - D_{02} - \bar{q}_v = 0 \qquad (6.65)$$

$$D_0 = \text{Bi}_{*L}\frac{h_*}{h_{*L}}, \quad D_1 = D_0\left(g_1\frac{\Phi U_\infty}{\text{Re}U} - g_2\frac{\Phi^2 U_\infty U'}{\text{Re}^2 U^3}\right), \quad D_2 = D_0\left(\frac{\Phi U_\infty}{\text{Re}U}\right) \qquad (6.66)$$

This equation may be solved in the same way as others: by the series close to the beginning of the plate and numerically for the rest of the plate. In the simplest case of self-similar flows, it is possible to introduce the universal functions using Equation (6.65). Some numerical results obtained in Reference [18] for incompressible fluid show that the dependency between the nonisothermicity coefficient and the temperature head variation qualitatively is the same as for gradientless flows: In the case of compressible gradient flows, the problem becomes complicated due to the fact that Dorodnizin-Stewartson variables do not uncouple the thermal and dynamic boundary layer equations.

Example 6.13: A High Heated Radiating Plate [4]

With regard to radiation, Equation (6.63) takes the following form:

$$\frac{d^2\vartheta}{d\zeta^2} - \frac{\text{Bi}_{*L1}C_x}{\sqrt{C}}\zeta^{-r_1/s_1}\left(\vartheta - \vartheta_{ad1} + \sum_{k=1}^{\infty}g_{1k}\zeta^k\frac{d^k\vartheta}{d\zeta^k}\right)$$

$$- \frac{\text{Bi}_{*L2}C_x}{\sqrt{C}}\zeta^{-r_2/s_2}\left(\vartheta - \vartheta_{ad2} + \sum_{k=1}^{\infty}g_{2k}\zeta^k\frac{d^k\vartheta}{d\zeta^k}\right) - N\vartheta^4 + H + \bar{q}_v = 0 \qquad (6.67)$$

$$N = \frac{\sigma(\varepsilon_1 + \varepsilon_2)T_{\infty1}^3 L^3}{\lambda_w\Delta}, \qquad H = \frac{\sigma(\varepsilon_1 + \varepsilon_2\vartheta_\infty^4)T_{\infty1}^3 L^2}{\lambda_w\Delta},$$

$$\vartheta = \frac{T_w}{T_\infty}, \vartheta_{ad} = \frac{T_{ad}}{T_{\infty1}}, \vartheta_\infty = \frac{T_{\infty2}}{T_{\infty1}} \qquad (6.68)$$

The solution of this equation near $\zeta = 0$ can be presented as in other cases in series with exponents $i/s_1 s_2$. The first $(2s_1 s_2 - r_2 s_1)$ coefficients, except a_0 and $a_{s_1 s_2}$, are zero. The rest of the coefficients are determined by an equation similar to Equation (6.49):

$$(r_1/s_1 - 1)(r_1/s_1 - 2)a_{2s_1 s_2 - r_1 s_2} + (\vartheta_{ad2} - a_0)\text{Bi}_{*L2} - \text{Bi}_{*L1}\vartheta_{ad1} = 0, \ i = 2s_1 s_2 - r_1 s_2$$

$$\text{Bi}_{*L1}[1 + g_{11}(j_1 - 1) + g_{21}(j_1 - 1)(j_1 - 2) + \cdots + g_{k1}(j_{1-1})\cdots(j_1 - k) + \cdots]a_{s_1 s_2(j_1 - 1)}$$

$$+ \text{Bi}_{*L1}[1 + g_{21} \times (j_2 - 1) + g_{22}(j_2 - 1)(j_2 - 2) + \cdots + g_{k2}(j - 1)\cdots$$

$$(j_2 - k) + \ldots] a_{s_1 s_2 (j_2 - 1)} - (i/s_1 s_2)[(i/s_1 s_2) - 1] a_i - b_{1[(i/s_1 s_2) - 2]} + N d_{i - 2 s_1 s_2} - H = 0,$$

$$j_1 = (i + r_2 / s_2 - s_1 s_2) / s_1 s_2, \quad j_2 = (i + r_1 / s_1 - s_1 s_2) / s_1 s_2 \quad i > 2 s_1 s_2 - r_1 s_2 \quad (6.69)$$

Here, d_n are coefficients of series in variable $\zeta^{n/s_1 s_2}$ presenting ϑ^4.

Equation (6.67) is nonlinear, and the universal functions could not be used. Therefore, this equation, as others of this type, should be solved numerically using as a starting value at $\zeta > 0$ a result obtained by series. In the case of a compressible fluid, a stagnation temperature T_{ad} is used as a scale, and the plate front temperature is usually at this temperature, and one of the boundary conditions is $\vartheta(0) = 1$. The other boundary condition is obtained from the fact that as distance from origin on a semi-infinite plate grows, the process of heat transfer diminishes because of increasing boundary layer thickness. As a result, the plate temperature tends to some asymptotic constant value $\vartheta_{w.as.}$. In this process, the derivatives of this asymptotic temperature become zero, $\zeta \to 1$, and from Equation (6.67) one gets a boundary condition in the form of an algebraic expression:

$$\frac{C_x}{\sqrt{C}} [\mathrm{Bi}_{*L1}(1 - \vartheta_{as}) + \mathrm{Bi}_{*L2}(\vartheta_{as} - \vartheta_\infty)] - N \vartheta_{as}^4 + H + \bar{q}_v = 0 \qquad (6.70)$$

Problem 6.4

Air flows $(\mathrm{Re} = 5 \cdot 10^4)$ over one side of a thin $(\Delta/L = 1/600)$ radiating $(N = H = 0.07)$ plate $(\lambda/\lambda_w = 0.135 \cdot 10^4)$ with uniform internal heat sources $(\bar{q}_v = 5.1)$. Another side of the plate is isolated. The front end is at stagnation temperature. This problem is solved by Sohal and Howell using numerical integration of the integro–differential equation [19] (Example 6.25).

Here is a simpler solution in series of variable $\zeta^{1/2}$ (laminar flow). In the case of a plate streamlined on one side, Equations (6.67) and (6.70) become

$$\frac{d^2 \vartheta}{d\zeta^2} - \mathrm{Bi}_{*L} \zeta^{-r/s} \left(\vartheta - 1 + \sum_{k=1}^{\infty} g_k \zeta^k \frac{d^k \vartheta}{d\zeta^k} \right) - N \vartheta^4 + H + \bar{q}_v = 0,$$

$$\mathrm{Bi}_{*L}(1 - \vartheta_{as}) - N \vartheta_{as}^4 + H + \bar{q}_v = 0 \qquad (6.71)$$

Solution of the first Equation (6.71) using the other Equation (6.71) and relation $\vartheta(0) = 1$ as boundary conditions is presented in a series. The coefficients $a_1 = a_3 = 0$, because the first $(2s - r) = 3$ coefficients, except a_0 and $a_s = a_2$, should be zero. The other coefficients are determined in terms of $a_0 = 1$ and a_2. Using Equation (6.69) gives $a_4 = -\bar{q}_v/2$, $a_5 = 0.432 \mathrm{Bi}_{*L} a_2$, $a_6 = 0.667 a_2 N$, $a_7 = -0.115 \mathrm{Bi}_{*L} \bar{q}_v$, $a_8 = 0.079 \mathrm{Bi}_{*L}^2 a_2 + 0.167 N(3a_2^2 - \bar{q}_v)$. Then, estimating $\mathrm{Bi}_{*L} = \mathrm{Nu}_{*L} \lambda L/\lambda_w \Delta = 0.535$ leads to the following series:

$$\vartheta = 1 + a_2 \zeta - 2.55 \zeta^2 + 0.231 a_2 \zeta^{5/2} + 0.0467 a_2 \zeta^3 - 0.314 \zeta^{7/2}$$

$$+ (0.0226 a_2 + 0.0351 a_2^2 - 0.0596) \zeta^4 + \cdots$$

TABLE 6.3

Variation of Temperature along a Highly Heated Radiated Plate (Problem 6.4)

$\zeta = x/L$	0	0.1	0.2	0.3	0.4	0.5	0.6	0.7	0.8	0.9	1.0
θ	1.00	1.31	1.58	1.82	2.01	2.18	2.31	2.45	2.57	2.68	2.79

Solving the second part of Equation (6.71), one gets the asymptotic temperature $\vartheta_{as} = 2.79$, and assuming that this condition is practically satisfied at the end of the plate at $\zeta = 1$, the value of coefficient $a_2 = 3.33$ is estimated. Results are given in Table 6.3. There is reasonable agreement between both data obtained by series and by solving an integro–differential equation in [19, Figure 5].

Example 6.14: Two Countercurrent Flowing Fluids Separated by a Thin Wall [4,20]

This problem is considered assuming that the thermal resistance of a plate is negligible in comparison with that of the fluids. This assumption considerably simplifies the problem that is more complicated than similar concurrent conjugate problems. At the same time, such a simplified problem retains the basic qualities of the same problem for the plate with finite resistance.

In the case of laminar flow after using Equation (3.32) with the first three terms, the equality of both heat fluxes leads to an equation determining the plate temperature:

$$[g_{22}\zeta^{3/2} + g_{12}\sigma(1-\zeta)^{3/2}]\vartheta'' + [g_{12}\zeta^{1/2} - g_{11}\sigma(1-\zeta)^{1/2}]\vartheta'$$
$$+ [\zeta^{-1/2} + \sigma(1-\zeta)^{-1/2}]\vartheta - \zeta^{-1/2} = 0 \tag{6.72}$$

Here, ϑ is defined by Equation (6.43) and $\sigma = h_{*1}/h_{*2}$. Boundary conditions $\vartheta(0) = 1$ and $\vartheta(1) = 0$ follow from the fact that at each end, the temperature of the plate is equal to the temperature of the fluid for which this end serves initially. Equation (6.72) is integrated numerically for different values of σ starting from some value of $\zeta = \zeta_i > 0$.

The solution is presented as a Cauchy's problem for two functions satisfying the following boundary conditions: $\theta_1(0) = \theta_1'(0) = \theta_2(0) = 1$ and $\theta_2'(0) = 0$. The values of ϑ for $\zeta < \zeta_i$ are calculated by series (6.45). Finally, the solution and coefficients of series are

$$\vartheta = \frac{\theta_2(1)\theta_1(\zeta) - \theta_1(1)\theta_2(\zeta)}{\theta_2(1) - \theta_1(1)},$$

$$a_0 = 1, \ a_1 = a_3 = 0, \ a_2 = \vartheta'(0), \ a_4 = \frac{a_2 g_{11} - 1}{2g_{21}}, \ a_5 = -\frac{4a_2(1 + g_{12})}{15 g_{21} \sigma} \tag{6.73}$$

Results obtained for $Pr > 0.5$ for which g_k are independent of Prandtl number are plotted in Figure 6.19a. It is seen that in the case of countercurrent flows, the temperature along the interface changes significantly. As a result, in the case of equal thermal resistances ($\sigma = 1$), the heat flux is

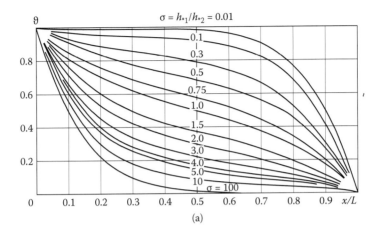

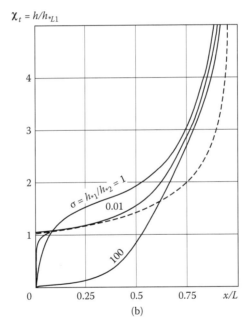

FIGURE 6.19

Heat transfer characteristics for two countercurrent flows separated by a plate with neglected thermal resistance: (a) variation of the temperature head along the plate, (b) variation of the nonisothermicity coefficient along the plate, ---- isothermal plate.

about 30% bigger than that calculated with the isothermal heat transfer coefficient. However, because the temperature head grows in flow directions on both sides of the interface, the distribution of the heat transfer coefficient does not have singularities (Figure 6.19b). At the same time, the results at the ends obtained by conjugate and traditional approaches differ in essence. The heat transfer coefficient at each of the ends obtained

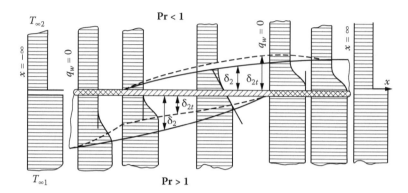

FIGURE 6.20
Scheme of the temperature profile deformation in countercurrent heat transfer.

in the latter problem is equal to that of one of the fluids because the heat transfer coefficient of another fluid $h_* \to \infty$. Therefore, the corresponding heat fluxes at the ends are finite. The heat fluxes at the ends obtained in the conjugate problem are zero. This follows from the fact that the temperature head at the ends tends to zero, as x, while the heat transfer coefficient tends to the infinite value as $x^{1/2}$. Thus, the heat flux tends to zero as $x^{1/2}$. Due to the conjugate conditions, the heat fluxes on the other sides tend to zero as well. As a result, the temperature profile on this side deforms close to end in the same way as in the case when flow impinges on an adiabatic wall (Section 3.5). The scheme of such profile deformation is plotted in Figure 6.20. This occurs because the fluids after the end of contacts arrive on an isolated surface.

It should be noted that in reality, the heat fluxes at the ends are not zero due to finite boundary layer thickness. In consequence, the deformation of velocity profile on the other side does not finish at the end but continues along the same part of the insulated surface.

The results presented here show that the effect of conjugation in the case of countercurrent flows is more significant than that in concurrent streams. In particular, in the latter case, the surface with zero resistance is isothermal (Example 6.5), and for the countercurrent flows even in this simplest case, the interface is considerably nonisothermal, and the effect of conjugation should be taken into account.

6.6 Reducing the Unsteady Conjugate Convective Heat Transfer Problem to an Equivalent Heat Conduction Problem

The methods presented above can also be used to obtain solutions to some classes of unsteady convection heat transfer problems. This holds in all cases when the unsteadiness is caused by thermal regime variation under the unchanging steady velocity distribution. In particular, thermal unsteadiness does not affect the velocity field in the thermal systems with incompressible fluids. Consequently, the methods of solving steady heat transfer problems are applicable to the corresponding unsteady problems if the quasi-steady-state approximation is valid. Because in such cases the unsteady effects in fluid are neglected, the relations obtained above for steady heat transfer are also valid for unsteady processes. Consequently, the method of reducing the conjugate problem to a heat conduction problem is applicable as well.

In the early work [21], Pomeranzev shows by qualitative analysis that an unsteady conjugate problem can be considered as a quasi-state problem if the thermal capacity of a fluid is small in comparison with that of a wall, as, for example, for a nonmetallic fluid–metal wall. Later, Perelman et al. [22] suggested a parameter that follows from the same reasoning by comparing the times of a propagation of a heat impulse in the wall and in a fluid (Example 2.6).

Here this approximate approach is studied using the exact solution of an unsteady thermal boundary layer equation obtained in Section 3.6.

6.6.1 Validity of Quasi-Steady Approximation — The Luikov Number as a Conjugate Criterion in Unsteady Heat Transfer

Consider a symmetric flow past a plate with variable surface temperature $T_w(t)$. In the case of a thermally thin body, the average across its thickness Equation (2.7) is

$$\frac{1}{\alpha_w} \frac{\partial T_{va}}{\partial t} - \frac{\partial^2 T_{av}}{\partial x^2} + \frac{q_w}{\lambda_w \Delta} = 0 \qquad (6.74)$$

Knowing that the conjugation effect in the case of unsteady heat transfer is basically determined by the term containing the first derivative with respect to time (Section 5.1), the following equation is obtained from Equation (6.74) after substituting for q_w series (3.83):

$$\frac{1}{\alpha_w} \frac{\partial T_{av}}{\partial t} - \frac{\partial^2 T_{va}}{\partial x^2} + \frac{h_*}{\lambda_w \Delta}\left[T_w - T_\infty + g_{01}\frac{x}{U}\frac{\partial T_w}{\partial t} + \cdots\right] = 0 \qquad (6.75)$$

The quasi-steady approximation is applicable when the term with a derivative of fluid temperature $\partial T_w/\partial t$ is negligible in comparison with the term with a derivative of solid temperature $\partial T_{av}/\partial t$. Thus, one of the following inequalities should be satisfied:

$$\frac{1}{\alpha_w} \frac{\partial T_{av}}{\partial t} \gg g_{01}\frac{h_* x}{\lambda_w U \Delta}\frac{\partial T_w}{\partial t} \quad \text{or} \quad \frac{\partial T_{av}}{\partial t} \gg g_{01}\frac{x}{\Delta}\frac{Nu_*}{Pe}\frac{\rho c_p}{\rho_w c_{pw}}\frac{\partial T_w}{\partial t} \qquad (6.76)$$

It follows from the last inequality that the ratio of physical properties (capacity ratio) as well as the other parameters in the right part of this inequality played a significant role in unsteady heat transfer processes.

It seems right to name this ratio the Luikov number $Lu = \rho c/\rho_w c_{pw}$, because A. V. Luikov was the leader of the group of scientists who first investigated the conjugate heat transfer [1,14,22–33]. Then, the last inequality takes the following form:

$$\frac{\partial T_{av}}{\partial t} \gg g_{01} \frac{x}{\Delta} \frac{Nu_* Lu}{Pe} \frac{\partial T_u}{\partial t} \tag{6.77}$$

Averaging the product of parameters along the body, one gets

$$\frac{\partial T_{av}}{\partial t} \gg \frac{g_{01}}{2} \frac{L}{\Delta} \frac{Nu_* Lu}{Pe} \frac{\partial T_u}{\partial t} \tag{6.78}$$

This inequality shows that the validity of quasi-steady approximation depends not only on a combination of several parameters but also on the relation between the time derivatives of solid and fluid. These derivatives are determined by sources of unsteadiness: the unsteady internal heat generations, the time variable boundary conditions, the alternating temperature of an incoming stream of fluid, and so forth. If the time scales of all sources are close to each others' time scales, the magnitude of time derivatives of the body $\partial T_{av}/\partial$ and fluid $\partial T_w/\partial t$ are of the same order. In such cases, the product of Nusselt, Peclet, and Luikov numbers along with relative body thickness Δ/L determine the validity of the quasi-steady approximation.

For the boundary layer heat transfer problems, the inequality (6.78) is usually satisfied due to high values of the Peclet number, especially if the body is thin and the Luikov number is less than one. However, in a case with conditions reversed, when the body is thick ($\Delta/L \approx 1$) and the Peclet number is relatively small, the result largely depends on the Luikov number. Because the validity is not obvious, the satisfaction of inequality (6.78) should be checked. If some of the unsteady sources produce much faster than other unsteady sources, the magnitudes of both time derivatives are different and the situation becomes uncertain. In this case, the control of inequality (6.78) is necessary as well. In these cases, when the applicability of the quasi-steady approach is not obvious, the result often depends on the Luikov number. This holds, for example, in systems with the same fluids and different bodies when the Nusselt and Peclet numbers are fixed.

A similar analysis as applied to Equation (6.76) shows that the Luikov number usually determines the coupled rate of the body–fluid. This is because at the known unsteadiness regime and other conditions given in each problem, the relation of both time derivatives of the body and fluid depends basically on the Luikov number. This is confirmed by the fact that the capacity ratio (Luikov number) is used as a criterion of the conjugation rate in solved problems (see Examples 6.17, 6.21, 7.8, and 7.17).

6.6.2 Universal Eigenfunctions for Unsteady Conjugate Heat Transfer Problems [34]

Consider two fluids at temperatures $T_{\infty1}(0)$ and $T_{\infty2}(0)$ flowing past a thermally thin plate with the internal heat sources $q_v(x,y)$ and the ends heated by fluxes $q_0(0)$ and $q_L(L)$. Let the steady-state heat transfer exist at $t < 0$, and at $t = 0$ the sources start to vary in time as $q_v(x,y,t)$, heat fluxes as $q_0(0,t)$ and $q_L(L,t)$, and the temperatures of fluids as $T_{\infty1}(t)$ and $T_{\infty2}(t)$. What distributions of the temperature $T_w(x,t)$ and the heat flux $q_w(x,t)$ are established on the plate after the initial transients died out?

In this case, the average across plate thickness Equation (2.7) has the form

$$\frac{1}{\alpha_w}\frac{\partial T_{vv}}{\partial t} - \frac{\partial^2 T_{av}}{\partial x^2} + \frac{q_{w1}+q_{w2}}{\lambda_w \Delta} - \frac{(q_v)_{av}}{\lambda_w} = 0 \qquad (6.79)$$

Considering the heat transfer in the fluids as a quasi-steady process, the following equation is obtained after substituting Equation (3.32) for heat fluxes into Equation (6.79):

$$\frac{\partial \theta}{\partial Fo} - \frac{\partial^2 \theta}{\partial \zeta^2} + Bi_{*L1}\zeta^{-r_1/s_1}\left(\theta + \sum_{k=1}^{\infty} g_{k1}\zeta^k \frac{\partial^k \theta}{\partial \zeta^k}\right)$$

$$+ Bi_{*L2}\zeta^{-r_2/s_2}\left[\theta - \theta_{\infty}(Fo) + \sum_{k=1}^{\infty} g_{k2}\frac{\partial^k \theta}{\partial \zeta^k}\right] + \frac{dT_{\infty1}}{dFo} - \bar{q}_v = 0 \qquad (6.80)$$

$$\theta = \frac{T_w - T_{\infty1}(t)}{T_{\infty2}(0)-T_{\infty1}(0)}, \qquad \theta_{\infty} = \frac{T_{\infty2}(t)-T_{\infty1}(t)}{T_{\infty2}(0)-T_{\infty1}(0)}, \qquad Fo = \frac{\alpha_w t}{L^2} \qquad (6.81)$$

where the Biot number is defined by Equation (6.43). Equation (6.80) describes an unsteady heat transfer in symmetrical or asymmetrical flow past one or two sides of a thermally thin plate at different flow regimes of Newtonian or non-Newtonian power fluids. Similar equations can be derived considering other unsteady heat transfer problems, for example, those that are studied above in the case of steady-state heat transfer.

Partial differential Equation (6.80) may be solved using a series of eigenfunctions. Consider the case when the exponent r/s and coefficients g_k are the same for both sides of the plate. Equation (6.80) simplifies in this case and takes the form

$$\frac{\partial \theta}{\partial Fo} - \frac{\partial^2 \theta}{\partial \zeta^2} + Bi_{\Sigma}\zeta^{-r/s}\left(\theta + \sum_{k=1}^{\infty} g_k\zeta^k \frac{\partial^k \theta}{\partial \zeta^k}\right) - \sigma\theta_{\infty}(Fo) + \frac{dT_{\infty}}{dFo} - \bar{q}_v = 0 \qquad (6.82)$$

where $Bi_\Sigma = Bi_{*L1} + Bi_{*L2}$, $\sigma = Bi_{*L1}/Bi_{*L2}$. Using this equation with the first two derivatives, Equation (6.82) is multiplied by function E to transform it to a Sturm-Liouville equation:

$$\frac{\partial}{\partial \zeta}\left(E_1 \frac{\partial \theta}{\partial \zeta}\right) + E_2\theta = E\left[\frac{\partial \theta}{\partial Fo} - \sigma\theta_\infty(Fo) + \frac{dT_\infty}{dFo} - \bar{q}_v\right] \qquad (6.83)$$

$$E = (1 - g_2 Bi_\Sigma \zeta^{2-r/s})^{\{g_1 s/[g_2 (2s-r)]\}-1}, \quad E_1 = E(1 - g_2 Bi_\Sigma \zeta^{2-r/s})\; E_2 = -E Bi_\Sigma \zeta^{-r/s} \qquad (6.84)$$

Eigenfunctions of the corresponding homogeneous problem are determined by the following equation:

$$(E_1 W_n')' + E_2 W_n = -\omega_n E W_n, \qquad W_n(0) = W_n(1) = 0 \qquad (6.85)$$

where ω_n are eigenvalues. Because $g_2 < 0$ and $r/s < 1$, the functions E and E_1 in the interval of variation $0 \le \zeta \le 1$ have no singularities, while $E_2 \to \infty$ when $\zeta \to 0$. However,

$$J = \int \left|\frac{E_2}{E_1}\right| d\zeta = Bi_\Sigma \int \frac{\zeta^{-r/s}}{1 - g_2 Bi_\Sigma \zeta^{2-r/s}} d\zeta \qquad (6.86)$$

is finite for any distance within the interval $0 \le \zeta \le 1$. This follows from the fact that as $\zeta \to 0$, $J \to Bi_\Sigma \zeta^{1-r/s}/(1-r/s)$ and, hence, $J \to 0$. Because the integral (6.86) is finite, the Sturm-Liouville problem in question is regular, and the classic results are valid for it.

Equation (6.85) can be refined if one determines heat flux using integral form (Equation 3.40) instead of series (3.32). Then, instead of Equations (6.82) and (6.85), one finds

$$\frac{\partial \theta}{\partial Fo} - \frac{\partial^2 \theta}{\partial \zeta^2} + Bi_\Sigma \zeta^{-r/s}\left[\int_0^\zeta f\left(\frac{\xi}{\zeta}\right)\frac{d\theta}{d\xi}d\xi + \theta(0., Fo)\right] - \sigma\theta_\infty(Fo) + \frac{dT_\infty}{dFo} - \bar{q}_v = 0 \qquad (6.87)$$

$$(E_1 W_n)' + E_2\left[\int_0^\zeta f\left(\frac{\xi}{\zeta}\right)\frac{dW_n}{d\xi}d\xi + W_n(0)\right] = -\omega_n E W_n, \qquad W_n(0) = W_n(1) = 0 \qquad (6.88)$$

The first four eigenvalues (Table 6.4) and eigenfunctions (Figures 6.21 and 6.22) are calculated for laminar and turbulent flows. This is done by numerical integration of Equation (6.83) and then the results are refined using

TABLE 6.4

Eigenvalue ω_n for Unsteady Conjugate Heat Transfer

	Laminar Flow				Turbulent Flow			
Bi_Σ	ω_1	ω_2	ω_3	ω_4	ω_1	ω_2	ω_3	ω_4
1	1.944	12.88	43.20	92.70	1.250	11.40	41.15	90.60
2	4.000	15.89	46.27	96.16	2.530	12.93	42.80	92.32
10	20.16	39.05	73.38	125.9	1266	25.15	56.04	106.3

Equation (6.88) to compute the correction $\varepsilon(\zeta)$ according to Equation (6.9). The solution in the vicinity of $\zeta = 0$ is presented as in steady-state problems by series (6.45).

The solution of Equation (6.83) or Equation (6.87) determining the sought temperature distribution is presented in a series of eigenfunctions. In the case

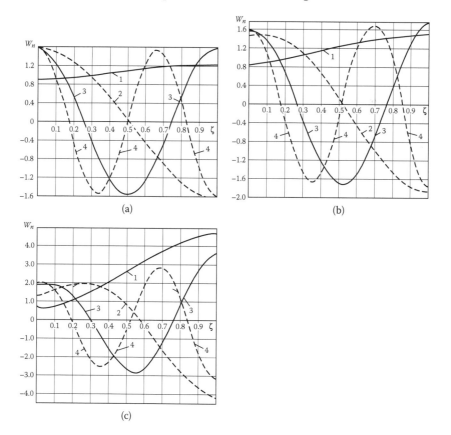

(a)

(b)

(c)

FIGURE 6.21

Eigenfunctions $W_n(\zeta)$ for unsteady conjugate heat transfer. Laminar flow. Pr > 0.5, (a) $Bi_\Sigma = 1$. (b) – $Bi_\Sigma = 2$, (c) $Bi_\Sigma = 10$, 1, 2, 3, 4-function numbers.

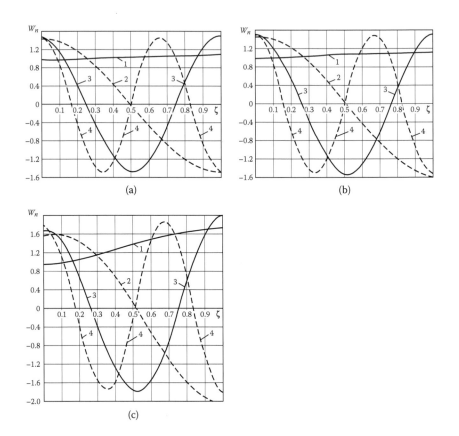

FIGURE 6.22
Eigenfunctions $W_n(\zeta)$ for unsteady conjugate heat transfer. Turbulent flow. Pr = 0.7, (a) $\mathrm{Bi}_\Sigma = 1$. (b) – $\mathrm{Bi}_\Sigma = 2$, (c) $\mathrm{Bi}_\Sigma = 10$, 1, 2, 3, 4-function numbers.

of thermally isolated ends of the plate, this solution has the following form:

$$\theta = \sum_{n=1}^{\infty} N_n(\mathrm{Fo})W_n(\zeta)$$

(6.89)

$$N_n(\mathrm{Fo}) = C_n\exp(-\omega_n\mathrm{Fo}) - \int_0^{\mathrm{Fo}} \exp\{[-\omega_n(\mathrm{Fo}-\tilde{\mathrm{Fo}})]\{a_n[T'_\infty(\tilde{\mathrm{Fo}}) - \sigma\theta_\infty(\tilde{\mathrm{Fo}})] - b_n(\tilde{\mathrm{Fo}})\}\}d\tilde{\mathrm{Fo}}$$

where the decomposing coefficients of 1 and $\bar{q}_v(\zeta, \mathrm{Fo})$ are determined as

$$a_n = \int_0^1 E(\zeta)W_n(\zeta)d\zeta, \qquad b_n = \int_0^1 E(\zeta)W_n(\zeta)\bar{q}_v(\zeta, \mathrm{Fo})d\zeta$$

(6.90)

Coefficients C_n should be found similar using the initial temperature distribution $\theta(\zeta, 0)$.

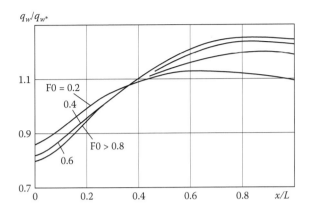

FIGURE 6.23
Variation of a ratio of unsteady heat fluxes obtained by conjugate and traditional approaches. Linearly varying in time temperature of laminar inflow, $Bi_\Sigma = 1$.

If the ends of the plate are heated using fluxes $q_0(0, Fo)$ and $q_L(1, Fo)$, the problem is reduced to the case with the isolated ends by the following substitution:

$$\tilde{\theta} = \theta + q_0(0, Fo)(\zeta - \zeta^2/2) + q_L(1, Fo)\zeta^2/2 \tag{6.91}$$

For a simple case wherein there are no sources and the temperature of the inflowing stream varies linearly with time, $T_\infty'(Fo) = const$, $\theta_\infty = 0$, and $\theta(\zeta, 0) = 0$, one gets

$$N_n(Fo) = const(a_n/\omega_n)[1 - \exp(-\omega_n Fo)] \tag{6.92}$$

It is seen from Figure 6.23 that the heat flux is less at the beginning and greater at the end of the plate than that obtained by traditional calculation. In this case, the temperature head grows in the flow direction, and, therefore, the difference is only $\pm(20 - 25)\%$.

As follows from Table 5.1, the coefficient $g_{01} = 2.4$ at the derivative with respect to time is about four times greater than $g_{10} = 0.63$ at the derivative with respect to the coordinate. This means that the conjugation effect is considerable in the case of unsteady heat transfer than that in steady or quasi-steady heat transfer.

The eigenfunctions $W_n(\zeta)$ are independent of boundary conditions of a particular unsteady problem and are, therefore, similar to functions (Equation 6.52) for steady heat transfer and are universal in this respect. The technique of using these functions for solving a particular unsteady problem is similar to that of the considered above case of steady heat transfer.

6.7 Integral Transforms and Similar Methods

The basic idea of any transform is the hope that the transformed equation can be solved more easily. For example, the Laplace transform converts the ordinary equation in an algebraic expression, and the partial equation is transformed in an ordinary differential equation. Thus, in this case, the transformed subsidiary equations are simpler and, in general, can be solved more readily. However, here arises another problem due to the fact that the most difficult procedure (except the simplest tabulated cases) is to inverse the solution from the Laplace space back to physical variables.

The Laplace transform as applied to conjugate heat transfer problems is extensively used by Sucec. Some of his results are considered in Chapter 2, because these solutions are found using linear or uniform (slug) velocity distribution across the boundary layer. Presented here are some of his other works carried out without those approximations. Many other studies are performed using the integral transform. For example, Rizk et al. [35] used slug approximation and this approach to study heat transfer from a small strip, and Travelho and Santos [36] investigated the unsteady heat transfer in a circular duct.

Some conjugate heat transfer problems are solved by applying the subsidiary equations in different ways or using the solution of a similar problem to a given one. In contrast to integral transforms, those approaches are approximate. Examples of such solutions are also considered in this section.

Example 6.15: A Parallel Plate Duct with Isolated Outer Surfaces and a Periodically Varying Inlet Temperature of Laminar Flow [37]

In the case of a fully developed velocity profile and neglecting an axial conduction, the energy balances on the fluid and on the wall lead to the following problems:

$$\frac{\partial \theta}{\partial t} + u_m \left(\frac{y}{R} + \frac{y^2}{2R^2} \right) \frac{\partial \theta}{\partial x} = \alpha \frac{\partial^2 \theta}{\partial y^2}, \qquad x = 0, t > 0, 0 \le y \le R, \theta = \vartheta_0 \sin \omega t \qquad (6.93)$$

$$y = 0, t > 0, x > 0, \rho_w c_{pw} \Delta \frac{\partial \theta}{\partial t} - \lambda \frac{\partial \theta}{\partial y} = 0, \qquad y = R, t > 0, x > 0, \frac{\partial \theta}{\partial y} = 0, \qquad (6.94)$$

Here θ, ϑ_0, R and u_m are temperature excess, amplitude, a half height of a duct, and a mass average velocity. For the periodic state, no initial condition is needed. In the improved quasi-state approach (Example 2.2), the heat flux is defined as

$$q_w = \int_0^{\bar{x}} G(\bar{x} - \xi) \frac{\partial \theta}{\partial \xi} (\xi, \tau) d\xi \qquad \bar{x} = \alpha x / R^2 u_m \qquad \tau = t - x / u_m \qquad (6.95)$$

for the second time domain $\tau \geq 0$ (after the transient first time domain). The kernel $G(\bar{x})$ is related to the solution of the case of steady-state heat transfer in a duct with isothermal walls. In contrast to the solution given in Example 2.2 with linear velocity distribution in a boundary layer, here the kernel is determined using the expression for heat flux found with the actual velocity profile. In this case for the wall temperature, one gets

$$G(\bar{x}) = \frac{\lambda}{R} \sum_{n=0}^{\infty} G_n \exp\left(-\frac{\mu_n^2 \bar{x}}{8}\right),$$

$$\frac{\partial \theta_w}{\partial(\omega\tau)} + \frac{1}{\bar{\Delta}} \int_0^{\bar{x}} \left\{ \sum_{n=0}^{\infty} G_n \exp\left[-\frac{\mu_n^2}{8}(\bar{x}-\xi)\right] \right\} \frac{\partial \theta_w}{\partial \xi} d\xi = 0, \qquad \bar{\Delta} = \frac{\rho_w c_{pw} R \omega \Delta}{\lambda} \qquad (6.96)$$

where μ_n are eigenvalues. The Laplace transform of this equation with respect to $\bar{x}$ yields

$$\frac{d\tilde{\theta}}{d(\omega\tau)} + \frac{1}{\bar{\Delta}} \sum_{n=0}^{\infty} \frac{G_n}{s+(\mu_n^2/8)}(s\theta_w - \vartheta_0 \sin \omega\tau) = 0 \qquad (6.97)$$

Solving this equation and retaining only the periodic unsteady-state portion gives

$$\frac{\tilde{\theta}}{\vartheta_0} = \frac{\sigma^2 s \sin \omega\tau}{1+\sigma^2 s^2} - \frac{\sigma \cos \omega\tau}{1+\sigma^2 s^2}, \qquad \sigma = \frac{1}{\bar{\Delta}} \sum_{n=0}^{\infty} \frac{G_n}{s+(\mu_n^2/8)} \qquad (6.98)$$

Expending the last equation for small s and taking into account that $\mu_0 < \mu_1 < \mu_2 ...$, one gets

$$\sigma = \frac{1}{\bar{\Delta}} \sum_{n=0}^{\infty} \frac{G_n}{s+(\mu_n^2/8)} \approx \frac{1}{\bar{\Delta}} \left[\sum_{n=0}^{j} \frac{G_n}{s+(\mu_n^2/8)} + \sum_{n=j+1}^{\infty} \frac{G_n}{\mu_n^2/8} \right] \text{ (for large } \bar{x}) \qquad (6.99)$$

After using this result, expression (6.98) becomes a fraction of polynomials in s (see Equation 6.100) and can be inverted by standard procedure. To check the accuracy, the calculation for $j = 0, 1$ and 2 for $\bar{\Delta} = 1, 2$ and 10 in the case of slug velocity u_m are performed. Comparing the results with exact solution [38], complete agreement is found for $j = 2$ and $\bar{x} \geq 0.1$. Then the temperature (for $j = 0$ and similar for $j = 1$ and 2), heat flux, and bulk mean temperature are defined:

$$\frac{\theta_w}{\vartheta_0} = \frac{Q_1(s)\sin \omega\tau - Q_2(s)\cos \omega\tau}{P(s)},$$

$$q_w = -\frac{\bar{\Delta}\lambda}{R} \frac{\partial \theta_w}{\partial(\omega\tau)}, \qquad \frac{\theta_B}{\vartheta_0} = \sin \omega\tau + \bar{\Delta} \int_0^{\bar{x}} \frac{\partial \theta_w}{\partial(\omega\tau)} d\xi \qquad (6.100)$$

The usual quasi-state and finite-difference solutions are obtained as well for comparison.

The basic result of this study is that an improved quasi-state approach based on the actual velocity profile and corresponding heat flux can be used with acceptable accuracy within and beyond the thermal entrance region of a duct. At the same time, the improved quasi-state approach based on the linear velocity profile given in Example 2.2 is restricted to the entrance region only, while the standard quasi-state approach is acceptably accurate basically in a thermally developed region. The last result is expected, because from general consideration, it follows that the quasi-state approximation usually fails at the beginning of the process (Section 6.6.1).

Sucec considered two other problems in the same formulation. In the first problem, the unsteadiness is caused by a sudden change of ambient temperature [39] and in the second by a sinusoidal variation with axial position generation in the wall [40]. In both cases, only the fastest part of the process is studied. Using this fact, the governing equation (Equation 6.93) in the first problem is simplified, and then after applying the Laplace transform, an exact solution is obtained. The same equation (Equation 6.93) in the second problem should be considered together with energy balance for generation. For this more complicated problem, the solution is found by the approximate method of a two integral equation. Theoretically, this method is similar to that which the author calls an improved quasi-steady approach outlined in this example. Details can be found in the literature [41]. Comparison of the results obtained in both problems with finite-difference and usual quasi-steady solutions lead to basically the same conclusion as just discussed in Example 6.15: the simple usual quasi-steady approach fails in the fast, beginning part of the process, and the more complicated methods are much better in this region.

Example 6.16: Alternative Quasi-Steady Approach

The quasi-steady approach similar to that discussed in Example 6.15 is suggested by Karvinen [42], who used as a subsidiary equation the same expression for steady-state heat flux but for a tube:

$$q_w(\bar{x}) = \frac{4\lambda}{d_1} \int_0^{\bar{x}} \sum_{n=0}^{\infty} G_n \exp[-\mu_n^2(\bar{x}-\xi)]dT_w(\xi) \tag{6.101}$$

where G_n and μ_n are eigenfunctions and eigenvalues and $\bar{x} = \lambda x / \rho c_p d_1^2 u_m$. This expression is applied to obtain a differential equation determining the lumped wall temperature:

$$\rho_w c_{pw} d_1 \Delta \frac{\partial T_w}{\partial t} + 8\lambda \int_0^{\bar{x}} \sum_{n=0}^{\infty} G_n \exp[-\mu_n^2(\bar{x}-\xi)]dT_w(\xi) + d_2 h_\infty(T_w - T_\infty) \tag{6.102}$$

where d_1, d_2, Δ, and h_∞ are tube diameters, wall thickness, and heat transfer coefficient to the surroundings. As an example, the case of a step change in inlet temperature is considered for the channel and for a tube. The wall temperature is calculated numerically, approximating the integrand by a series. Numerical results are in agreement with experimental data obtained by the author.

Example 6.17: A Parallel Plate Duct with an Outer Wall Subjected to Convection with Environment and Turbulent Flows with a Periodically Varying Inlet Temperature [43]

Considering quasi-steady approximation and thermally thin radial lamping walls, the authors transform the original complex system of equations and boundary conditions to a simpler periodic problem assuming the solution in the following form (notations in Example 6.15):

$$\bar{u}(\bar{r})\frac{\partial\theta(\bar{r},\bar{x})}{\partial\bar{x}} = \frac{\partial}{\partial\bar{r}}\left[\varepsilon(\bar{r})\frac{\partial\theta(\bar{r},\bar{x})}{\partial\bar{r}}\right] - i\bar{\omega}\theta(\bar{r},\bar{x})$$

$$\frac{\partial\theta}{\partial\bar{r}}(1,\bar{x}) + \left(Bi + \frac{i\bar{\omega}}{Lu}\right)\theta(1,\bar{x}) = b\frac{\partial^2\theta(1.\bar{x})}{\partial\bar{x}^2} \tag{6.103}$$

$$\Theta(\bar{r},\bar{x},\tau) = \theta(\bar{r},\bar{x})\exp(i\bar{\omega}t), \qquad 0<\bar{r}<1, \bar{x}>0 \qquad \bar{\omega}=\frac{\omega R^2}{\alpha}$$

$$\theta(\bar{r},0) = 1, \, 0\le\bar{r}\le1 \quad \frac{\partial\theta}{\partial\bar{r}}(0,\bar{x}) = 0, \, \bar{x}>0,$$

$$\Theta = \frac{T(\bar{r},\bar{x},\bar{t})-T\infty}{\vartheta_0}, \bar{x}=\frac{x}{r_1 Pe}, \qquad \bar{r}=\frac{r}{R}, \quad \bar{t}=\frac{\alpha t}{R^2}, \quad \bar{\Delta}=(r_2-r_1)/r_1 \tag{6.104}$$

$$\varepsilon = 1 + \frac{V_{tb}}{\alpha}(Pr = Pr_{tb} = 1), \, Lu = \rho c_p/\rho_w c_{pw}, \, b = \frac{\bar{\Delta}}{Pe^2}\left(\frac{\lambda_w}{\lambda}\right),$$

$$Bi = \frac{h_\infty rR}{\lambda}, \, Pe = \frac{u_{max}R}{\alpha}$$

where ε is eddy diffusivity, Lu is the Luikov number (Section 6.6.1), θ is quasi-steady temperature, and $\bar{u} = u/u_{max}$.

Solution of this simplified problem strongly depends on the evaluation of the complex eigenvalues and eigenfunctions of the corresponding nonclassical Sturm-Liouville problem in the complex domain. For such a problem, no known solution is available. To find an approximate solution, the generalized integral transform is used by considering a subsidiary problem that is related to the classical steady Graetz problem:

$$\frac{d}{d\bar{r}}\left[\varepsilon(\bar{r})\frac{d\psi(\mu_i,\bar{r})}{d\bar{r}}\right] + \mu_i^2\bar{u}(\bar{r})\psi(\mu_i,\bar{r}) = 0,$$

$$\frac{d\psi(\mu_i,\bar{r})}{d\bar{r}} = 0, \bar{r}=0 \quad \frac{d\psi(\mu_i,\bar{r})}{d\bar{r}} + Bi\psi(\mu_i,\bar{r}) = 0, \bar{r}=1 \tag{6.105}$$

The eigenfunctions set of this system yields the pair of integral transform and inversion:

$$\theta(\bar{r},\bar{x}) = \sum_{i=1}^{\infty}\frac{\psi(\mu_i,\bar{r})}{N_i^{1/2}}\theta(\bar{r}),$$

$$\theta_{.i.a}(\bar{x}) = \int_0^1\bar{u}(\bar{r})\frac{\psi(\mu_i,\bar{r})}{N_i^{1/2}}\theta(\bar{r},\bar{x})d\bar{r}, \quad N_i = \int_0^1\bar{u}(\bar{r})\psi^2(\mu_i,\bar{r})d\bar{r} \tag{6.106}$$

Applying these relations to transform Equation (6.103) after some standard procedures and algebraic manipulations given in Reference [43], leads to expressions for the wall and fluid bulk temperatures and wall heat flux in polar form in terms of amplitudes and phase lags:

$$\Theta_w(\bar{x}) = A_w(\bar{x})\exp[-i\phi_w(\bar{x})], \qquad A_w(\bar{x}) = \{[\mathrm{Re}\,\Theta_w(\bar{x})]^2 + [\mathrm{Im}\,\Theta_w(\bar{x})]^2\}^{1/2},$$

$$\phi_w(\bar{x}) = \tan^{-1}\frac{\mathrm{Im}\,\Theta_w(\bar{x})}{\mathrm{Re}\,\Theta_w(\bar{x})} \tag{6.107}$$

Numerical results show the effects of the parameters: conjugation (Lu), wall axial conduction (b), and Biot number (Bi). For the smallest values of Lu $\approx 5 \cdot 10^{-4}$, the oscillations in the fluid temperature are damped within a short distance from the duct inlet because of the larger thermal capacitances of the wall. For the larger values of Lu $\approx 5 \cdot 10^{-3}$, the thermal wave penetrates farther downstream due to the smaller thermal capacitance of the wall and, consequently, requires a longer length for the same amount of heat to be stored at the wall. The amplitudes of the wall temperature flattened when parameter b increases. This occurs due to improved heat diffusion along the wall, especially very close to the inlet where the thermal gradients are larger. The amplitudes decay slower when the Reynolds number increases from 10^4 to 10^5. In the case of $b = 0$, the wall temperature amplitudes decay faster as the value of Lu decreases. The effect of Lu on the amplitude of the bulk fluid temperature is similar to that for the wall temperature, and the effect of axial wall conduction on bulk fluid temperature turns out to be little. Similar trends are observed with higher Biot number, because an increasing Biot number decreases the wall amplitudes as the external thermal resistance becomes smaller. The smaller is the value of Lu, the larger the heat flux amplitudes. This occurs due to significant attenuation in the wall temperature. On the other hand, with the larger Lu, the axial wall conduction leads to an increase in the heat flux amplitudes. This effect is more pronounced at the inlet and is negligible for smaller values of Lu.

These results are obtained in the case of fixed dimensionless frequency $\bar{\omega} = 0.1$ and indicate, in particular, that for systems of gases flowing inside metal walls, the effect of conjugation cannot be neglected in regions close to the inlet.

6.8 Solutions in Asymptotic Series in Eigenfunctions

Asymptotic solutions are obtained for small and large axial coordinates in different conjugate heat transfer problems. Luikov et al. in an early work [28] gave the solution of heat transfer problems for a plate in a compressible fluid and for flows in the channel and tube. They used the Fourier sine transform, reducing the system of governing equations to an integral equation with a solution presented by an asymptotic series. Stein et al. [44] presented an asymptotic analysis of conjugate heat transfer from a flush-mounted source in an incompressible fluid flow with linear velocity distribution across the boundary layer. Wang et al. [45] obtained an asymptotic solution of conjugate heat transfer between a laminar impingent jet and a disk with given temperature or heat flux distribution on its surface. The asymptotic conjugate analysis of the drying process on a continuously moving porous plate is given by Grechannyy et al. [46,47]. They studied the effect of the temperature and concentration heads distribution on heat and mass transfer from such a plate to flow of fluid entrained by it moving.

Several authors presented asymptotic solutions of conjugate heat transfer in the channels and tubes. Davis and Gill [48] considered heat transfer between the Poiseuille-Couette flow in a parallel plates channel studying the effects of axial conduction by eigenfunctions method. The method of eigenfunction is also used by Hickman [49], who obtained an asymptotic solution for the Nusselt-Graetz problem taking into account the effect of the wall but neglecting the axial conduction. Lee and Ju [50] solved a conjugate problem for a duct in the case of high Prandtl numbers, and Papoutsakis and Ramkrishna [51] presented a solution of a more complicated case of a low Prandtl number when axial conduction in a fluid should be taken into account. These solutions that are based on the asymptotic expansion of the Graetz method are efficient for high abscissas but converge slowly for small values of x. Pozzi and Lupo [52] have provided a solution of a conjugate heat transfer problem for a duct, which is efficient for small values of x.

Example 6.18: Heat Transfer between a Flush-Mounted Source on an Infinite Slab and an Incompressible Fluid Flow [44]

In this case, the thermal boundary layer develops inside the dynamic boundary layer starting from the front edge of the heated strip. Therefore, it is reasonable to use a linear velocity distribution across the thermal boundary layer. Because the axial conduction of a fluid is taken into account, both equations for body and for fluid become elliptic, which makes the solution complicated. The governing mathematical model is

$$cy\frac{\partial T}{\partial x} = \alpha\left(\frac{\partial^2 T}{\partial x^2} + \frac{\partial^2 T}{\partial y^2}\right), \qquad \frac{\partial^2 T_s}{\partial x^2} + \frac{\partial^2 T_s}{\partial y^2} = 0 \qquad (6.108)$$

$$T_s \to 0, |x_s| \to \infty, T \to 0, x^2 + y^2 \to \infty, \ \lambda\left.\frac{\partial T}{\partial y}\right|_{y=0} = \lambda_w\left.\frac{\partial T_s}{\partial y}\right|_{y=0}, \ |x| > \frac{L}{2}, \left.\frac{\partial T_s}{\partial y}\right|_{y=-\Delta} \qquad (6.109)$$

where L is the heat source length and the last equation corresponds to an insulated slab.

The Fourier integral transforms the problem in the upper half-plane:

$$\frac{\partial^2 \tilde{T}_s}{\partial y^2} - s^2 \tilde{T}_s = 0, \quad is\, cy\tilde{T} = \alpha\left(\frac{\partial^2 \tilde{T}}{\partial y^2} - s^2 \tilde{T}\right),$$

$$\frac{\partial \tilde{T}_s}{\partial y}(s, -\Delta) = 0, \quad \tilde{T}_s(s,0) = \tilde{T}(s,0) = \tilde{T}_w(s) \tag{6.110}$$

The solution of the first equation is a simple task. The second is transformed to an Airy equation by introducing a new variable instead of y. Solutions of both equations are as follows:

$$\tilde{T}_s(s,y) = \tilde{T}_w(s)\frac{\cosh[s(y+\Delta)]}{\cosh(s\Delta)}, \quad \tilde{T}(s,\zeta) = C_1(s)Ai(\zeta) + C_2(s)Bi(\zeta) \tag{6.111}$$

$$\zeta = (iPs)^{1/3}y + (iP)^{-2/3}s^{4/3}, \quad P = Pe/L^2, \quad Pe = cL^2/\alpha$$

where $Ai(\zeta)$ and $Bi(\zeta)$ are Airy functions. Analysis [44] shows that $C_2(s) = 0$. Then, after returning back to variable y and using the last condition (6.110), one obtains

$$\tilde{T}(s,y) = \tilde{T}_w \frac{Ai[(iPs)^{1/3}y + (iP)^{-2/3}s^{4/3}]}{Ai[(iP)^{-2/3}s^{4/3}]} \tag{6.112}$$

Both solutions for $\tilde{T}_s(s,y)$ and $\tilde{T}(s,y)$ contain unknown $\tilde{T}_w(s)$, which is defined from condition (6.109), equating it to a heat source. Then, the Fourier transform gives

$$-q(x) = \frac{\partial T}{\partial y}\bigg|_{y=0} = \lambda_w \frac{\partial T_s}{\partial y}\bigg|_{y=0},$$

$$\tilde{T}_w = -\frac{\tilde{q}(s)LAi}{\lambda(is\,LPe)^{1/3}\dfrac{Ai'\,[(iPe)^{-2/3}(sL)^{4/3}]}{Ai[(iPe)^{-2/3}(sL)^{4/3}]} - sL\lambda_w \tanh(s\Delta)} \tag{3.113}$$

There are no known techniques to perform an inverse Fourier transform of this expression; thus, authors performed a detailed investigation to get an asymptotic solution for large x. It is found that the leading contribution to the surface temperature is given by

$$T_w(x) \cong \frac{3^{1/6}Q\Gamma(1/3)}{2\pi(x\lambda)^{2/3}(c\rho c_p)^{1/3}} - \frac{Q_{m1}\Gamma(1/3)}{3^{5/6}\pi\lambda^{2/3}x^{5/3}(c\rho c_p)^{1/3}} + O(x^{-2/3}) \tag{3.314}$$

Here Q and Q_{m1} are the total and the first moment amounts of heat released by the heat source. This term can be classified [44] into a contribution from pure convection, from the interaction of convection and the conduction in a solid, and from the interaction of convection and conduction in the fluid. It is also found that downstream of the heat source, the two leading terms come from pure convection. The asymptotic results are in agreement with the author's numerical solution.

Example 6.19: Heat Transfer between Two Fluids Separated by a Thin Wall Flowing Concurrently or Countercurrently in the Double Pipe — Conjugate Graetz Problem [51]

The flow velocities are fully developed. One fluid (1) is flowing in a tubular space of the double pipe, while the other (2) flows in an annual space exchanging the energy with the surroundings. Axial conduction in a fluid and in a wall are taken into account.

The energy equations for fluids and the boundary conditions are as follows:

$$-\frac{1}{r}\frac{\partial}{\partial r}\left[\lambda r\frac{\partial T}{\partial r}\right] - \lambda\frac{\partial^2 T}{\partial r^2} + \rho c_p u_x(r)\frac{\partial T}{\partial x} = 0, \quad u_{x1} = U_1\left(1-\frac{r^2}{r_1^2}\right),$$

$$u_{x2} = \pm U_2\left[1-\left(\frac{r}{R}\right)^2 + (\bar{\Delta}^2-1)\frac{\ln(r/R)}{\ln\bar{\Delta}}\right] \tag{6.115}$$

$$-\lambda_2\frac{\partial T}{\partial r}\bigg|_{r=R} = h_e[T(x,R)-T_0], \quad -\lambda_1\frac{\partial T}{\partial r}\bigg|_{r=r_1} = h[T(x,r_1)-T(x,r_1+\Delta)] \tag{6.116}$$

$$-\lambda_1\frac{\partial T}{\partial r}\bigg|_{r=r_1} = -\lambda_2\frac{r_1+\Delta}{r_1}\frac{\partial T}{\partial r}\bigg|_{r=r_1+\Delta}, \quad r=0, \frac{\partial T}{\partial r} = 0 \tag{6.117}$$

Here U is the characteristic velocity, h and h_e are heat transfer coefficients for the resistance of separated and outer walls, respectively, and $\bar{\Delta} = (r_1+\Delta)/R$, r_1, R, r_2 are the radiuses of the inner and outer tubes and width of the annular of the double pipe. The plus at u_{x2} is for the concurrent and the minus is for the countercurrent problems.

To decompose the energy equation, two functions are introduced:

$$S_1(x,r) = \int_0^r\left(-\lambda_1\frac{\partial T}{\partial x} + \rho_1 c_{p1} u_{x1} T\right)2\pi\zeta d\zeta,$$

$$S_2(x,r) = S_1(x,r) + \int_{r_1+\Delta}^r\left(-\lambda_2\frac{\partial T}{\partial z} + \rho c_p u_{x2} T\right)2\pi\zeta d\zeta \tag{6.118}$$

Using these functions reduces the solution for each of the fluids to the same two problems:

$$\frac{\partial S}{\partial x} = -2\pi r\left(-\lambda\frac{\partial T}{\partial r}\right), \quad \frac{\partial S}{\partial r} = 2\pi r\left(-\lambda\frac{\partial T}{\partial x} + \rho c_p u_x T\right),$$

$$S_1(x,r), 0 < r < r_1, \quad S_2(x,r), r_1+\Delta < r < R$$

Introducing the dimensionless variables and using Equation (6.116) leads to a system:

$$\frac{\partial \theta}{\partial \xi} = \mathrm{Pe}^2 u(\eta)\theta - \frac{\mathrm{Pe}^2}{2\eta}\frac{\partial J}{\partial \eta}, \quad \frac{\partial J}{\partial \xi} = 2\eta\frac{\partial \theta}{\partial \eta}.$$

$$J(\xi,0) = 0, \frac{\lambda_2}{\lambda_1}J(\xi,\bar{\Delta}) = J[\bar{\Delta} - (\Delta/R)] \tag{6.119}$$

$$\frac{\partial J}{\partial \xi}(\xi,1) = 2\frac{\partial \theta}{\partial \eta}(\xi,1) = -2\mathrm{Bi}_e\theta(\xi,1),$$

$$\xi = \frac{x}{R\mathrm{Pe}}, \quad \eta = \frac{r}{R}, \quad \theta = \frac{T - T_2}{T_0}, \quad \mathrm{Pe} = \frac{U_2 R}{\alpha_2}, \quad \mathrm{Bi} = \frac{hR}{\lambda_1} \tag{6.120}$$

$$\frac{\partial J}{\partial \xi}[\xi,\bar{\Delta} - (\Delta/R)] - [\bar{\Delta} - (\Delta/R)]\frac{\partial \theta}{\partial \eta}[\xi,\bar{\Delta} - (\Delta/R)]$$
$$= -2[\bar{\Delta} - (\Delta/R)] \ \mathrm{Bi} \ [\theta[\xi,\bar{\Delta} - (\Delta/R)] - \theta(\xi,\bar{\Delta})] \tag{6.121}$$

$$J(\xi,\eta) = \left[\int_0^\eta -\frac{1}{\mathrm{Pe}}\frac{\partial \theta}{\partial \xi} + u\theta\right]2\pi\zeta d\zeta,$$

$$u(\eta) = u_1(\eta) = \frac{\mathrm{Pe}_1}{\mathrm{Pe}_2}\left[1 - \frac{\eta^2}{[\bar{\Delta} - (\Delta/R)]^2}\right] \quad 0 < \eta < [\bar{\Delta} - (\Delta/R)] \tag{6.122}$$

$$J(\xi,\eta) = \frac{\lambda_1}{\lambda_2}J[\bar{\Delta} - (\Delta/R)]\left[\int_b^\eta -\frac{1}{\mathrm{Pe}}\frac{\partial \theta}{\partial \xi} + u\theta\right]2\pi\zeta d\zeta,$$

$$u(\eta) = u_2(\eta) = \pm\left[1 - \eta^2 + (\bar{\Delta}^2 - 1)\frac{\ln \eta}{\ln \bar{\Delta}}\right] \quad \bar{\Delta} < \eta < 1 \tag{6.123}$$

Here $J(\xi,\eta)$ is dimensionless function $S(x,r)$, and T_0 is a characteristic temperature.

The numerical calculation is performed for the flow in an annualar space around a solid cylinder with a uniform heat source. Two versions are considered: with heat transfer through the outer walls and with insulated outer walls. Solutions are presented in the usual Graetz-type form of asymptotic series of the eigenfunctions. It is known that such series are efficient for high abscissas but converge slowly for small values of x. Authors conclude that in their examples the series converge unusually fast at Pe = 5 but did not point out the reason for this. They found that unlike the single-stream Graetz problem, the effect of axial heat conduction in the fluid cannot be ignored, even for Peclet numbers larger than 40 to 50, and that Peclet numbers significantly higher than 100 will be required for axial conduction in the fluid to become insignificant.

Example 6.20: Temperature Field in the Entrance of a Plane Duct with a Fully Developed Velocity Profile [52]

The goal of this paper is to give an example of a solution that requires only few terms of the usual eigenfunctions series to obtain good accuracy for small abscissas. Ignoring the axial conduction, the temperatures of a thin wall and of fluid and boundary conditions in dimensionless variables are given as

$$T_s = T_w + (T_0 - T_w)\frac{y-1}{\Delta/R}, \qquad (1-y^2)\frac{\partial\theta}{\partial x} = \frac{\partial^2\theta}{\partial y^2},$$

$$\theta(0,y) = \frac{\partial\theta}{\partial y}\bigg|_{y=0} = 0, \qquad \theta_w - 1 = -\Lambda\frac{\partial\theta}{\partial y}\bigg|_{y=1} \qquad (6.124)$$

where T_0 and T_i are outer wall and fluid inlet temperatures; $\theta = (T - T_i)/(T_0 - T_i)$; R is the half-height of the duct; the reference lengths for dimensionless x and y are R and RPe, respectively; $Pe = u_{max}R/\alpha$; and $\Lambda = \lambda\Delta/\lambda_w R$ is the coupled parameter.

The Laplace transform of Equation (6.124) yields

$$\frac{\partial^2\tilde{\theta}}{\partial y^2} - s(1-y^2)\tilde{\theta} = 0, \qquad \frac{\partial\tilde{\theta}}{\partial y}\bigg|_{y=0} = 0, \qquad \tilde{\theta}(s,1) - \frac{1}{s} = -\Lambda\frac{\partial\tilde{\theta}}{\partial y}\bigg|_{y=1} \qquad (6.125)$$

The solution of this problem for the transformed fluid and surface temperatures gives

$$\tilde{\theta}(s,y) = \exp\left[\frac{(1-y^2)i\sqrt{s}}{2}\right]\frac{F_1(s,y)}{F_1(s,1)}\tilde{\theta}(s,1),$$

$$\tilde{\theta}(s,1) = \frac{F_1(s,1)}{s}[(1-i\sqrt{s}\Lambda)F_1(s,1) + \Lambda(s+i\sqrt{s})F_2(s,1)] \qquad (6.126)$$

where the two confluent hypergeometric functions $F(\gamma,\delta,x)$ are

$$F_1(s,y) = F\left(\frac{1-i\sqrt{s}}{4},\frac{1}{2},i\sqrt{s}y^2\right), \quad F_2(s,y) = F\left(\frac{5-i\sqrt{s}}{4},\frac{3}{2},i\sqrt{s}y^2\right) \qquad (6.127)$$

It is known that to get a solution for small values of x, one should consider an asymptotic solution in the Laplace space for $s \to \infty$. Using the integral formula for $F(\gamma,\delta,x)$,

$$\frac{\Gamma(\delta-\gamma)\Gamma(\delta)}{\Gamma(\gamma)}F(\gamma,\delta,x) = \int_0^1 \exp[x\zeta + (\gamma-1)\log\zeta + (\gamma-\delta-1)\log(1-\zeta)]d\zeta \qquad (6.128)$$

and stationary phase method [53], a transformed surface temperature is obtained:

$$\tilde{\theta}(s,1) = \sum_{k=0}^{m} a_k r^k \Big/ s \sum_{k=0}^{m} b_k r^k, \qquad r = s^{-1/3} \tag{6.129}$$

If r_{mk} is m roots of the polynomial in the denominator, this equation can be written as follows:

$$\tilde{\theta}(s,1) = \sum A_{mk} \frac{r_{mk}^2 s^{-1/3} - r_{mk} s^{-2/3} + s^{-1}}{r_{mk}^3 (r_{mk}^{-3} + s)} \tag{6.130}$$

The inverse transform of this function gives the surface temperature in physical space:

$$\theta_w = \sum_{k=1}^{\infty} A_{mk} \left\{ 1 + \exp\left(-\frac{x}{r_{mk}^3}\right) \left[\frac{3x^{1/3} F(1/3, 4/3, x/r_{mk}^3)}{r_{mk} \Gamma(1/3)} \right. \right.$$

$$\left. \left. - \frac{3x^{2/3} F(2/3, 5/3, x/r_{mk}^3)}{2 r_{mk}^2 \Gamma(2/3)} - 1 \right] \right\} \tag{6.131}$$

In the cases of $\Lambda \le 0.01$ and $\Lambda \ge 1$, this result can be simplified to get, respectively,

$$\theta_w = \frac{a_1}{b_1} \left[\frac{1 - \Lambda x^{-1/3}}{a_1 \Gamma(2/3)} + \frac{\Lambda^2 x^{-2/3}}{a_1^2 \Gamma(1/3)} \right],$$

$$\theta_w = \frac{1}{\Lambda} \sum_{n=0}^{\infty} c_n \frac{x^{(1+n)/3}}{\Gamma[(4+n)/3]} \tag{6.132}$$

The first expression is obtained using the asymptotic formula for the hypergeometric function, and the second is a result of the inversion of series given the solution in the Laplace space. Constant coefficients a, b, c are given in Reference [52].

The fluid bulk mean temperature and the local Nusselt number are defined as

$$\theta_{bl}(x) = \frac{3}{2\Lambda} \int_0^x [1 - \theta_w(\zeta)] d\zeta, \qquad \mathrm{Nu} = \frac{4(\partial\theta/\partial y)_{y=1}}{\theta_w - \theta_{bl}} \tag{6.133}$$

The accuracy given by Equation (6.131) for small and large Λ as well as by the first Equation (6.132) for $\Lambda \le 0.01$ and by the second Equation (6.132) for $\Lambda \ge 1$ is shown by comparison of the values of Nusselt numbers computed with the data adopted from Reference [54] which are calculated using 120 terms of the eigenfunction series.

Example 6.21: The Pipe with Hydrodynamically and Thermally Fully Developed Flow and a Thick Wall with Sufficient Heat Conduction and Capacity [55]

For the time domain $t_{max} < x/2u_m$, the equations and associated conditions are

$$\frac{\partial T}{\partial t} = \alpha \frac{1}{r} \frac{\partial}{\partial r}\left(r\frac{\partial T}{\partial r}\right) - \mu\left(\frac{\partial v}{\partial r}\right)^2 \quad 0 < t < t_{max}, 0 < r < R$$

$$\frac{\partial T_s}{\partial t} = \alpha_w \frac{1}{r} \frac{\partial}{\partial r}\left(r\frac{\partial T_s}{\partial r}\right), \quad R < r < R_0 \qquad\qquad (6.134)$$

$$t = 0, \ T = T_s = T_i; \ t > 0, \ r = 0, \ \frac{\partial T_s}{\partial r} = 0; \ r = R, \ T = T_s, \ \lambda\frac{\partial T}{\partial r} = \lambda_w\frac{\partial T_s}{\partial r},$$

$$r = R_0, \ q = -\lambda_w\frac{\partial T_s}{\partial r} \quad \text{ or } \quad T_s = T_w$$

Using dimensionless variables, one puts these equations and conditions in a unified form:

$$\frac{\partial\theta}{\partial \bar{t}} = f(\bar{r})\frac{1}{\bar{r}} \frac{\partial}{\partial \bar{r}}\left(\bar{r}\frac{\partial\theta}{\partial \bar{r}}\right) + \sigma(\bar{r}), \ \bar{t} > 0, \ 0 < \bar{r} < \bar{R}_0 \quad f(\bar{r}) = \left(\frac{1}{k}\right),$$

$$\sigma(\bar{r}) = \left(\begin{array}{c} c\bar{r}^2 \\ 0 \end{array}\right) \begin{array}{c} 0 \leq \bar{r} \leq 1 \\ 1 \leq \bar{r} \leq \bar{R}_0 \end{array} \qquad\qquad (6.135)$$

$$\bar{t} = 0, \ \theta = 0; \ \bar{t} > 0, \ \bar{r} = 0, \ \frac{\partial\theta}{\partial \bar{r}} = 0, \ \bar{r} = 1, \ \theta^- = \theta^+, \ \left(\frac{\partial\theta}{\partial \bar{r}}\right)^- = \Lambda\left(\frac{\partial\theta}{\partial \bar{r}}\right)^+,$$

$$\bar{r} = \bar{R}_0, \ \frac{\partial\theta}{\partial \bar{r}} = -\frac{1}{\Lambda} \quad \text{ or } \quad \theta = \theta_w \qquad\qquad (6.136)$$

$$\bar{r} = \frac{r}{R}, \ \bar{t} = \frac{\alpha t}{R^2}, \ \Lambda = \frac{\lambda_w}{\lambda}, \ k = \frac{\alpha_w}{\alpha}, \ a) \ \theta = \frac{T - T_i}{T_w - T_R}, \ c = \frac{16\mu u_m}{\lambda(T_w - T_R)},$$

$$b) \ \theta = \frac{\lambda(T - T_i)}{qR}, \ c = \frac{16\mu u_m}{qR} \qquad\qquad (6.137)$$

Thus, system (Equation 6.134) is transformed into one equation with a single domain with discontinuous properties. In Equations (6.134) and (6.135), R_0 is an external pipe radius, T_R is a reference temperature, $(^-)$ and $(^+)$ denote the fluid and the wall sides at the interface, c is defined using formula $u = 2u_m(1 - r^2/R^2)$ and a different definition of θ and is related to boundary conditions on the outer surface of the wall: (a) for a temperature and (b) for a heat flux. The eigenvalue problem associated with Equations (6.135) and (6.136) is as follows:

$$f(\bar{r})\frac{1}{\bar{r}}\frac{\partial}{\partial \bar{r}}\left(\bar{r}\frac{\partial G}{\partial \bar{r}}\right)=-g^2G \quad \bar{r}=0, \quad \frac{\partial G}{\partial \bar{r}}=0, \quad \bar{r}=1, \quad G^-=G^+,$$

$$\left(\frac{\partial G}{\partial \bar{r}}\right)^- = \Lambda\left(\frac{\partial G}{\partial \bar{r}}\right)^+, \quad \bar{r}=\bar{R}_0, \quad \frac{\partial G}{\partial \bar{r}}=0 \quad or \; G=0 \tag{6.138}$$

The solution of this problem can be expressed using the Bessel functions:

$$G_1 = e_1 J_0(m\bar{r}) + e_2 Y_0(m\bar{r}), \quad 0 \le \bar{r} \le 1;$$

$$G_2 = f_1 J_0(m\bar{r}/\sqrt{k}) + f_2 Y_0(m\bar{r}/\sqrt{k}), \quad 1 \le \bar{r} \le \bar{R}_0 \tag{6.139}$$

To satisfy the condition for $\bar{r}=0$, it must be $e_2=0$. To satisfy the last boundary condition at $\bar{r}=\bar{R}_0$ for $\partial G/\partial \bar{r}=0$, the remaining three conditions yield

$$\mathbf{Ax}=0 \qquad \mathbf{x}=(e_1,f_1,f_2)^T$$

$$\mathbf{A} = \begin{bmatrix} -J_0(g_n) & J_0(g_n/\sqrt{k}) & Y_0(g_n/\sqrt{k}) \\ -(\sqrt{k}/\Lambda)J_1(g_n) & J_1(g_n/\sqrt{k}) & Y_1(g_n/\sqrt{k}) \\ 0 & J_1(g_n\bar{R}_0/\sqrt{k}) & Y_1(g_n\bar{R}_0/\sqrt{k}) \end{bmatrix} \tag{6.140}$$

For $G=0$ at $\bar{r}=\bar{R}_0$, these three conditions give a similar expression for $\mathbf{A}$ in which functions $J_0(g_n\bar{R}_0/\sqrt{k})$ $Y_0(g_n\bar{R}_0/\sqrt{k})$ are substituted for the last two functions. The solution of homogeneous Equation (6.140) exists only if det $\mathbf{A}=0$. The last equation gives

$$G_{1n}=J_0(g_n\bar{r}), \qquad G_{2n}=A_n J_0(g_n\bar{r}/\sqrt{k})+B_n Y_0(g_n\bar{r}/\sqrt{k}) \tag{6.141}$$

$$A_n = \frac{J_0(g_n)Y_1(g_n/\sqrt{k})-(\sqrt{k}/\Lambda)J_1(g_n)Y_0(g_n/\sqrt{k})}{J_0(g_n/\sqrt{k})Y_1(g_n/\sqrt{k})-J_1(g_n/\sqrt{k})Y_0(g_n/\sqrt{k})} \tag{6.142}$$

$$B_n = \frac{(\sqrt{k}/\Lambda)J_1(g_n)J_0(g_n/\sqrt{k})-J_0(g_n)J_1(g_n/\sqrt{k})}{J_0(g_n/\sqrt{k})Y_1(g_n/\sqrt{k})-J_1(g_n/\sqrt{k})Y_0(g_n/\sqrt{k})} \tag{6.143}$$

The function (6.141) does not form an orthogonal set, because the derivative of G_n is discontinuous at $\bar{r}=1$. However, these functions can be made orthogonal by the weight function that can be found using the method of Yeh [56]. Applying this approach and the condition of orthogonality, the authors find the weight functions for both eigenfunctions G_n:

$$\sum_{i=1}^{2}\int_0^{\bar{r}_i} G_n(\bar{r})G_m(\bar{r})F_i\bar{r}d\bar{r}=0, \text{ if } n\ne m, \ne 0 \text{ if } n=m, \qquad F_1=1, F_2=\Lambda \tag{6.144}$$

Solutions for the temperature distribution $\theta(\bar{r}, \bar{t})$ along the interface in the cases of the outer wall temperature (a) or heat flux (b) boundary conditions are obtained as

$$\theta = \theta_w + \sum_{n=1}^{\infty} G_n(\bar{r})\left[c_n \exp(-g_n^2 \bar{t}) + \frac{V_n}{g_n^2} \right],$$

$$\theta = W_0 \bar{t} - \frac{1}{\Lambda \bar{R}_0}\left(\frac{\bar{r}^2}{2} - \frac{1}{4} \right) + \sum_{n=1}^{\infty} G_n(\bar{r})\left[c_n \exp(-g_n^2 \bar{t}) + \frac{W_n}{g_n^2} \right] \tag{6.145}$$

$$V_n = \frac{c \displaystyle\int_0^1 G_{1n}(\bar{r}) \bar{r}^3 d\bar{r}}{\displaystyle\int_0^1 G_{1n}(\bar{r}) \bar{r} d\bar{r} + \frac{\Lambda}{k} \int_0^{\bar{R}_0} G_{2n}^2(\bar{r}) \bar{r} d\bar{r}},$$

$$c_n = -\theta_w \frac{\displaystyle\int_0^1 G_{1n}(\bar{r}) \bar{r} d\bar{r} + \frac{\Lambda}{k} \int_0^{\bar{R}_0} G_{2n}^2(\bar{r}) \bar{r} d\bar{r}}{\displaystyle\int_0^1 G_{1n}^2(\bar{r}) \bar{r} d\bar{r} + \frac{\Lambda}{k} \int_0^{\bar{R}_0} G_{2n}^2(\bar{r}) \bar{r} d\bar{r} - \frac{V_n}{g_n^2}} \tag{6.146}$$

$$W_n = \frac{\displaystyle\int_0^1 \left(c\bar{r}^2 + \frac{2}{\Lambda \bar{R}_0} \right) G_{1n}(\bar{r}) \bar{r} d\bar{r} + \frac{\Lambda}{k} \int_0^{\bar{R}_0} \frac{2k}{\Lambda \bar{R}_0} G_{2n}(\bar{r}) \bar{r} d\bar{r}}{\displaystyle\int_0^1 G_{1n}^2(\bar{r}) \bar{r} d\bar{r} + \frac{\Lambda}{k} \int_0^{\bar{R}_0} G_{2n}^2(\bar{r}) \bar{r} d\bar{r}},$$

$$c_n = -\frac{1}{2 \Lambda \bar{R}_0} \frac{\displaystyle\int_0^1 G_{1n}(\bar{r}) \bar{r}^3 dr + \frac{\Lambda}{k} \int_0^{\bar{R}_0} G_{2n}(\bar{r}) \bar{r}^3 d\bar{r}}{\displaystyle\int_0^1 G_{1n}^2(\bar{r}) \bar{r} d\bar{r} + \frac{\Lambda}{k} \int_0^{\bar{R}_0} G_{2n}^2(\bar{r}) \bar{r} d\bar{r} - \frac{W_n}{g_n^2}} \tag{6.147}$$

The numerical results are computed using 50 terms of the series. These results show the following:

1. The temperature decreases with increasing values of the conjugation parameter $\sqrt{k}/\Lambda = \sqrt{\rho c_p \lambda/(\rho c_p \lambda)_w} = \sqrt{\text{Lu}(\lambda/\lambda_w)}$, where Lu is the Luikov number (Section 6.6.1), and it reaches a steady state in case (a); the temperature also decreases with increasing values of the conjugation parameter and time, but it does not reach a steady state in case (b).

2. The greater the conjugation parameter is, the greater the effect of fluid properties, so that at a given time, the interface temperature is closer to the initial condition $\theta = 0$.
3. The temperature increases with the radius and with increasing dissipation.
4. Despite the dissipation, in the case of heat extraction from the outer wall and of high conjugation parameter, the heat flux reversal occurs at the interface.

6.9 Superposition and Other Methods

Example 6.22: Heat Transfer between Two Laminar or Turbulent Concurrent or Countercurrent Flows Separated by a Thin Plate [57]

Here the method of superposition is used to solve the problem considered in Example 6.5. Substituting Equation (1.38) for heat flux into Equation (6.11) for a thin plate gives

$$-\int_0^{\bar{x}_1} f\left(\frac{\xi}{\bar{x}_1}\right)\frac{d\theta_1}{d\xi}d\xi = H\frac{\bar{x}_1^{1-m_1}}{\bar{x}_2^{1-m_2}}\int_0^{\bar{x}_2} f\left(\frac{\xi}{\bar{x}_2}\right)\frac{d\theta_2}{d\xi}d\xi,$$

$$\theta_2(\bar{x}_2)-\theta_1(\bar{x}_1) = \frac{G}{\bar{x}_1^{1-m_1}}\int_0^{\bar{x}_1} f\left(\frac{\xi}{\bar{x}_1}\right)\frac{d\theta_1}{d\xi}d\xi \qquad (6.148)$$

$$\theta = \frac{T-T_{\infty 2}}{T_{\infty 1}-T_{\infty 2}}, \quad \bar{x} = \frac{Ux}{\nu}, \quad H = \frac{K_2\lambda_2\,\mathrm{Pr}_2^{n_2}(U_2 v_1)^{m_2}}{K_1\lambda_1\,\mathrm{Pr}_1^{n_1}(U_1 v_2)^{m_1}},$$

$$G = \frac{K_1\lambda_1 U_1^{m_1}\,\mathrm{Pr}_1^{n_1}}{\lambda_w v_1^{m_1}}, \quad f\left(\frac{\xi}{x}\right) = \left[1-\left(\frac{\xi}{x}\right)^{C_1}\right]^{-C_2}$$

Here, $h_* = K(\lambda/\bar{x})\mathrm{Re}_x^m\,\mathrm{Pr}^n$, $\bar{x}_2 = \bar{x}_1$ for concurrent and $\bar{x}_2 = \mathrm{Re}_{L1}-\bar{x}_1$ for countercurrent flows.

To solve these equations, the interval $0 < \bar{x} < \mathrm{Re}_{L1}$ is broken up into N small subintervals $\Delta_{i,j}$. If $s_{i,k}$ denotes $d\theta/dx$ on the kth subinterval, Equation (6.148) for the concurrent flows and wall temperatures $\theta_1(\bar{x}_{1,j})$ and $\theta_2(\bar{x}_{1,j+1})$ have the forms

$$-\sum_{k=1}^{j} s_{i,k}\int_{\bar{x}_{1,k-\Delta_k}}^{\bar{x}_{1,k}} f\left(\frac{\xi}{\bar{x}_{1,j+1}}\right)d\xi = H\frac{\bar{x}_{1,j}^{1-m_1}}{\bar{x}_{2,j+1}^{1-m_2}}\sum_{1}^{j+1} s_{2,k}\int_{\bar{x}_{2,k-\Delta_k}}^{\bar{x}_{2,k}} f\left(\frac{\xi}{\bar{x}_{2,j+1}}\right)d\xi$$

$$\theta_1(\bar{x}_{1,j}) = 1+\sum_{k=1}^{j} s_{1,k}\Delta_k \qquad (6.149)$$

$$-\sum_{k=1}^{j+1} s_{2,k}\Delta_k - \sum_{1}^{j} s_{1,k}\Delta_k - 1 = \frac{G}{\bar{x}_1^{1-m_1}} \int_{\bar{x}_{2,k-\Delta_k}}^{\bar{x}_{2,k}} f\left(\frac{\xi}{\bar{x}_{1,j}}\right) d\xi,$$

$$\theta_2(\bar{x}_{2,j+1}) = \sum_{k=1}^{j+1} s_{2,k}\Delta_k \tag{6.150}$$

Solving the system of $2N$, those linear algebraic equations, one then calculates temperatures. For the concurrent flows, the computing is simpler because in such a case, $s_{i,k}$ is calculated recursively, while the matrix is required for the case of countercurrent flows.

Using the numerical results, the effect of parameters H and G on the heat transfer intensity is presented. The first parameter is a measure of the relative heat transfer conductance of the streams, and the second is a measure of the conductance of the plate relative to that of one of the streams. The following are determined:

1. For any values of H, a decrease in G or an increase in $\bar{x} = \mathrm{Re}_x$ results in limit $\theta = 1/(1+H)$.
2. For fixed H and G, heat flux decreases as Re_{x1} increases because the boundary layer becomes thicker. As $\mathrm{Re}_x \to 0$, the heat flux approaches unity.
3. For given values of G and Re_x, an increase in H results in an increased heat flux. However, when H grows, the rate of that increase decreases. Some other results of this work are considered in Example 2.5.

The heat transfer characteristics in this study are determined by three parameters that can be substituted by two $H = Bi_2/Bi_1$ and $\bar{x}_1 G = Bi_1$ containing only a Biot number for each flow [7].

Example 6.23: Two Quiescent Fluids Separated by a Vertical Thin Plate [58]

Although formulation of this problem is similar to the previous one, the solution of it is different because the problem is nonlinear due to natural convection when velocity and energy boundary layer equations are coupling. Therefore, the superposition method is not applicable, and using the same condition (Equation 6.11) for a thin plate, the approximate formula for heat flux in the case of natural convection derived in Reference [59] is employed. As a result, a system defining the plate temperatures is obtained:

$$C(\mathrm{Pr}_1)\frac{\lambda_1 \Delta}{\lambda_w L} Ra_1^{1/4}\theta_{w1}^{5/3}(\bar{x}_1)\left[\int_0^{\bar{x}_1}\theta_{w1}^{5/3}(\xi)d\xi\right]^{-1/4}$$

$$= \bar{q} = C(\mathrm{Pr}_2)\frac{\lambda_2 \Delta}{\lambda_w L} Ra_2^{1/4}\theta_{w2}^{5/3}(\bar{x}_1)\left[\int_0^{\bar{x}_2}\theta_{w2}^{5/3}(\xi)d\xi\right]^{-1/4} \tag{6.151}$$

$$1-\theta_{w1}+\theta_{w2} = C(\mathrm{Pr}_1)\frac{\lambda_1\Delta}{\lambda_w L}Ra_1^{1/4}\theta_{w1}^{5/3}(\bar{x}_1)\left[\int_0^{\bar{x}_1}\theta_{w1}^{5/3}(\xi)d\xi\right]^{-1/4},$$

$$\theta_w = \frac{T_{wi}-T_{\infty i}}{T_{\infty 1}-T_{\infty 2}}, \quad \bar{q} = \frac{q\lambda_w}{\Delta(T_{\infty 1}-T_{\infty 2})}$$

Here, $\bar{x} = x/L$, $\xi_2 = 1-\xi_1$ and constant $C(\mathrm{Pr})$ is given in Reference [60]. These equations show that the required plate temperatures are dependent on two parameters $H = (\lambda_1/\lambda_w)(\Delta/L)Ra_1^{1/4}$ and $G = (\lambda_1/\lambda_2)(\Delta/L)$ $(Ra_1/Ra_2)^{1/4}$. Equation (6.151) is solved numerically by iterations using 40 intervals for ξ between 0 and 1. Results computed are in reasonable agreement with the author's experimental data and show the following:

1. When thermal resistances of both fluids are equal, there are sharp temperature variations near the top and bottom of the plate.
2. In this case, the heat flux is symmetrical about the midpoint of the plate and practically constant over about 80% of the central part of it.
3. The more the parameter G departs from unity, the more fluid and thermal resistances differ and the greater is the asymmetry of the flux.
4. The effect of the thermal interaction of the two natural streams is only moderate. The heat transfer coefficients are about 12% higher than those obtained using a constant temperature wall.

Example 6.24: Heat Transfer between Two Fluids Separated by a Vertical Thin Plate Local Similarity Approach [61]

Introducing similarity variables transforms the governing system containing Equations (2.1), (2.2), and (2.3) (without unsteady, pressure gradient, and dissipation terms) to the form

$$\mathrm{Pr}\,F'''+FF''-\frac{2}{3}F'^2+(1+\mathrm{Pr})\Theta = \frac{4}{3}\xi\left(F'\frac{\partial F}{\partial\xi}-F''\frac{\partial F}{\partial\xi}\right),$$

$$\Theta''+F\Theta' = \frac{4}{3}\xi\left(F'\frac{\partial\Theta}{\partial\xi}-\Theta'\frac{\partial F}{\partial\xi}\right) \qquad (6.152)$$

$$F(\xi,\eta) = \left(\frac{1+\mathrm{Pr}}{\mathrm{Pr}}\right)^{1/4}\frac{\psi(x,y)}{(4/3)x^{3/4}}, \quad \Theta(\xi,\zeta) = \frac{|T(x,y)-T_\infty|}{T_{\infty 1}-T_{\infty 2}},$$

$$\xi = x, \quad \zeta = \frac{y}{(4/3x)^{1/4}}, \quad \eta = \left(\frac{\mathrm{Pr}}{1+\mathrm{Pr}}\right)^{1/4}\zeta$$

These equations refer to both fluids, $T_{\infty 1} > T_{\infty 2}$ and $x_2 = 1-x_1$. The boundary conditions are simple: $\eta = 0, F = F' = 0, \eta = \infty, F' = \Theta = 0$, while conjugate conditions are as follows:

$$\frac{\Theta_1'(\xi_1,0)}{(4\xi_1/3)^{1/4}} = \frac{\lambda_2}{\lambda_1}\left[\frac{Ra_2\,Pr_2(1+Pr_1)}{Ra_1\,Pr_1(1+Pr_2)}\right]^{1/4}\frac{\Theta_2'(\xi_2,0)}{(4\xi_2/3)^{1/4}}, \Theta_1(\xi_1,0)+\Theta_2(\xi_2,0)-1$$

$$= \frac{\Theta_2'(\xi_2,0)}{(4\xi_2/3)}\frac{\Delta\lambda_2}{L\lambda_1}\left(\frac{Ra_2\,Pr_2}{1+Pr_2}\right)^{1/4} \tag{6.153}$$

The basic idea of using governing equations in the form of Equation (6.152) is that when F and Θ are independent of ξ, the right-hand parts of equations become zero, and these take a self-similar form. Therefore, the right-hand parts of Equation (6.152) are a measure of the departure of these equations from similarity. The method of solution of such equations is as follows. The first solution is obtained assuming that the right-hand parts of Equation (6.152) are zero. Each of the following approximations is performed calculating these parts and satisfying conjugate conditions using the data of the previous step and finite differences for estimating the derivatives. The details are given in References [61] and [62]. However, it should be noticed that this method of solving the system of the equation required much more computation time then that applied by Viskanta and Abrams [57] (Example 6.22).

The conclusions obtained in this study basically agree with those formulated in Reference [58]. However, there are some differences. In particular, in this analysis, the temperatures near the ends of the plate change more gradually and the predicted temperatures at the top of the cool sides and at the bottom of the warm sides are higher. Another discrepancy is in the values of average Nusselt numbers which in some cases reach 17%. It is worth mentioning that according to Lock and Ko [61], the departure of the velocity and temperature distribution in this conjugate problem from corresponding similar results obtained in this nonconjugate problem is significant.

Example 6.25: A Radiating Plate with an Internal Source and an Insulated Bottom in Laminar or Turbulent Flows [19]

Substituting heat flux

$$q_w = \left(\lambda\frac{d^2T_w}{dx^2}+q_v\right)\Delta-\varepsilon\sigma\left(T_w^4-T_\infty^4\right), \tag{6.154}$$

into formula (Equation 1.50) after introducing dimensionless variables yields the governing equations in the following form:

$$\theta = 1+C\frac{\lambda_w}{\lambda}\,\bar{x}^{-r}\int_0^{\bar{x}}\left[1-\left(\frac{\xi}{x}\right)^{C_1}\right]^{1-C_2}\left(\Delta\frac{d^2\theta}{d\xi^2}-\frac{\theta^4}{\Lambda}+\frac{1+\bar{q}_v}{\Lambda}\right)d\xi, \qquad C=K\,Pr^{-n}\,Re^{-m}$$

$$\bar{x}=\frac{x}{L},\quad \bar{\Delta}=\frac{\Delta}{L},\quad \theta=\frac{T}{T_\infty},\qquad \Lambda=\frac{\lambda_w}{\varepsilon\sigma T_\infty^3 L},\qquad \bar{q}_v=\frac{q_v\Delta}{\varepsilon\sigma T_\infty^4} \tag{6.155}$$

Expanding the influence function in binominal series and using constants

$$K = 0.623, 3.32, \quad n = 1/3, \ 3/5, \quad m = 1/2, \ 4/5 \quad r = 1/2, \ 4/5, \quad C_1 = 3/4, 9/10,$$
$$C_2 = 1/3, 1/9$$

for laminar and turbulent flows, respectively, leads to the following two equations:

$$\theta = 1 + C\frac{\lambda_w}{\lambda}\overline{x}^{1/2}\left\{3.53\left(\frac{1+\overline{q}_v}{\Lambda}\right)\right.$$

$$\left. + \sum_{i=1}^{j}\left(\overline{\Delta}\frac{d^2\theta_i}{d\xi^2} - \frac{\theta_i^4}{\Lambda}\right)\left[\frac{1}{j-1} + \frac{8}{21}\frac{i^{7/4}-(i-1)^{7/4}}{(j-1)^{7/4}} + \frac{2}{9}\frac{i^{5/2}-(i-1)^{5/2}}{(j-1)^{5/2}} + \cdots\right]\right\} \quad (6.156)$$

$$\theta = 1 + C\frac{\lambda_w}{\lambda}\overline{x}^{1/5}\left\{9.83\left(\frac{1+\overline{q}_v}{\Lambda}\right) + \sum_{i=1}^{j}\left(\overline{\Delta}\frac{d^2\theta_i}{d\xi^2} - \frac{\theta_i^4}{\Lambda}\right)\right.$$

$$\left. \times \left[\frac{1}{j-1} + \frac{80}{171}\frac{i^{19/10}-(i-1)^{19/10}}{(j-1)^{19/10}} + \frac{160}{567}\frac{i^{14/3}-i^{14/3}}{(j-1)^{14/3}} + \cdots\right]\right\} \quad (6.157)$$

These equations are solved using iterations. The following are some basic conclusions:

1. The high values of plate conductivity (parameter Λ) result in an appreciable decrease of the plate temperature at small $\overline{x}$, but by increasing $\overline{x}$, this effect reduces.
2. Increasing internal heat generation or enlarging the plate thickness at low conductivity as well as reducing the radiation part of heat (parameter $\overline{q}_v$) yields an increase in the plate temperature.
3. Neglecting radiation in turbulent flow does not cause severe errors. The maximum error in plate temperature is about 45.5% in a laminar flow, and the corresponding error in a turbulent flow is only 5.5%.

6.10 Green's Function and the Method of Perturbation

Example 6.26: Heat Transfer from a Small Heated Strip of Length 2*a* Embedded in an Infinite Plate Streamlined by an Incompressible Laminar Flow [2]

Because the small heated strip is considered, the assumption of a linear velocity distribution across the thermal boundary layer is reasonable (Example 6.18). The equations governing the temperature field in a fluid and a body are as follows:

$$cy\frac{\partial T}{\partial x} = \alpha\frac{\partial^2 T}{\partial y^2}, \qquad \frac{\partial^2 T_s}{\partial x^2} + \frac{\partial^2 T_s}{\partial y^2} = 0 \qquad (6.158)$$

where cy is the velocity distribution. The boundary and the first conjugate conditions are as usual. The second conjugate condition takes into account the heat flux from the strip:

$$-\lambda \frac{\partial T}{\partial y}\bigg|_{y=0} - \lambda_w \frac{\partial T_s}{\partial y}\bigg|_{y=0} = q_0 f(x) \tag{6.159}$$

Here $f(x)$ is a top-hat function equal to unity on $-a < x < a$ and equal to zero elsewhere.

The temperature along the interface may be stated as an integral of an unknown heat flux and Green's function:

$$T_w = \int_{-\infty}^{\infty} q_w(\xi) G(x - \xi) d\xi \tag{6.160}$$

Here Green's function $G(x - \xi)$ is the influence function of type (Equation 1.50) which determines the temperature change at position x due to an infinitesimal heat flux located at position ξ. Applying relation (6.160) to conjugate condition (6.159) and transforming the integral to a sum yields the system of algebraic equations:

$$\sum_{j=1}^{m} [\phi(x_i, x_j) + \phi_w(x_i, x_j)] q^+(x_j) = \sum_{j=1}^{m} \phi_w(x_i, x_j) f_j \tag{6.161}$$

Here $\phi(x_i, x_j)$ and $\phi_w(x_i, x_j)$ are influence functions for the fluid and body, and $q^+ = -(\lambda/q_0)(\partial T/\partial y)_{y=0}$ is the dimensionless heat flux. The set of m linear algebraic equations determines m unknown heat fluxes. Typically, $m = 32$ variable-length surface elements are needed to adequately describe the desired unknown $q^+(x_j)$.

The numerical results are obtained for a wide range of conjugate Peclet number $\Lambda = (\lambda/\lambda_w) \mathrm{Pe}_c^{1/3}$ and body thickness Δ. The parameter $\Lambda = (\lambda/\lambda_w) \mathrm{Pe}_c^{1/3}$ characterizing the relationship between thermal resistances of the fluid and body is a variant of the Biot number (Section 6.1). Distribution of the temperature, heat flux, and average Nusselt number are obtained. The results show that as Λ increases, the temperature decreases everywhere, and at $\Lambda = 100$, the temperature is essentially zero upstream of the heated zone. The heat flux entering the fluid grows as Λ increases, and at $\Lambda = 100$ approaches 1 on the heated region, which means that most of the heat flows directly to the fluid. Numerical results give the correlation for the average Nusselt number:

$$\mathrm{Nu} = -\frac{2a}{T - T_\infty} \left(\frac{\partial T}{\partial y}\right)_{y=0} = 1.5\Lambda - 1.3 \left(\frac{\Delta/a}{\Delta/a + 0.6}\right) \left(\frac{\Lambda}{\Lambda + 2.2}\right) \tag{6.162}$$

This formula is applicable for $0.01 \leq \Delta/a \leq 100$ and $1 \leq \Lambda \leq 100$.

Another example of using Green's function is given in Reference [63] where it is shown how an operator defined in terms of Green's function can be applied to derive the solution of conjugate heat transfer problems.

Example 6.27: Heat Transfer between a Translating Liquid Drop and Another Immiscible Liquid of Infinite Extent under the Influence of an Electric Field at a Low Peclet Number [64]

Problems with low Peclet number are required to consider the full energy equation, because in such a case, the term $\partial^2 T/\partial x^2$ is of the same magnitude as the term $\partial^2 T/\partial y^2$. Thus, in this case, the system of governing equations is

$$Pe\left(u\frac{\partial\theta}{\partial x}+v\frac{\partial\theta}{\partial y}\right)=\frac{\partial^2\theta}{\partial x^2}+\frac{\partial^2\theta}{\partial y^2},\quad \frac{\partial\hat{\theta}}{\partial Fo}+Pe\left(u\frac{\partial\hat{\theta}}{\partial x}+v\frac{\partial\hat{\theta}}{\partial y}\right)=\frac{\partial^2\hat{\theta}}{\partial x^2}+\frac{\partial^2\hat{\theta}}{\partial y^2},$$

$$\theta=\frac{T-T_\infty}{T_0-T_\infty},\quad Pe=\frac{U_\infty R}{\alpha} \tag{6.163}$$

where overhead denotes dispersed phase, coordinates and velocities are referred to a drop radius R and its terminal velocity U_∞, respectively. The heat transfer within the drop is treated as transient because the steady state cannot be attained unless both phases are in thermal equilibrium. The electrodynamic problem in the creeping regime (very low motion) was solved by superposing Taylor and Hadamard-Rybczynski flows [65].

The approximate analytical solution of this complicated problem is obtained in the form of perturbation series considering the Peclet number as a small parameter, $\varepsilon=Pe$. Such a series describes changes occurring due to the small parameter in the unperturbed solution, which corresponds to the problem with $Pe=0$. This last problem refers to the case when heat transfer is provided by conduction without convection.

Two series are employed to get the solution for inner (drop) and outer domains:

$$\theta^{in}(r,\varphi)=\sum_{k=0}^\infty g_k(\varepsilon)\theta_k^{in}(r,\varphi),\quad \theta^{out}(\tilde{r},\varphi)=\sum_{k=0}^\infty G_k(\varepsilon)\theta_k^{out}(\tilde{r},\varphi),\quad \tilde{r}=\varepsilon r \tag{6.164}$$

These solutions must be matched at the interface asymptotically using the usual conjugate conditions. The temperature is computed up to and including the first order in the Peclet number. To examine the influence of the external field upon the total transport rates, higher-order terms for the host phase are also determined. Some conclusions are as follows:

1. According to first-order solution, the effect of an electric field alters the temperature and heat flux inside and outside the droplet, but the net heat transfer rate remains unchanged.
2. Unlike the case of thermal transport at high Peclet numbers, where the electric field has a decisive role even with a relatively low strength, no pronounced consequences are observed before the field is increased, and the applied voltage is sufficient.

3. The usage of an electric field is only effective at high values of its strength.
4. The electrodynamic couplings may carry a direct application in combustion where a change in temperature distribution inside a heterogeneous droplet would enhance the likelihood of a secondary atomization.

References

1. Luikov, A. V., 1974. Conjugate heat transfer problems, *Int. J. Heat Mass Transfer* 17: 257–265.
2. Cole, K. D., 1997. Conjugate heat transfer from a small heated strip, *Int. J. Heat Mass Transfer* 40: 2709–2719.
3. Dorfman, A. S., 1985. A new type of boundary condition in convective heat transfer problems, *Int. J. Heat Mass Transfer* 28: 1197–2003.
4. Dorfman, A. S., 1982. *Heat Transfer in Flow around Nonisothermal Bodies* (in Russian). Mashinostroenie, Moscow.
5 Dorfman, A. S., 1985. Errors arising in convective heat transfer problem when using conditions of the third kind, *High Temperature*, 23: 430–434.
6. Lipvezkaya, O. D., 1977. Conjugate heat transfer between two turbulent flowing fluids separated by thin plate (in Russian), *Teplophysika I Teplotechnika* 33: 75–79.
7. Dorfman, A. S., 1970. Heat transfer from liquid to liquid in a flow past two sides of a plate, *High Temperature* 8: 515–520.
8. Dorfman, A. S., 1972. Calculation of the thermal fluxes and the temperatures of the surface of a plate with heat transfer between fluids flowing around the plate, *High Temperature* 10: 293–298.
9. Dorfman, A. S., 1985. Combined heat transfer over the initial segment of a plate in a flow, *Heat Transfer-Soviet Research* 18: 52–74.
10. Van Dyke, M. D., 1964. *Perturbation Methods in Fluid Mechanics*. Academic Press, New York.
11. Dorfman, A. S., and Davydenko, B. V., 1980. Conjugate heat exchange for flows past elliptical cylinders, *High Temperature* 18: 275–280.
12. Schlichting, H., 1979. *Boundary Layer Theory*. McGraw-Hill, New York.
13. Sokolova, I. N., 1957. Plate temperature streamlined by supersonic flow (in Russian), *Theoretical Aerodynamic Investigations*, pp. 206–221, ZAGI, Oborongis, Moscow.
14. Luikov, A. V., Perelman, T. L., Levitin, R. S., and Gdalevich, L. B., 1970. Heat transfer from a plate in a compressible gas flow, *Int. J. Heat Mass Transfer* 13: 1261–1270.
15. Dorfman, A. S., 1975. Temperature-distribution singularities on the separation surface during heat transfer between a plate and the liquid flowing around it, *High Temperature* 13: 97–100.
16. Shvets, Y. I., Dorfman, A. S., and Didenko, O. I., 1975. Some characteristics of heat transfer between two moving fluids separated by a wall containing heat sources, *Heat Transfer-Soviet Research* 7: 25–31.

17. Loytsyanskiy, L. G., 1962. *The Laminar Boundary Layer*. Fizmatgiz Press, Moscow [Engl. trans., 1966].
18. Shvets, Y. I., Dorfman, A. S., and Didenko, O. I., 1975. Conjugate heat transfer from thin plate in gradient flow past it (in Russian), *Teplofiizika i Teplotechnika* 28: 23–27.
19. Sohal, M. S., and Howell, J. R., 1973. Determination of plate temperature in case of combined conduction, convection and radiation heat exchange, *Int. J. Heat Mass Transfer* 16: 2055–2066.
20. Shvets, Y. I., Dorfman, A. S., and Didenko, O. I., 1975. Heat transfer between two countercurrently flowing fluids separated by a thin wall, *Heat Transfer-Soviet Research* 7: 32–39.
21. Pomeranzev, A. A., 1960. Heating of the plate by supersonic flow (in Russian), *J. Engn. Physics* 3: 39–46.
22. Perelman, T. L., Levitin, R. S., Gdalevich, L. B., and Khusid, B. M., 1972. Unsteady-state conjugated heat transfer between a semi-infinite surface and incoming flow of a compressible fluid-II. Determination of a temperature field and analysis of result, *Int. J. Heat Mass Transfer* 15: 2563–2573.
23. Perelman, T. L.,1961. On conjugated problems of heat transfer. *Int. J. Heat Mass Transfer* 3: 293–303.
24. Perelman, T. L., 1961. One problem for equations of mixed type (in Russian), *J. Engn. Physics* 4: 121–125.
25. Perelman, T. L., 1961. On asymptotic expansion of solutions of one type of integral equations (in Russian), *Prikl. Math. Mech.* 25: 1145–1147.
26. Kumar, I., and Bartman, A. B., 1968. Conjugate heat transfer from radiated plate to laminar boundary layer of compressible fluid (in Russian). In *Teplo- i Massoperenos*, 9: 181–198, Nauka i Technika, Minsk.
27. Kumar, I., 1968. On conjugate heat transfer problem for laminar boundary layer with injection, *J. Engn. Physics* 14: 781–791.
28. Luikov, A. V., Aleksashenko, V. A., and Aleksashenko, A. A., 1971. Analytical methods of solution of conjugated problems in convective heat transfer, *Int. J. Heat Mass Transfer* 14: 1047–1056.
29. Gdalevich, L. B., and Khusid, B. M., 1971. Conjugate unsteady-state heat transfer of a thin plate in an incompressible fluid flow, *J. Engn. Physics* 20: 1045–1052.
30. Perelman, T. L., Levitin, R. S., Gdalevich, L. B., and Khusid, B. M., 1972. Unsteady conjugate heat transfer between a semi-infinite surface and incoming flow of a compressible fluid. I-Reduction to the integral relation, *Int. J. Heat Mass Transfer* 15: 2551–2561.
31. Perelman, T. L., 1961. Heat transfer from streamlined plate with heat sources to laminar boundary layer (in Russian), *J. Engn. Physics* 4: 54–61.
32. Luikov, A. V., 1974. Conjugate convective heat transfer problems, *Int. J. Heat Mass Transfer* 17: 257–265.
33. Kopeliovich, B. L., Perelman, T. L., Levitin, R. S., Khusid, B. M., and Gdalevich, L. B., 1976. Unsteady conjugate heat exchange between a semi-infinite surface and a stream of compressible fluid flowing over it (in Russian), *J. Engn. Physics* 30: 337–339.
34. Dorfman, A. S., 1977. Solution of the external problem of unsteady state convective heat transfer with coupled boundary conditions, *Int. Chem. Engn.* 17: 505–509.

35. Rizk, T. A., Kleinstreuer, C., and Ozisik, M. N., 1992. Analytic solution to the conjugate heat transfer problem of flow past a heated block, *Int. J. Heat Mass Transfer* 35: 1519–1525.

36. Travelho, J. S., and Santos, W. F. N., 1998. Unsteady conjugate heat transfer in a circular duct with convection from ambient and periodically varying inlet temperature, *ASME J. Heat Transfer* 120: 506–511.

37. Sucec, J., and Sawant, A. M., 1984. Unsteady conjugated forced-convection heat transfer in a parallel plate duct, *Int. J. Heat Mass Transfer* 27: 95–101.

38. Sparrow, E. M., and DeFarias, F. N., 1968. Unsteady heat transfer in ducts with time varying inlet temperature and participating walls, *Int. J. Heat Mass Transfer* 11: 837–853.

39. Sucec, J., 1987. Exact solution for unsteady conjugated heat transfer in the thermal entrance region of a duct, *ASME J. Heat Transfer* 109: 295–299.

40. Sucec, J., 2002. Unsteady forced convection with sinusoidal duct wall generation: The conjugate heat transfer problem, *Int. J. Heat Mass Transfer* 45: 1631–1642.

41. Sucec, J., and Radley, D., 1990. Unsteady forced convection heat transfer in a channel, *Int. J. Heat Mass Transfer* 33: 683–690.

42. Karvinen, R., 1988. Transient conjugated heat transfer to laminar flow in a tube or channel, *Int. J. Heat Mass Transfer* 31: 1326–1328.

43. Guedes, R. O. C., Ozisik, M. N., and Cotta, R. M., 1994. Conjugated periodic turbulent forced convection in parallel plate channel. *ASME J. Heat Transfer* 116: 40–46.

44. Stein, C. F., Johansson, P., Bergh, J., Lofdahl, L., Sen, M., and Gad-el-Hak, M., 2002. An analytical asymptotic solution to a conjugate heat transfer problem, *Int. J. Heat Mass Transfer* 45: 2485–2500.

45. Wang, X. S., Dagan, Z., and Jlji, L. M., 1989. Conjugate heat transfer between a laminar impinging liquid jet and solid disk, *Int. J. Heat Mass Transfer* 32: 2189–2197.

46. Grechannyy, O. A., Dolinsky, A. A., and Dorfman, A. S., 1988. Flow, heat and mass transfer in the boundary layer on a continuously moving porous sheet, *Heat Transfer-Soviet Research* 20: 52–64.

47. Grechannyy, O. A., Dolinsky, A. A., and Dorfman, A. S., 1988. Effect of non-uniform distribution of temperature and concentration differences on heat and mass transfer from and to a continuously moving porous plate, *Heat Transfer-Soviet Research* 20: 355–368.

48. Davis, E. J., and Gill, W. N., 1970. The effects of axial conduction in the wall on heat transfer with laminar flow, *Int. J. Heat Mass Transfer* 13: 459–470.

49. Hickman, H. J., 1974. An asymptotic study of the Nusselt-Graetz problem, Part I: Large x behavior, *J. Heat Transfer* 96: 354–358.

50. Lee, W. C., and Ju, Y. H., 1986. Conjugate Leveque solution for Newtonian fluid in a parallel plate channel, *Int. J. Heat Mass Transfer* 29: 941–947.

51. Papoutsakis, E., and Ramkrishna, D., 1981. Conjugated Graetz problems, *Chem. Engn. Sci.* 36: 381–391.

52. Pozzi, A., and Lupo, M., 1989. The coupling of conduction with forced convection in a plane duct, *Int. J. Heat Mass Transfer* 32: 1215–1221.

53. Bender, C. M., and Orsag, S. A., 1978. *Advanced Mathematical Methods for Scientists and Engineers.* McGraw-Hill, New York.

54. Shah, R. K., and London, A. L., 1978. *Laminar Flow Forced Convection in Ducts.* Academic Press, New York.

55. Olek, S., Elias, E., Washolder, E., and Kaizerman, S., 1991. Unsteady conjugated heat transfer in laminar pipe flow, *Int. J. Heat Mass Transfer* 34: 1443–1450.

56. Yeh, C., 1980. An analytical solution to fuel-and-cladding model of the rewetting of a nuclear fuel rod, *Nucl. Engn. Des.* 61: 101–112.

57. Viskanta, R., and Abrams, M., 1971. Thermal interaction of two streams in boundary layer flow separated by a plate, *Int. J. Heat Mass Transfer* 14: 1311–1321.

58. Viskanta, R., and Lankford, D. W., 1981. Coupling of heat transfer between two natural convection systems separated by a vertical wall, *Int. J. Heat Mass Transfer* 24: 1171–1177.

59. Raithby, G. D., and Hollands, K. G. T., 1975. A general method of obtaining approximate solutions to laminar and turbulent natural convection problems. In *Advances in Heat and Mass Transfer,* edited by T. F. Irvine, Jr., and J. P. Hartnett. 11: 265–315. Academic Press, New York.

60. Churchill, S. W., and Ozoe, H., 1973. A correlation for laminar free convection from a vertical plate, *ASME J. Heat Transfer* C95: 540–541.

61. Lock, G. S. H., and Ko, R. S., 1973. Coupling through a wall between two free convective systems, *Int. J. Heat Mass Transfer* 16: 2087–2096.

62. Hayday, A. A., Bowlus, D. A., and McGraw, R. A., 1967. Free convection from a vertical plate with step discontinuities in surface temperature, *Int. J. Heat Mass Transfer* 89: 244–250.

63. Friedly, J. C., 1983. Transient response of linear conjugate heat transfer problem. American Society of Mechanical Engineers and American Institute of Chemical Engineers, Heat Transfer Conference. Seattle, WA, July 24–28, 8 p.

64. Nguyen, H. D., and Chung, J. N., 1992. Conjugate heat transfer from a translating drop in an electric field at low Peclet number, *Int. J. Heat Mass Transfer* 35: 443–456.

65. Chang, L. S., Carleson, T. E., and Berg, J. C., 1982. Heat and mass transfer to a translating drop in an electric field, *Int. J. Heat Mass Transfer* 25: 1023–1030.

7

Numerical Methods for Solving Conjugate Convective Heat Transfer Problems

7.1 Analytical and Numerical Methods

The finite-difference method was developed long before it became a powerful tool for solving differential equations (after the use of computers came to the forefront) [1]. Since the understanding and technique of the finite-difference method seems to be simpler than that of analytical methods, it was believed that the time of analytical methods was over.

Although the techniques of analytical and numerical approaches indeed are different, both methods are based on the same fundamental principles. The distinction between both approaches is only that these basic principles are applied in the former to infinite-small differences, while in the latter they are used for small but finite size values. For example, both analytical and numerical derivatives are determined by the same principle. However, to calculate an analytical derivative, one needs to have some knowledge, and even with that knowledge, sometimes calculations are not easy. At the same time, to obtain the finite-difference derivative using the difference between function values at two grid points is not a problem. This feature of numerical methods makes them seem no less complicated techniques than the analytical approach.

Thus, many researchers, especially young people without experience, start to produce programs for solving different differential equations, trying to study complex contemporary problems. However, it soon became clear that only deeply understanding the properties of each problem and carefully checking and testing the program may lead to the proper numerical solution. Only an investigator who adopted a corresponding part of current knowledge can possess the complex technique of numerical solution, which just seems to be simple, and then interpret the obtained results. Otherwise, an insufficiently considered and prepared program can give an unrealistic outcome.

After the applications of computers were expanded, analytical methods not only retained their importance, but gained new functions as well. In particular, despite the many recommendations and rules for preparing and qualitatively checking numerical programs, the best way to test

231

and control accuracy is to compare the result obtained by some software program with results obtained by applying the proper simple analytical solution [2].

Some cases when analytical solutions are especially useful are listed below:

1. The formulae for the finite-difference derivatives are typically obtained using a Taylor series. For a grid point i located midway between points $i-1$ and $i+1$, using the first three terms of the Taylor series, the following two expressions can be obtained:

$$f_{i-1} = f_i - (x_i - x_{i-1})\left(\frac{df}{dx}\right)_i + \frac{1}{2}(x_i - x_{i-1})^2\left(\frac{d^2 f}{dx^2}\right)_i + \cdots \quad (7.1)$$

$$f_{i+1} = f_i + (x_i - x_{i-1})\left(\frac{df}{dx}\right)_i + \frac{1}{2}(x_i - x_{i-1})^2\left(\frac{d^2 f}{dx^2}\right)_i + \cdots \quad (7.2)$$

By adding and subtracting these equations, one obtains formulae for the first two derivatives:

$$\left(\frac{df}{dx}\right)_i = \frac{f_{i+1} - f_{i-1}}{x_{i+2} - x_{i-1}}, \quad \left(\frac{d^2 f}{dx^2}\right)_i = \frac{f_{i-1} - 2f_i + f_{i+1}}{(x_i - x_{i-1})^2} \quad (7.3)$$

These formulae may be used only if the function in question is analytic. If there are some grid points where the function is singular (e.g., if at these points one or more derivatives become infinite), Equation (7.3) cannot be applied. Such a case is considered in Section 6.4, where it is shown that the wall temperature distribution on a thermally thin plate near $x = 0$ is not an analytical function of the coordinate x. It is rather presented as a series in integer powers of variable $x^{1/s}$, where s is the denominator of the exponent in the relation for an isothermal heat transfer coefficient. For laminar and turbulent flows, this variable is $x^{1/2}$ and $x^{1/5}$, respectively. Some examples and methods for deriving corresponding variables for other cases are given in Section 6.4.2.

It is obvious that the derivative with respect to coordinate x is proportional to $x^{-1/2}$ or to $x^{-4/5}$ for laminar or turbulent flow, or to $x^{(1/s)-1}$ for any other value of s. Hence, this derivative becomes infinite at $x = 0$ for laminar or turbulent flow and for other cases in which an exponent r/s in the relation for an isothermal heat transfer coefficient is less than unity. However, if one introduces a new variable $z = x^{1/s}$, the temperature distribution turns into an analytical function, the corresponding derivative with respect to z becomes finite, and Equations (7.3) can be used.

Another similar situation takes place for the velocity and temperature boundary layer equations in Prandtl-Mises form (Equation 1.3). These equations have a singularity at the body surface where the stream function is zero, $\psi = 0$. This follows from the fact that the longitudinal velocity near the surface is proportional to the transverse variable: $u \sim cy$. This relation holds true because on the surface $u = 0$, but $du/dy \neq 0$ due to the friction. Then, according to continuity equation (1.1), $du/dx \sim dv/dy$, and hence, $v \sim cy^2$. Using these estimations and knowing that both convective terms in the thermal boundary equation (1.1) are of the same order $u(dT/dx) \sim v(dT/dy)$, one gets that near the surface $y(dT/dy) \sim 1$, $\exp T \sim cy$ and $T \sim cy$. At the same time, it follows from the definition (Equation 1.3) of the stream function that near the surface $\psi \sim uy$, and since $u \sim cy$ for the stream function, one obtains $\psi \sim cy^2$. From this, it follows that $y \sim \psi^{1/2}$ and, finally, $u \sim c\psi^{1/2}$ and $T \sim c\psi^{1/2}$.

Therefore, near the surface, both derivatives of the velocity and of the temperature with respect to the independent variable ψ become infinite. Due to this singularity, Equation (7.3) cannot be applied. However, introducing a new variable $z = \psi^{1/2}$ solves this problem again. Thus, analytical analysis leads to understanding what variable should be used to overcome the singularities of boundary layer equations, in particular, and in the case of numerical solutions as well.

2. The other difficulty that usually arises in preparing a program for numerical solution is attaining the proper distribution of the grid points inside a considered domain. This distribution should correspond to the distribution of the studied function gradients. In achievement of this conformity, one can get significant information by analyzing a field of an interested function in the analytical solution of a similar problem. In particular, it is known from exact solutions of the boundary layer problems that the maximum values of the velocity and temperature gradients are at the wall. These maximum gradients gradually decrease as the distance from the wall grows until they become zero at $y \to \infty$.

On the basis of this information, one creates a grid with corresponding distribution point or applies a special variable, for instance, $z = \Delta/[1 - (y/\delta)]$, where Δ is a size of a mesh at the wall, which increases as the distance from the wall grows.

3. Analytical solutions are useful as well when one needs to choose a function for approximately describing the solution behavior between the grid points. It is obvious that this function should be as close as possible to actual dependence. Because this actual function is unknown, the best way to find one of the closest to it is to examine a proper analytical solution. In particular, for the case of a

numerical solution of a boundary layer equation, it is reasonable to apply the polynomial or some other approximate distribution that are usually used in integral methods (Section 1.4).

4. Apparently, the most important function of analytical solutions is that they can be used as a reference in checking and testing the software. Comparing the computation results of the same problem with an exact analytical solution, one determines not only the usefulness of the program in general, but estimates as well the deviations of obtained values from exact data.

The examples presented here demonstrate the role of analytical results in developing the software for numerical solutions. At the same time, there is no doubt of the significance of the numerical approaches in obtaining the digital data from analytical solutions. Therefore, there is no reason to oppose the analytical and numerical methods; rather, it is reasonable to consider both approaches as a united, combined method of investigation and solution of contemporary theoretical and practical problems. In fact, the numerical and analytical methods are two means supplementing each other. The former is a powerful technique for obtaining an approximate solution of almost any complex problem using a known mathematical model, and the latter provides the possibility of finding some exact solutions of relatively simple problems, investigating general properties of studied phenomena, and developing models on a basis of these data. Because any contemporary problem is a challenge, only by using both methods in combination could one expect to obtain adequate results.

The author hopes that this book is a convincing example of the fruitfulness of such a complex approach. If the exact and highly accurate approximate solutions of the nonisothermal and conjugate heat transfer problems presented in Chapters 3 through 6 were used for checking and testing numerical methods, in particular, those described in this and the following chapters, then the author would believe that the goal of his work is achieved.

7.2 Approximate Analytical and Numerical Methods for Solving Differential Equations

Approximate methods for solving differential equations were developed and widely used many years before they became a basis of modern numerical methods after computers start to become common, powerful tools for calculation. However, before the computer era, these methods were used only for the entire computation domain of interest; therefore, they were applied as analytical approaches. The use of computers makes it possible to divide the computation domain into small subdomains and apply the same approximate methods for each one. It is obvious that this pattern vastly increases the possibilities and the accuracy of these simple analytical approaches and converts them, in essence, into contemporary numerical methods.

Numerical methods differ in their methods of discretization of the computation domain and in the analytical approximate methods for solving the differential equations for each subdomain. Depending on the first procedure, the numerical methods can be classified into three basic groups: the finite-difference method (FDM), finite-element method (FEM), and boundary element method (BEM). The old one, the finite-difference method [1] usually uses uniform grids for discretization and calculates in the points of these grids the numerical derivatives by formula (7.3), which are a finite-difference version of the usual derivatives in calculus. The two other techniques compute the values of the studied function in each usually irregular distributed subdomain using different approximate methods, as previously these were applied analytically for entire domains.

The distinction between these two last approaches results from various numbers of subdomains that are needed to obtain a solution. The finite-element method required employing the subdomains of the whole field of an investigated function, but the boundary-element method used only a part of the whole number of subdomains located just on the boundaries.

There are also two of the most widely used modern modifications of the finite-difference method (CVFDM) and finite-element method (CVFEM) which differ from earlier versions by using the control-volume (CV) formulation for deriving the equations for grid points or elements (discretization equations [2]). In that approach, the discretization equations are obtained as a result of integrating the differential equations over each control volume. Here the basic idea is that such an equation expresses the conservation laws for finite volume just as the differential equation expresses these laws for an infinitesimal volume.

The modified finite-difference method (CVFDM), for simplicity, is called the finite-difference method (FDM), but the modified finite-element method is called the finite-volume method (FVM) because the volume coincides as an element. There are thus three commonly used types of finite-difference approaches: FDM, FVM, and BEM.

The distinctions between different analytical approximate methods employed in numerical methods are convenient to describe using the weighted residual approach (see, for example, References [2], [3], or [4]). The weighted residual method is a generalized well-known method of moments that was widely used before the computer era, in particular, in boundary layer theory by the name integral method (Section 1.4). The concept of the weighted residual method can be explained as follows.

Let there be a need to find an approximate solution of a differential equation $F(u) = 0$ subjected to a given boundary condition. First, the given boundary condition should be converted to a homogeneous one. Then, a function

$\tilde{u} = f(x)$ is chosen that exactly satisfies the boundary condition but contains one or more unknown parameters, for instance, a polynomial with undefined coefficients or some other function. Substituting this function into the differential equation yields a residual $R = F(\tilde{u})$ because $\tilde{u}$ is an approximate solution and, hence, does not satisfy the equation under consideration. Multiplying this residual by some weighting function w and integrating over the considered domain S, one tries to minimize an average error:

$$\int_S wR\,dx = \int_S wF(\tilde{u})\,dx = 0 \tag{7.4}$$

Applying this relation and choosing a series of weighting functions, one obtains as many algebraic equations as are required to determine the unknown parameters. Substituting these parameters into function $\tilde{u} = f(x)$ completed the formation of an approximate solution. Various approximate methods applied in numerical solutions to obtain the required system of algebraic equations differ from each other by classes of these weighting functions.

For example, the method of moments results from a series of weighting functions: $1, x, x^2, \dots$ The well-known integral method is a case of this method when only the first one is used (i.e., $w = 1$).

Example 7.1: One-Dimensional Conduction Problem for a Plane Wall

The governing equation and boundary conditions are:

$$\lambda \frac{d^2T}{dx^2} + q = 0 \qquad x = 0 \quad T = T_0 \quad x = L \quad T = T_L \tag{7.5}$$

where T_0 and T_L are temperatures of surfaces and q is uniform heat generation. To solve the problem using the method of moments, one introduces a new variable ϑ that transforms the boundary condition to a homogeneous one to get the problem in the following form:

$$\vartheta = T - T_0(1-\xi) - T_L\xi, \quad \frac{d^2\vartheta}{d\xi^2} + \bar{q} = 0 \quad \xi = 0, \quad \xi = 1, \quad \vartheta = 0 \quad \xi = \frac{x}{L}, \quad \bar{q} = \frac{qL^2}{\lambda} \tag{7.6}$$

For using the first two moments, one should choose a function with two parameters that satisfy boundary conditions. For instance, $\vartheta = a_1\xi(\xi-1) + a_2\xi^2(\xi^2-1)$. Substituting this approximate solution into the governing equation leads to residual $R = 2a_1 + (12\xi^2 - 2) + \bar{q}$. Then, using the first two moments as weighting functions $w = 1, \xi$, one obtains the following from Equation (7.4):

$$\int_0^1 [2a_1 + (12\xi^2 - 2)a_2 + \bar{q}]\,d\xi = 0, \qquad \int_0^1 [2a_1 + (12\xi^2 - 2)a_2 + \bar{q}]\xi\,d\xi = 0, \tag{7.7}$$

which determine coefficients $a_1 = -\bar{q}/2$ and $a_2 = 0$. Thus, the desired solution is

$$T = \frac{qL^2}{2\lambda}\frac{x}{L}\left(1-\frac{x}{L}\right)+T_0\left(1-\frac{x}{L}\right)+T_L\frac{x}{L} \tag{7.8}$$

In this particular case, the obtained result is an exact solution.

Similar solutions can be obtained using other approximate methods. The different results arise from various series of weighting functions used to obtain the required system of algebraic equations. Thus, for Galerkin's method, the weighting functions are the same as the functions chosen for satisfying the boundary conditions so that Equation (7.7) becomes

$$\int_0^1 [2a_1 + (12\xi^2 - 2)a_2 + \ddot{q}]\xi(\xi-1)d\xi = 0,$$

$$\int_0^1 [2a_1 + (12\xi^2 - 2)a_2 + \ddot{q}]\xi^2(\xi^2 - 1)d\xi = 0 \tag{7.9}$$

Solving these equations yields the same solution (Equation 7.8). An analogous solution can be obtained by using the point collocation method with Dirac delta weighting functions or the subdomain collocation method. In the last case, a calculation domain is divided into several subdomains. For instance, for two domains, instead of Equation (7.7), one obtains a system

$$\int_0^{1/2} [2a_1 + (12\xi^2 - 2)a_2 + \ddot{q}]d\xi, \qquad \int_{1/2}^1 [2a_1 + (12\xi^2 - 2)a_2 + \ddot{q}]d\xi, \tag{7.10}$$

which leads to the same result (Equation 7.8).

A special case is the subdomains method when the domain under consideration is divided into numbers of subdomains, and one assumes that the weighting function is $w = 1$ for one of the subdomains at a time and $w = 0$ for all others. Physically, this approach implies that the average residual error is zero over each small domain. In particular, the control-volume formulation pertains to this type of method. In this case, the system of algebraic equations is obtained by integrating the governing differential equation over each of the subdomains. Applying this approach to the same simple one-dimensional conduction equation as in (7.5), one gets

$$\int_a^b \lambda \frac{d^2T}{dx^2}dx + \int_a^b qdx = 0, \qquad \lambda\left[\left(\frac{dT}{dx}\right)_b - \left(\frac{dT}{dx}\right)_a\right] + \int_a^b qdx = 0$$

$$\frac{T_{i+1} - T_i}{x_{i+1} - x_i} - \frac{T_i - T_{i-1}}{x_i - x_{i-1}} + \bar{q}(x_b - x_a) = 0 \tag{7.11}$$

Here, a and b denote the midway points between x_i and x_{i-1} and x_{i+1} and x_i, respectively. In deriving the last Equation (7.11) from the first, it is assumed that the temperature changes between grid points linearly.

This example shows the typical way to use the control-volume approach in finite-difference numerical methods. This way and formula (7.3) obtained via the Taylor series are the two most similar techniques for deriving finite-difference equations. The former is preferable because the control-volume approach provides an exact satisfaction of integral conservation of mass, momentum, and energy over each subdomain and, hence, for the whole calculation domain.

Another specific case, which is called weak formulation, is employed in the finite-element and boundary-element numerical methods. In the finite-difference method, an approximate solution is obtained by satisfying the differential equation at the grid points, in finite-element and boundary-element methods the approximate solution is found by distribution of the error of this solution over each of the subdomains.

To introduce the basic concepts of the finite-element and boundary-element approaches, consider again the one-dimensional conduction equation (7.5) for domain $(0, 1)$. By multiplying this equation by some weighting function w after integrating, the following equation is obtained:

$$\int_0^1 \left(\frac{d^2T}{dx^2} + \frac{q}{\lambda} \right) w dx = 0 \quad x = 0, x = 1, T = 0 \tag{7.12}$$

Integrating by parts, two times, setting $u = w, dv = (d^2T/dx^2)\, dx$ for the first time and $u = dw/dx, dv = (dT/dx)dx$ for the second, leads to two expressions:

$$\int_0^1 \left(-\frac{dT}{dx}\frac{dw}{dx} + \frac{q}{\lambda}w \right) dx + \left[\frac{dT}{dx}w \right]_0^1 = 0,$$

$$\int_0^1 \left(T\frac{d^2w}{dx^2} + \frac{q}{\lambda}w \right) dx + \left[\frac{dT}{dx}w \right]_0^1 - \left[T\frac{dw}{dx} \right]_0^1 = 0 \tag{7.13}$$

The first expression is the starting statement for the finite element method, and the second is the starting statement for the boundary element method. These expressions are named weakening expressions because the process of integrating by parts reduces the requirement of continuity of the studied function.

The basic idea of using the second equation (7.13) is that such a form of relation makes it possible to replace the search for an approximate solution by choosing a proper weighting function to get the boundary problem in which only boundary conditions should be satisfied. There are two ways that this can be done [3,4]:

1. *Selecting such a weighting function that satisfies the governing equation in its homogenous form.* Satisfying the homogeneous version of a given equation yields such weighting function

$$\frac{d^2w}{dx^2} = 0, \quad w = a_1 x + a_2, \quad \frac{dw}{dx} = a_1 \tag{7.14}$$

The second, Equation (7.13) becomes

$$\int_0^1 \frac{q}{\lambda} w dx + \left[\frac{dT}{dx} w\right]_0^1 - \left[T \frac{dw}{dx}\right]_0^1 = 0,$$

$$\int_0^1 \frac{q}{\lambda}(a_1 x + a_2) dx + \left(\frac{dT}{dx}\right)_1 (a_1 + a_2) - \left(\frac{dT}{dx}\right)_0 a_2 = 0 \tag{7.15}$$

Because the last equation should be satisfied for any arbitrary a_1 and a_2, one gets

$$\int_0^1 \frac{q}{\lambda} x dx + \left(\frac{dT}{dx}\right)_1 = 0, \qquad \int_0^1 \frac{q}{\lambda} dx + \left(\frac{dT}{dx}\right)_1 - \left(\frac{dT}{dx}\right)_0 = 0 \tag{7.16}$$

These equations give the desired values of boundary derivatives at $x = 0$ and $x = 1$ and then the solution of the problem:

$$\left(\frac{dT}{dx}\right)_1 = -\frac{q}{2\lambda}, \qquad \left(\frac{dT}{dx}\right)_0 = \frac{q}{2\lambda} \qquad T = \frac{q}{2\lambda} x(1-x) \tag{7.17}$$

2. *Using a function, usually the Dirac delta function, to satisfy the governing equation.* In this case, one assumes that the delta function satisfies the governing equation regardless of boundary conditions and obtains

$$\frac{d^2 w}{dx^2} = -\delta_i, \qquad \delta_i = \begin{cases} 1, & x = x_i \\ 0, & x \neq x_i \end{cases} \qquad w = \begin{cases} x & x \leq x_i \\ x_i & x > x_i \end{cases} \tag{7.18}$$

Then the second equation (7.13) gives

$$-\int_0^1 T \delta(x) dx + \int_0^1 \frac{q}{\lambda} w dx + \left(\frac{dT}{dx}\right)_1 w_1 - \left(\frac{dT}{dx}\right)_0 w_0 = 0 \tag{7.19}$$

Because the first integral equals T_i and $w_0 = 0$, this equation determines the desired function:

$$T_i = \frac{q}{\lambda}\left(\int_0^{x_i} x dx + \int_{x_i}^1 x_i dx\right) + \left(\frac{dT}{dx}\right)_1 \qquad w_1 = \frac{q}{\lambda}\left(\frac{x_i^2}{2} + x_i - x_i^2\right) + \left(\frac{dT}{dx}\right)_1$$

$$x_i = \frac{q}{\lambda}\left(x_i - \frac{x_i^2}{2}\right) + \left(\frac{dT}{dx}\right)_1 x_i \tag{7.20}$$

Using the boundary condition $T_i = 0$ at $x_1 = 1$, one finds an unknown derivative at $x_i = 1$ and then the same solution (7.17):

$$\left(\frac{dT}{dx}\right)_1 = -\frac{q}{2\lambda}, \qquad T = \frac{q}{2\lambda} x(1-x) \tag{7.21}$$

Transforming the second, Equation (7.13), using the delta function as done in (2) is a commonly used approach in the boundary element method. After the first term is modified in this way, the second equation (7.13) becomes

$$T_i = \int_0^1 \frac{q}{\lambda} w dx + \left[\frac{dT}{dx} w \right]_0^1 - \left[T \frac{dw}{dx} \right]_0^1 \tag{7.22}$$

This is a common form of equation in the boundary element approach from which it is clear that the unknown function in this case is defined only by boundary conditions. In contrast to this, the finite element method is based on the first equation (Equation 7.13) from which follows that the unknown function is defined by information about the whole domain.

Although the presentation of different numerical methods is done using the simple one-dimensional equations, the described procedures and features in principle are valid in more complicated cases for various two- and three-dimensional problems [3,4].

7.3 Difficulties in Computing Convection-Diffusion and Flow

There are special difficulties and procedures in applying numerical methods to convective heat transfer and flow problems. The basics of these difficulties are discussed below [2,5].

7.3.1 The Control-Volume Finite-Difference Method

7.3.1.1 *Computing Pressure and Velocity*

The main difficulty in solving the Navier-Stokes equations

$$\rho \left(\frac{\partial u}{\partial t} + u \frac{\partial u}{\partial x} + v \frac{\partial u}{\partial y} \right) = -\frac{\partial p}{\partial x} + \mu \left(\frac{\partial^2 u}{\partial x^2} + \frac{\partial^2 u}{\partial y^2} \right),$$

$$\rho \left(\frac{\partial v}{\partial t} + u \frac{\partial v}{\partial x} + v \frac{\partial v}{\partial y} \right) = -\frac{\partial p}{\partial y} + \mu \left(\frac{\partial^2 v}{\partial x^2} + \frac{\partial^2 v}{\partial y^2} \right) \tag{7.23}$$

is the unknown pressure. Although there is no special equation for pressure, it is indirectly determined by the continuity equation. Because the velocity field can be calculated only by knowing the pressure distribution, the continuity equation can be satisfied only when the velocities are computed using a proper pressure. Thus, to satisfy the continuity equation, the pressure should be known.

One well-known method to overcome the difficulty in determining pressure is to eliminate it from governing Equations (7.23). This can be achieved by introducing the stream function and cross-differentiating. Because two equations (Equation 7.23) contain the derivatives of pressure with respect to different coordinates, the cross-differentiating leads to mixed derivatives in each of the equations which can then be eliminated. As a result, one obtains an equation for the stream function and after solving it determines the velocity components. Unfortunately, this method for two-dimensional equations cannot be used for a three-dimensional problem, because in this case, a stream function does not exist.

The direct method for pressure calculation is presented in Reference [2]. It is based on determining a pressure difference for a control volume using the common procedure of defining the values of dependent variable for grid points. However, as shown in Reference [2], such a procedure fails in the case of using standard three-point control volume with a midway located main grid point. It turns out that such a standard approach yields only uniform zero pressure and hence cannot be applied.

A reasonable real pressure distribution can be obtained by using special control volumes that form a staggered grid [2]. In such a control volume model, the velocity components and corresponding pressure are calculated for the main points located on the control-volume faces that are set midway between two adjacent points normal to the velocity component direction. Thus, the x-directed component u is calculated at the y-directed faces, and, vice versa, the y-directed component v is calculated at the x-directed faces. This means that the u-component is calculated on the faces from the left and from the right of the main point, while the v-component is calculated on the faces located above and below the main point. The location of the mean point with respect to two adjacent other points is not important. It is important only that the main point is on the control volume face. For example, in the grid with control volumes having different length, it is immaterial that the main points lying on the faces are not in the midst of two neighboring points. Using the staggered grid eliminates the difficulties of calculating the pressure field, but a corresponding computer program becomes more complicated because it must record all information about the location of the velocity components and must perform tiresome interpolations.

On the basis of the above discussed difficulties and peculiarities of the numerical method of defining the velocity components and pressure fields, a special calculation procedure is developed which is named the SIMPLE (Semi-Implicit Method for Pressure-Linked Equations). It was published by Patankar and Spalding in their paper [6] and in two others [7,8] and is described by Patankar in his book [2].

This iterative procedure begins by requiring the guessing of the pressure field. Then, using the finite-difference technique and guessed pressure field, the momentum equations (7.23) for velocity components are solved. The second iteration is performed by applying just found values of velocity components to calculate the pressure difference between adjacent grid points. This gives the new pressure field that is used to get new velocity components. These iterations are carried on until the continuity equation is satisfied. To control the process of satisfaction, the special equation is derived. The continuity equation is integrated over the control volume to obtain for compressible fluid the following equation:

$$\frac{(\rho - \rho^0)\Delta x \Delta y}{\Delta t} + [(\rho u)_{(i+1/2),j} - (\rho u_{(i-1/2),j})]\Delta y + [(\rho v)_{i,(j+1/2)} - (\rho v)_{i,(j-1/2)}]\Delta x = 0$$

(7.24)

where ρ^0 is the density in the previous step of time. This equation is written for the main point (i, j) using four neighboring points: $(i+1), j$ and $(i-1), j$ points from the left and from the right and $i, (j+1)$ and $i, (j-1)$ points above and below the main point.

Because the density ρ is usually calculated from the equation of state, and hence, it depends on pressure and temperature, the first term in Equation (7.24) should be calculated by an additional iterative process. By substituting calculation results in Equation (7.24), one estimates the residual and decides to stop iterations when the residual becomes sufficiently small. It is obvious that in the case of incompressible fluid, the first term in Equation (7.24) should be omitted as well as the density ρ in the other terms.

To improve the process of convergence, some revised versions were developed. Among the most used versions are SIMPLER (SIMPLE Revised) [2], SIMPLEC (SIMPLE Consistent) [9], and SIMPLEM (SIMPLE Modified) [10]. A noniterative procedure, PICO (Pressure-Implicit with Splitting of Operators) was also developed [11]. The comparison shows that the differences in the efficiency of these approaches are generally modest. The analysis of available studies reveals that family SIMPLE is a reliable, practical calculating tool for a variety of applications and for solving current, complicated technological and theoretical problems [11].

7.3.1.2 Computing Convection–Diffusion

Consider a steady one-dimensional equation with only the convection and diffusion terms. Integration of such an equation over the control volume yields

$$\frac{d}{dx}(\rho cuT) = \frac{d}{dx}\left(\lambda \frac{dT}{dx}\right), (\rho cuT)_{i+1/2} - (\rho cuT)_{i-1/2} = \left(\lambda \frac{dT}{dx}\right)_{i+1/2} - \left(\lambda \frac{dT}{dx}\right)_{i-1/2}$$

(7.25)

Using the piecewise-lineal profile for approximation between grid points and Equation (7.3) for the first derivative, the corresponding finite-difference equation is obtained:

$$\frac{1}{2}(\rho u c)_{i+1/2}(T_{i+1} + T_i) - \frac{1}{2}(\rho u c)_{i-1/2}(T_i + T_{i-1}) = \frac{\lambda_{i+1/2}(T_{i+1} - T_i)}{x_{i+1} - x_i} - \frac{\lambda_{i-1/2}(T_i - T_{i-1})}{x_i - x_{i-1}}$$

(7.26)

where $i + 1/2$ and $i - 1/2$ denote points located midway between points $i+1$ and i and between points i and $i-1$, respectively.

It can be shown that this fine-looking central-difference scheme leads to unrealistic results [2]. To see this, rearrange the last expression to get two others:

$$T_i \left[\frac{(\rho c u)_{i+1/2}}{2} - \frac{(\rho c u)_{i-1/2}}{2} + \frac{\lambda_{i-1/2}}{x_i - x_{i-1}} + \frac{\lambda_{i+1/2}}{x_{i+1} - x_i} \right]$$

$$= T_{i+1} \left[\frac{\lambda_{i+1/2}}{x_{i+1} - x_i} - \frac{(\rho c u)_{i+1/2}}{2} \right] + T_{i-1} \left[\frac{\lambda_{i-1/2}}{x_i - x_{i-1}} + \frac{(\rho c u)_{i-1/2}}{2} \right] \quad (7.27)$$

$$a_i = a_{i+1/2} + a_{i-1/2} + (\rho c u)_{i+1/2} - (\rho c u)_{i-1/2} \qquad a_i = a_{i+1} + a_{i-1}$$

Here a stands for the expressions in brackets. The last relation follows from the previous one because according to the continuity law, $(\rho u)_{i+1/2} = (\rho u)_{i-1/2}$.

Consider now a simple example. Let us say, for instance, that we have

$$\lambda_{i+1/2}/(x_{i+1/2} - x_i) = \lambda_{i-1/2}/(x_i - x_{i-1}) = 1/2, \quad (\rho c u)_{i+1/2} = (\rho c u)_{i-1/2} = 3/2$$

$$T_{i+1} = 200, \quad T_{i-1} = 100$$

(7.28)

Then $T_i = 75$, but if $T_{i+1} = 100$, $T_{i-1} = 200$, then $T_i = 225$. These results are clearly unrealistic because T_i cannot fall outside its neighbors T_{i+1} and T_{i-1}. It is obvious that such unrealistic results are possible in any case if $|\rho c u|$ exceeds $2\lambda/\Delta x$.

There are some possibilities of overcoming this difficulty. The simplest is the upwind scheme. Because the difficulty arises when (ρu) is negative and exceeds $2\lambda/\Delta x$, in the upwind formulation the dependent variable is defined as $(\rho c u T)_{i+1/2} = T_i \| (\rho c u)_{i+1/2}, 0 \| - T_{i+1} \| -(\rho c u)_{i+1/2}, 0 \|$, where an operator $\| A, B \|$ denotes the greater of A and B. This means that the value at the faces of the control volume (between i and $i \pm 1$ midway points) are determined instead of the central-difference approach as follows: $T_{i+12} = T_i$ if $(\rho c u)_{i+1/2} > 0$. The value of $T_{i-1/2}$ is defined similarly.

The exact solution of the governing equation (7.25) with boundary conditions $x = 0, T = T_0$ $x = L, T = T_L$ is

$$\frac{T - T_0}{T_L - T_0} = \frac{\exp \text{Pe}(x/L) - 1}{\exp \text{Pe} - 1}, \qquad \text{Pe} = \frac{\rho c u L}{\lambda} \qquad (7.29)$$

The Peclet number here is the ratio of the power of convection and diffusion. Analysis shows [2] that for large values of $|Pe|$, the temperature at the middle is nearly equal to that at the upwind boundary, and this is the assumption used in the upwind scheme. However, in this scheme, the boundary temperature remains the same for all values of Peclet number. The other imperfection is that for large $|Pe|$, at the middle the derivative dT/dx is almost zero so that the diffusion is almost absent. At the same time, in the upwind scheme, the diffusion is calculated always by applying a linear profile that overestimates it for large Peclet numbers.

The exponential scheme follows all the properties of the exact solution. To present this scheme, consider the governing equation (7.25) in the form of total flux J of convection and diffusion:

$$\frac{dJ}{dx} = 0 \quad or \quad J_{i+1/2} - J_{i-1/2} = 0 \quad J = \rho cuT - \lambda \frac{dT}{dx} \tag{7.30}$$

Applying the exact solution (7.29) as a profile between grid points and substituting the total flux, one obtains the governing equation for the exponential scheme:

$$(\rho cu)_{i+1/2} \left(T_i + \frac{T_i - T_{i+1}}{\exp Pe_{i+1/2} - 1} \right) - (\rho cu)_{i-1/2} \left(T_{i-1} + \frac{T_{i-1} - T_i}{\exp Pe_{i-1/2} - 1} \right) \tag{7.31}$$

Despite the advantages of the exponential scheme, it is not widely used because for two- and three-dimensional problems, this scheme can be applied only as an approximate approach, and because the exponential scheme is expensive to compute, there is no reason to use it. Two easily computed schemes with the qualitative behavior of the exponential scheme are usually used instead of the exponential scheme.

The hybrid scheme reasonably approximates the exponential scheme. To see this, consider the dimensionless coefficient a_{i+1} at temperature T_{i+1} in Equation (7.31) as a function of the Peclet number (Δx is substituted for L):

$$\bar{a}_{i+1} = \frac{a_{i+1}\Delta x}{\lambda_{i+1/2}(\rho cu)_{i+1/2}} = \frac{Pe_{i+1/2}}{\exp Pe_{i+1/2} - 1} \qquad Pe = \frac{\rho cu \Delta x}{\lambda} \tag{7.32}$$

Analysis of this exact result shows [2] that specific properties of the function $\bar{a}_{i+1} = f(Pe_{i+1/2})$ are as follows:

$$Pe \to \infty, \bar{a}_{i+1} \to 0; \quad Pe \to -\infty, \bar{a}_{i+1} \to -Pe; \quad Pe = 0, \bar{a}_{i+1} \to 1 - Pe/2 \tag{7.33}$$

The hybrid scheme approximates these properties by three straight lines:

$$Pe < -2, \bar{a}_{i+1} = 0; \quad -2 \le Pe \le 2, \bar{a}_{i+1} = 1 - Pe/2; \quad Pe > 2, \bar{a}_{i+1} = 0 \tag{7.34}$$

Thus, the hybrid scheme works in the same way as the exponential one: it is like the central-difference scheme for the range $-2 \le Pe_{i+1/2} \le 2$, and outside of this range it reduces the diffusion to zero, similar to the upwind scheme. That is the reason why this scheme is named the hybrid scheme.

A better approximation to exponential behavior provided the power law scheme that is constructed as follows [2]:

$$Pe < -10, \bar{a}_{i+1} = -Pe; \; -10 \leq Pe < 0, \bar{a}_{i+1} = (1+0.1Pe)^5 - Pe;$$

$$0 \leq Pe \leq 10, \bar{a}_{i+1} = (1+0.1Pe)^5; \; Pe > 10, \bar{a}_{I=1} = 0$$

(7.35)

Simple calculation indicates that the power law scheme closely approximates the exact solution.

Thus, all three schemes are applicable for solving practical problems except the central-difference scheme that can only be used for low Peclet numbers with unreasonable fine grids.

7.3.1.3 False Diffusion

There are two types of false diffusion [2]. The first type is the common misunderstanding, when by comparison using the central-difference and upwind schemes, one concludes that the upwind scheme produces false diffusion. The other type is the real false diffusion that arises in situations when the calculation results show the presence of diffusion despite the diffusion coefficient being zero.

It can be shown using the Taylor expansion that the central-difference scheme has second-order accuracy, and the upwind scheme is of first-order accuracy. On the other hand, it follows from comparison of these two schemes that they would be equivalent if the diffusion coefficient in the upwind scheme were increased by $\rho c u \Delta x / 2$ for that coefficient in the central-difference scheme [2]. From these two true facts, one sometimes draws the wrong conclusion that the upwind scheme produces false diffusion. The central-difference scheme is better than the upwind one only in the case of small Peclet numbers when the Taylor expansion is applicable. For large Peclet numbers, the truncating Taylor series cannot be used for analyzing the convention-diffusion dependence, because in this case that dependence is of an exponential type. At the same time, as discussed above, the central-difference scheme leads to unrealistic results in the case of large Peclet numbers.

The real false diffusion occurs only in a multidimensional situation and cannot be observed in steady one-dimensional problems. For example, if two parallel two-dimensional streams of equal velocities and different temperatures come in contact, the diffusion process forms a mixed layer only in the case of the nonzero diffusion coefficient. However, when the diffusion coefficient is zero, such streams remain separated with temperature discontinuity at the interface despite the presence of a temperature gradient. Thus, if in such a situation the computer program shows a smeared profile in the cross-section, it is obvious that a real false diffusion is taking place.

In general, false diffusion arises when the flow is oblique to the grid lines and there is a gradient in the direction normal to the flow. Because false diffusion is most severe when the flow direction makes an angle 45° with the grid lines [2], the intensity of false diffusion can be reduced by adjusting the flow along the grid lines. The other way to reduce false diffusion is to use a small Δx and Δy because it is known that this results in small Peclet numbers when the central-difference scheme works perfectly.

7.3.1.4 A Case of a Moderate Mach Number

In the case of the incompressible fluid, density is constant and, as a result, only the velocity components depend on the pressure. Therefore in this case only, the correct pressure field yields a velocity distribution that satisfies the continuity equation. In the other limiting case of highly compressible flows, velocity components are almost independent of pressure, and the pressure influences only the density through the equation of state. Proceeding from these two facts, one concludes that in the case of flow with moderate Mach numbers, both velocity components and density should depend significantly on pressure. In this case, only the correct pressure distribution could give velocities (from the momentum equations) and density (from state equation) distribution that satisfy the continuity equation.

The procedure of solving the momentum and energy equations remains the same as in the case of incompressible flow. The equation for the pressure is also the same (Equation 7.24) but without the first term. This term should be omitted because in this case both dependences of the velocities and density are taken into account, not in sequence manner as in the case of small Mach numbers, but rather simultaneously [5].

7.3.2   The Control-Volume Finite-Element Method [2,5]

The control-volume finite-element method is basically very close to the control-volume finite-difference method described above and should not be considered as a different method. The only advantage that distinguishes the former from the latter is the ability to use the irregular grids, because in the finite-difference approach, mainly uniform grids are employed. Such irregular, for example, triangular grids are more flexible and allow for local grid refinement [2].

The difficulties discussed in the last section are inherent in the finite-element method as well. These difficulties have been resolved, and a control-volume finite-element method similar to the control-volume finite-difference method was developed [2,5,12]. This method is characterized by the following features:

1. For the triangular grids, the values of dependent variables are calculated for the grid points found at the vertices of the triangles that

play a role of main points. The lines joined the centroid of each triangular element with midpoints on its sides dividing each element
into three equal areas, regardless of the form of triangle element.
These areas collectively construct the nonoverlapping contiguous
polygonal volume elements that are similar to these in the finite-
difference approach.

2. Many CFD (computational fluid dynamics) codes used the staggered
   grids that do not have the problems of central-difference schemes.
   However, the staggered grids cannot be used for nonorthogonal
   grids and unstructured meshes that are typical for the finite-
   element approach.

   Therefore, in the finite-element method, instead of the staggered grids, two other approaches are used [5,12]. One consists of
   unequal-order formulations for pressure and velocity components
   in which for the former are used a sparser grid and a lower-order
   interpolation than that used for the latter. In the second technique,
   the equivalent of the colocated momentum–interpolation scheme
   is applied. However, the first approach requires two sets of control
   volumes that make calculation awkward and the excessively fine
   grids in the case of high Reynolds numbers or pressure gradients.
   As the result, the co-located momentum–interpolation scheme has
   been adopted in computational practice.

The term "co-located" pertains to methods in which both the velocities
and the pressure are calculated at the same set of nodes located at the center
of the control volume, while in the staggered grids the pressure is calculated
at the center and the velocities are determined at the faces of the control volume. In the co-located schemes, the velocity and the pressure are obtained
using the following expressions:

$$\tilde{u}_{i+1/2} = \tilde{u}_{i-1/2}, \qquad \tilde{u}_i = \frac{\sum\limits_{nb} a_{nb} u_{nb}}{a_i} - \frac{\Delta V_i}{a_i}\left(\frac{dp}{dx}\right)_i \tag{7.36}$$

$$\left(\frac{dp}{dx}\right)_{i+1/2} = \frac{p_{i+1} - p_i}{x_{i+1} - x_i}, \qquad \left(\frac{dp}{dx}\right)_{i-1/2} = \frac{p_i - p_{i-1}}{x_i - x_{i-1}} \tag{7.37}$$

$$a_i u_i = \sum\limits_{nb} a_{nb} u_{nb} - \Delta V_i\left(\frac{dp}{dx}\right)_i \qquad (Au)_{i+1/2} = (Au)_{i-1/2}$$

where a_i and a_{nb} are the discretization coefficients, A is a face area, and ΔV_i
is the volume of the control volume. Indices refer to the main (i) and two
neighbor ($i-1/2$ and $i+1/2$) points, and indices $i-1$ and $i+1$ pertain to control volume faces.

Equations (7.36) and (7.37) are derived using the discretization momentum and continuity equations for the staggered grids. Because staggered grid equations do not have the problems of the central-difference schemes. Equation (7.36) obtained from these equations do not have these problems. Hence, Equation (7.36) can be applied instead of the staggered grids in general, and, in particular, can be used in the co-located schemes.

The methods of solving flow and heat transfer governing equations discussed above are based on single-grid solutions. These approaches are flexible, reliable, have potential, and are convenient and are working well. For many years, they have been employed usefully in science and industry. However, the convergence of solutions in the single-grid schemes was acceptable some decades ago when computer memory was not enough to use fine grids. The computational technique available now requires very fine grids (up to $(1-10) \cdot 10^6$ fine-volumes) to ensure accuracy and reduce the errors.

There are several ways to accelerate the convergence of the family SIMPLE methods. They include better formulation of the pressure-correction scheme; a more efficient approach to solving the linear equations of momentum and pressure/pressure correction; and a treatment of interequation coupling, such as among turbulence, chemical species, and momentum/continuity equations. However, because the entire set of equations is highly nonlinear and coupled, any one of the equations can slow the convergence of the entire set [5].

The poor convergence of iterative schemes arises from the low-frequency components of the error, because the high-frequency components converge rapidly during initial iterations. On the basis of that fact, the idea of multigrid techniques arises. This approach consists of eliminating the poor converging part of error and increasing its convergence by putting it on the coarse grid. Then, the correct variant is interpolated back to the finer grid to correct the previously obtained poorly converged part of the error. A review of the studies of multigrid schemes is presented in Reference [5].

The other way to improve the convergence and reduce the time of mesh generation of SIMPLE-type methods is to use the meshless methods. The meshless methods apply the compact or global interpolation on nonordered irregular spatial domains employing special radial basis functions (RBFs) [13,14]. The primary feature of a radial basis function is that it depends only upon the radial distance between knots or nodal points. There are several global radial basis functions commonly used [13]. The interpolation with this function has been successfully used and demonstrated striking convergence. In particular, a scheme with multiquadratic radial basis functions (MQ-RBFs) shows exponential convergence. For example, it was shown that the solution of the three-dimensional Poisson's equation can be obtained with only 60 randomly distributed knots to the same accuracy as the finite-element method solutions with 71,000 linear elements. Several researchers successfully applied the MQ-RBF schemes for solution of various initial and boundary problems for different types of ordinary and partial differential equations [13].

Recently, the local collocation meshless method was proposed, and later this approach was developed to formulate using the MQ-RBF scheme a localized meshless collocation method [14]. Employing this technique, the procedure for solving the conjugate heat transfer problems was developed.

7.4   Numerial Methods of Conjugation

The following governing equations are used in the conjugate convective heat transfer problems for the fluid and the body domains, respectively:

For steady and unsteady flows in the fluid domain:

Continuity equation

Two- or three-dimensional Navier-Stokes and energy equations

Two- or three-dimensional dynamic and thermal boundary layer equations for high Reynolds numbers

Two or three-dimensional dynamic and thermal equations for low and moderate Reynolds numbers

One-dimensional velocity and temperature equations

For steady and unsteady conduction in the body domain:

Two- or three-dimensional unsteady equations

Two- or three-dimensional steady Laplace or Poisson (with internal heat generation) equations

One-dimensional unsteady or steady equations for a thin (negligible longitudinal conduction) or a thermally thin (negligible transverse conduction) body

Although many combinations of these equations are possible, the majority of the known investigations of the conjugate convection heat transfer are using for a flow (1) two- or three- (rare) dimensional steady or unsteady elliptic Navier-Stokes and energy equations, or (2) two- or three-dimensional steady or unsteady parabolic dynamic and thermal boundary layer equations, and for a body (3) two-dimensional steady Laplace or Poisson equations or (4) steady or unsteady (more often) one-dimensional conduction equations.

There are several methods with which to perform the conjugation between solution domains:

1. A direct approach in which the governing equations for both the fluid and body are conjugated by simultaneously solving one large set of equations.

For example, a method proposed by Patankar [2,15] is of that type. In Patankar's approach, the one generalized conservation expression is considered instead of both separated equations for the fluid and solid. If any dependent variable (for example, velocity or temperature) is defined by ϕ, this general differential equation is [2]

$$\frac{\partial}{\partial t}(\rho\phi) + div(\rho\mathbf{u}\phi) = div(\Gamma grad\phi) + S \tag{7.38}$$

where $\mathbf{u}$ is the velocity vector, Γ is the diffusion coefficient, and S is the source term.

The conjugate problem in this approach is solved using one large calculation domain including both domains for the fluid and body, their interface, and the outer boundary conditions. To ensure that the velocity is zero in the solid when the velocity field is calculated, one puts for the grid points in the body domain a very large value of diffusion coefficient Γ, while at the same time for the grid points in the fluid domain, the diffusion coefficient Γ is made equal to the real fluid viscosity. Such a procedure would provide zero velocity on the whole body domain, including the surface. As a result, the fluid would meet the correct boundary conditions on the surface. When the temperature field is computed, one specifies in the general equation the real values of diffusion coefficient Γ for the fluid and for the body, respectively, in their domains.

Thus, such a procedure gives the entire temperature field in the whole domain as a result of automatically matching the temperature distribution in the fluid and the body.

At the same time, in some cases, applying the direct conjugation involves difficulties, in particular, those that occur due to the mismatch in the structure of the coefficient matrices in BEM, FEM, and FDM solvers [16].

2. The iterative conjugation method consists of a separated solution of each set of equations for the fluid and for the body, respectively. An idea of such a strategy is that each solution of the body or the fluid equation produces the boundary condition for the other along the body–fluid interface. It starts from guessing one of the fields on the interface which is then used as a boundary condition for the other. For example, one begins assuming the temperature or heat flux distribution along the interface that is used as a boundary condition for solving the equation for the body. Then, the obtained variable distribution along the interface is applied as a boundary condition for solving the system of equations for the fluid, and so on. If the process would converge, the iterations would continue until the required accuracy criterion is achieved. This iterative approach is widely used in finite-difference and finite-element or finite-volume methods.

However, one of the disadvantages of these commonly applied numerical approaches is that these require an entire meshing of both the fluid and the body domains and use of numerical differentiating to get the heat fluxes. An alternative algorithm that does not require these procedures was developed by Divo et al. [16].

Although this approach, as with many others, uses the finite-volume method to mesh the whole fluid domain and applies the iterative procedure for conjugation, it employs for the body domain the boundary element method that requires discretization only of the boundaries. Besides, in this case, the interfacial heat fluxes needed to provide continuity in the conjugate problem are obtained when the body conduction is carried out; hence, numerical differentiating is avoided. Thus, this method is significantly simple compared to the typically used FDM, FEM, or FVM. Articles [16–18] present some solutions of conjugate heat transfer problems obtained by this method.

3. The superposition method gives the expression for heat flux on the gradientless streamlined plate in the case of an arbitrary continuous temperature distribution (Equation 1.38 from Chapter 1):

$$q_w = h_* \left[T_w(0) - T_\infty + \int_0^x \left[1 - \left(\frac{\xi}{x} \right)^{C_1} \right]^{-C_2} \frac{dT_w}{d\xi} d\xi \right]$$

where $C_1 = 3/4, C_2 = 1/3$ for laminar flows and $C_1 = 9/10, C_2 = 1/9$ for turbulent flows. The solution of the conjugate heat transfer problem can be obtained if one substitutes this relation for the heat flux in the conduction equation for the plate. In the case of the thermally thin plate, the conjugate problem after substituting Equation (1.38) reduces to integro–differential or nonlinear differential equations with a weak singularity. These equations are solved numerically using standard or special SIMPLER approaches. Many solutions of this type are carried out in 1961 through 1969 [19]. In one such solution given in Reference [19], the plate temperature is defined in the case of combined convection, conduction, and radiation heat transfer (Example 6.25).

The superposition method was also used by Barozzi and Pagliarini [20] for the case of fully developed laminar flow in a pipe. In this study, the superposition method (Duhamel integral) is used for fluid and the finite-element method is employed for numerically solving Laplace's equation (Example 7.3).

A general approach of this type that reduces a conjugate convective heat transfer problem to an equivalent heat conduction problem is suggested in References [21,22] and is outlined in Section 6.3.

This method is based on the exact solutions given in Chapter 3 for an arbitrary nonisothermal surface in the laminar incompressible and compressible flows. This technique is also applicable to unsteady laminar flow, continuously moving surface, and flow of non-Newtonian power fluids for which the analogous exact solutions are derived in Chapter 3. For the case of turbulent flow and some other situations, the highly accurate approximate solutions of the same type are obtained in Chapters 3 and 4.

In particular, the general forms of the Duhamel integral (Equation 1.38), like Equation (3.40) for laminar flow and similar formulae for turbulent flow and other cases, are applicable to flows with nonzero pressure gradient due to using Görtler's variable Φ (see Equation 3.3), which takes into account the presence of the pressure gradient. As shown in Chapter 3, the relations for the gradientless flow became valid for gradient flows if the variable Φ was substituted for the axial coordinate x. Another advantage of this approach is that there are differential formulae, like Equation (3.32), and analogous formulae that are equivalent to generalized Duhamel integrals and are used to simplify the conjugate problem, especially in the case of thin bodies.

The examples of analytical solutions applying this method are given in Chapter 6. The numerical method based on this approach was developed by Davydenko [23]. This method consists of a finite-difference solution of the Laplace equation using as a boundary condition the generalized Duhamel integral of type (3.40) based on employing the Görtler variable and, hence, applicable to the above-mentioned different cases, including the flows with nonzero pressure gradient. As an example, a heat transfer from elliptical cylinders to laminar flow is considered (Example 6.7).

7.5 Examples of Numerical Studies of the Conjugate Convective Heat Transfer in Pipes and Channels

Example 7.2: A Horizontal Channel Heated from below by q_w = Constant Fully Developed Laminar Flow [24]

The system of equation for incompressible flow consists of two-dimensional continuity and Navier-Stokes and energy equations with sources $S(x,y) = -dP/dx$ and $S(x,y) = -dP/dy + (\text{Gr}/\text{Re}^2)\theta$ for u and v velocity components, respectively. Here the term $(\text{Gr}/\text{Re}^2)\theta$ stands for buoyancy effects, $\theta = (T - T_\infty)\lambda/q_w H$, and H is a channel height. For the walls, Laplace's equation is solved. The conjugate and the boundary condition are different for the heated and for both insulated parts of the wall at the inlet and outlet of the channel. The total heat transfer coefficient that takes into account the resistances of the walls from the plexiglass and the insulation from the fiberglass is used by formulating the conjugate and boundary conditions.

It is assumed that the flow enters with a parabolic velocity profile and with ambient temperature. This assumption is verified by measurements. The exiting conditions are quite complex and depend basically on recirculation effects. According to some known data, it is accepted that $\partial u/\partial x = v = 0$, $\theta = 0$ if $u \leq 0$ (inflow) and $\partial \theta/\partial x = 0$ if $u \geq 0$ (outflow).

A finite-volume technique with different grid size for body and flow is employed. The Successive Over Relaxation code is used to solve the pressure equation, and the Tridiagonal Matrix Algorithm for the solution of a nonlinear coupled system of momentum, energy, and continuity equations. The grid density is increased until the two successive solutions of flows, thermal fields, and mass and energy conservation differ by less than 1%.

The numerical results are in agreement with the author's experiments. The basic conclusions are as follows:

1. In the case of low Reynolds number (Re = 9.48), the buoyancy causes two rolls with axis of rotation perpendicular to the flow direction. The upstream roll produces a recirculation zone, and the downstream roll entrains flow from outside. As the Reynolds number increases to Re = 29.7, these rolls become smaller, and flow entrainment is reduced. It is observed that the intensity of these and transverse rolls as well as oscillatory movements and turbulence depend on the channel cross-sectional sizes and heating rates. The study shows that the effect of wall conductivity on the roll location is small.

2. The longitudinal asymmetry of temperature profiles is caused by channel flow. The heated region temperature is higher than that of insulation which results in the generation of a thermal plume above the surface. In the case of an aluminum heated region, the temperature distribution was highly uniform which differs significantly from the case of a ceramic heated region when temperature uniformity is reduced due to poor thermal material diffusivity and to an increased heat transfer to the insulation.

3. Comparison of the results obtained for conjugate and nonconjugate approaches demonstrates the significance of conjugate modeling. This comparison is made for different heated region conductivities in the range corresponding to materials such as plexiglass, ceramics, stainless steel, and aluminum. In the case of a uniform heated surface, the nonconjugate model predicts a highly nonuniform temperature profile. In contrast, the conjugate model gives the highly uniform temperature profile. This occurs not only due to the redistribution of the thermal energy by itself, but also to the increase of the thermal energy loss to the insulation. These results are confirmed by analysis of the effect of the thermal conductivity of the heated and insulated regions and of the wall thickness on the ratio of average heated region temperature predicted by both nonconjugate and conjugate models.

4. The numerical and experimental results show that the conjugate effects are significant and usually should be taken into account, except in two cases: (1) for the thin walls, low insulation or high thermal conductivity of the heated regions, the surface can be considered as an isothermal one, and (2) at the low heated region thermal conductivity, the heat transfer may be modeled as $q_w = \text{const}$.

Example 7.3: A Pipe with a Fully Developed Laminar Flow Heated Symmetrically at the Outer Surface by Uniform Heat Flux [20]

The governing equations for the fluid and body and boundary and conjugate conditions in dimensionless variables are as follows:

$$u\frac{\partial\theta}{\partial x} = 4\left(\frac{\partial^2\theta}{\partial r^2} + \frac{1}{r}\frac{\partial\theta}{\partial r}\right), \qquad \theta(0,r) = \frac{\partial\theta}{\partial r}(x,0) = 0,$$

$$4\left(\frac{\partial^2\theta_s}{\partial r^2} + \frac{1}{r}\frac{\partial^2\theta_s}{\partial r^2}\right) + \frac{1}{\text{Pe}^2}\frac{\partial^2\theta_s}{\partial x^2}, \qquad \frac{\partial\theta_s}{\partial x}(0,r) = \frac{\partial\theta_s}{\partial x}(L,r) = 0 \qquad (7.39)$$

$$\frac{\partial\theta_s}{\partial r}(x+2\Delta) = \frac{\lambda}{\lambda_w(1+2\Delta)}, \quad \theta_s(x,1) = \theta(x,1), \quad q_w = \frac{\lambda_w}{\lambda}\frac{\partial\theta}{\partial r}(x,1)$$

Dimensionless variables are scaled using the pipe radius R, mean velocity U, Peclet number $\text{Pe} = 2RU\rho c/\lambda$, initial temperature T_e, outer heat flux q_0, and $\theta = (T - T_e)\lambda/q_0R$. $\theta_s = (T_s - T_e)\lambda_w/q_0R$. The solution procedure consists of applying the superposition method for the fluid equation in the form

$$\theta_w - \theta_{bl} = \frac{2q_w(0)}{\text{Nu}_{x,q}(x)} + 2\int_0^x \frac{dq_w(\xi)}{d\xi}\frac{d\xi}{\text{Nu}_{x,q}(x-\xi)},$$

$$\theta_{bl} = 8\int_0^x q_w(\xi)d\xi, \quad 2\frac{\partial\theta_w}{\partial r}(x,1) = \frac{\lambda}{\lambda_w}(\theta_w - \theta_{bl})\text{Nu}_x \qquad (7.40)$$

and the finite-element method for the body equation. Here θ_{bl} is the bulk temperature, and $\text{Nu}_{x,q}$ is the Nusselt number for the case $q_w = \text{const.}$, which for the flows in the ducts is given in Reference [25]. The last equation is an auxiliary relation between heat flux and temperature difference on the fluid–body interface that authors chose to guess temperature distribution to improve the iterations' convergence.

The iteration process begins by guessing the values at the right part of the last Equation (7.40) and computing the corresponding distribution of the derivative $(\partial\theta_w/\partial r)_{r=1}$ on the interface. Using these results and the finite-element method to solve the second, Equation (7.39) for the body, two other equations (Equation 7.40), and the last Equation (7.39) to find θ_w, θ_{bl} and q_w gives data for the second iteration that begins again from employing the auxiliary relation. Authors claim that using the auxiliary equation reduces the number of needed iterations. For the finite-element approach, the variational formulation is used, which is not popular in heat transfer studies [2].

The following basic results are formulated:

1. The temperature distribution on the interface shows that there is an isothermal region in the nearness of the inlet. In this case, the wall-to-fluid temperature difference is such that the heat is transferred from the fluid to the wall. This is similar to that observed in other studies in the case of uniform $q_w = \text{const.}$ heating.

   An almost isothermal temperature distribution exists along the heated section for the higher conductivities and wall thicknesses and for the lower lengths of the heated section and Peclet numbers. The plots for the same heated section length have a common point of intersection in the second part of this section that is shifted to the end of the pipe with increasing pipe length.

2. In the initial part of the section, the heat flux decreases from very high values to nominal magnitude $q_w = 1$. Then, it varies in two ways. The first is when it reaches the minimum and then goes back up to the nominal value but does not reach it. In this case, the temperature distribution inside the wall is close to one dimension, and the wall may have been considered as thermally thin.

   The other type is when the heat flux decreases monotonically along the heated section so that the axial component of the temperature gradient decreases sharply from the outer to the inner wall surface. In that case, the wall should be seen as thermally thick, and disregarding wall heat conduction in this case leads to large errors.

3. The dependence $\mathrm{Nu}_x - \mathrm{Nu}_{x*} = f(x)$ starts from zero at some axial position and close to the end decreases sharply, reaching the curve $\mathrm{Nu}_{x,q} = f(x)$ in such a way that Nu_x always remains lower than $\mathrm{Nu}_{x,q}$.

4. The increasing Peclet number reduces the effect of axial conduction much more than the corresponding decreasing wall-to-fluid conductivity ratio or the wall thickness. The other way to considerably reduce the effect of axial wall conductivity is to increase the pipe length.

Example 7.4: A Thick-Walled Parallel-Plate Channel with Poiseuille-Couette Incompressible Laminar Flow and a Moving Upper Wall [18]

The system of governing equations contains the energy equation for the fully developed velocity field of incompressible flow and Laplace's equation for the solid. The boundary conditions and fluid velocity profile are as follows:

$$\rho c_p u \frac{\partial T}{\partial x} = \lambda \frac{\partial^2 T}{\partial y^2} + \mu \left(\frac{\partial u}{\partial y} \right)^2, \quad x = 0, \ T = T_e, \ \frac{\partial T_s}{\partial x} = 0,$$

$$x = L, \ \frac{\partial}{\partial x} \left(\frac{T_w - T}{T_w - T_m} \right) = 0, \ \frac{\partial T_s}{\partial x} = 0, \quad y = H, \ T = T_e, \ y = -\Delta, \ q_s = q_0 \tag{7.41}$$

$$u = [6U_m - 2U - (6U_m - 3U)(y/H)](y/H)$$

The conjugate conditions consist of equalities of the temperatures and the heat fluxes of the fluid and the body on the interface. Here T_e is the entrance temperature, H is the channel height, T_m and U_m are the fluid mean temperature and velocity, and T and U are the temperature and velocity of the moving surface. The fluid temperature field is computed using the finite-difference method, while for the body, Laplace's equation is solved by applying the boundary-element approach. The advantages of such a technique are discussed in Section 7.4. The energy equation for fluid is modified so as to cluster the grid points close to the wall and to the moving surface.

To conjugate both solutions, the iterative procedure is used. The calculation is started by guessing the heat flux distribution on the interface and solving the Laplace equation to find the temperature distribution on the interface. Then, using these data, the energy equation for a fluid is solved. The new heat flux distribution is estimated using the approximate equation $q^{n+1} = (\omega_s q_s^n + \omega q^n)/(\omega_s + \omega)$, where ω is a weighting function. By numerical experiments, it is found that for relatively rapid convergence of interactions, the weighting functions ratio should satisfy the inequality $\omega/\omega_s < 0.03$.

The numerical calculation gives the following results:

1. The range of conductivity ratio of the body to fluid is from 80.6 to 8060. As the conductivity ratio increases, the interfacial temperature at the leading edge becomes higher and its changes become smaller. As this ratio becomes small, the wall temperature at the entrance approaches the entrance flow temperature. The effect on the Nusselt number with conductivity ratio variable is very little.

2. For the changes of dimensionless wall thickness from 0.0042 to 0.21, the effect is similar. When the wall thickness increases, the interfacial temperature becomes closer to the entrance fluid temperature, similar to that in the case of a smaller conductivity ratio. In contrast, when the wall thickness decreases, the conjugate problem gets closer to the case of constant heat flux with the only difference in the temperature at the entrance region. The thin wall case shows a higher dimensionless temperature that is caused by heat storage in the wall, despite it being thin.

3. The channel aspect ratio was varied from 0.001 to 0.127. In this case, both the interfacial temperature and heat flux are strongly influenced. When the ratio is small, the fluid flow is reduced, and the temperature increases more rapidly. In the case of a large aspect ratio, the temperature changes slightly, but the Nusselt number varies greatly.

4. The effect of the ratio of the free surface to the fluid mean velocities was studied for the values from 0.075 to 2.03. When the mean velocity increases with unchanged free surface velocity, the changes in interfacial temperature are small. This is because in this case the free surface temperature is low. When the mean fluid velocity is low, the convective heat transfer in the cross-channel direction is high, but in the opposite case, this part of heat transfer

is greatly reduced, and this caused the temperature to increase down the channel. This is confirmed by Nusselt number distribution because for the high velocity ratio, a low Nusselt number is observed.

Example 7.5: The Thick-Walled Pipe of an Arbitrary Shape of Cross-Section with Fully Developed Velocity [26]

Because the shape of a cross-section is arbitrary and the velocity is fully developed, the momentum equation has the relatively simple form, $\nabla^2 u = (1/\mu)(dp/dx)$, and the equations for the fluid and for the solid temperature fields and conjugate and the boundary conditions are like these in the other two-dimensional problems. Both equations for the fluid and solid are solved using the boundary element technique. The dual reciprocity boundary element method is used to solve the heat convective problem, and the conventional boundary element method is applied to solve the heat conduction equation. The details of transforming the governing equations and performing the procedure of these methods may be found in Reference [26].

The two separate solutions are conjugated using the numerically stable iterative procedure. To start iterations, both the heat transfer coefficient and the mean fluid temperature at the body–fluid interface are guessed.

Two cases, the uniform temperature or uniform heat flux imposed on the outer surface of the thick-walled pipe, are calculated and compared with an available analytic solution to test the suggested method.

Example 7.6: Heat Transfer between the Forced Convective Flow inside and the Natural Convective Flow outside of the Vertical Pipe [27]

The flow enters the pipe with the temperature T_e and fully developed velocity profile. It is assumed that $T_e > T_\infty$ so there is the energy exchange between the internal flow and the environment. For the internal flow, only the energy equation is solved, but while for the external flow, the full system of equations should be considered:

$$(1-r^2)\frac{\partial \theta}{\partial x} = \frac{i}{r}\frac{\partial}{\partial r}\left(r\frac{\partial \theta}{\partial r}\right), \quad x=0,\ \theta=1,\ r=0, \quad \frac{\partial(ru)}{\partial x}+\frac{\partial(rv)}{\partial r}=0$$

$$u\frac{\partial u}{\partial x}+v\frac{\partial u}{\partial r}=\frac{(Ra/Pe)\theta}{8\,Pr_o}+\frac{1}{r}\frac{\partial}{\partial r}\left(r\frac{\partial u}{\partial r}\right) \qquad u\frac{\partial \theta}{\partial x}+v\frac{\partial \theta}{\partial r}=\frac{1}{r}\frac{\partial}{\partial r}\left(r\frac{\partial \theta}{\partial r}\right), \qquad (7.42)$$

$$Ra = \frac{g\beta_o(T_e-T_\infty)D^3\,Pr_o}{v_o^2} \quad r\to\infty\ x\to 0,\ \theta,\ u\to 0$$

The conjugate conditions are $\theta_i = \theta_o$, $(\partial\theta/\partial r)_i = (\lambda_o/\lambda_i)(\partial\theta/\partial r)$, where $\theta = (T-T_\infty)/(T_e-T_\infty)$. The application of dimensionless variables here is scaled using R for r, RPe for x, RPe/v_o for u, R/v_o for v; subscripts o,i denote outer and inner flows. Because the wall is considered as thin with no thermal resistance, the variable r is less than one ($r \leq 1$) and more than one ($r \geq 1$) for inside and outside the pipe, respectively.

Numerical solutions of both the energy equation for inside flow and the full system of equations for outside flow are performed using the Patankar-Spalding approach [6] that is similar to SIMPLE described in [2] and discussed above (Section 7.3.1). The iterative procedure is applied to conjugate both the inside and outside solutions. To improve the convergence of iterations, the local heat transfer coefficient distribution on the interface is used. The procedure starts from solving the external problem using the boundary condition of uniform wall temperature. These results are used as a boundary condition for the inside pipe flow. Then, two equations (θ_{bl}-balk temperature)

$$\left(-\frac{\partial \theta}{\partial r}\right)_0^j = \frac{\mathrm{Nu}_i^j \lambda_i (\theta_{bl}^{j-1} - \theta_w^j)}{2(1 + S^j)\lambda_o}, \qquad T_{bl}^j = \frac{T_{bl}^{j-1} + S^j T_w^j}{1 + S^j}, \qquad S^j = 2\Delta x^j \mathrm{Nu}_i^j, \qquad (7.43)$$

follow from the heat balance and conjugate conditions, along with Equation (7.42) are employed to solve the outer problem. These data and relations for inner flow similar to the last one $(-\partial\theta/\partial r) = (\mathrm{Nu}_o/2)$ $(\lambda_o/\lambda_i)\theta_{wi}$ are used to solve the inner problem and so on until the convergence of the temperature and heat flux on the interface is achieved. Authors claim that due to using the heat transfer coefficient distribution instead of applying the temperature or heat flux distribution, the convergence of iterations is achieved during three to five iterations.

The following results are obtained:

1. The Nusselt number for the forced convection of inside flow is insensitive to the thermal boundary conditions. Data of conjugate values of Nusselt number for different λ_o/λ_i, Pr_o and $\mathrm{Ra/Pe}$ are bounded between those for uniform wall temperature and uniform heat flux. Near the inlet, the Nusselt number values tend to be closer to uniform wall temperature, and for the larger distance from the inlet, the Nusselt number values become closer to uniform heat flux.

2. The Nusselt number for the external flow is compared with the results for isothermal vertical cylinder Nu_{cy}. Analyzing the ratio $\mathrm{Nu}_o/\mathrm{Nu}_{cy}$ shows that the values obtained in the conjugate problem are always lower than those in the case of uniform temperature. Increasing both λ_o/λ_i and $\mathrm{Ra/Pe}$ leads to rapid decreasing of the temperature difference $(T_w - T_\infty)$ along the pipe, and finally results in a decrease of the Nusselt number. At the fixed values of these parameters, the variations of Pr_o from 0.7 (air) to 5 (water) do not practically affect the values of the Nusselt number.

3. The wall temperature decreases with an increase in the downstream distance along the pipe. This effect intensifies at large values of λ_o/λ_i and $\mathrm{Ra/Pe}$. A special case is the large values of λ_o/λ_i which may be regarded as corresponding to internal airflow and external water flow. It is clear that such a situation shows that the external thermal resistance is very small in comparison with that

of internal flow, and this results in a significant drop of the wall temperature upstream at the small distances from the entrance. A comparison of $\mathrm{Pr}_o = 0.7$ and $\mathrm{Pr}_o = 5$ indicates that this parameter does not play a significant role.

4. The bulk temperature ratio $(T_{bl} - T_\infty)/(T_e - T_\infty)$ determines the heat transfer effectiveness because it compares the heat transfer rate for the length of pipe between the entrance and a certain location to that for an infinitely long pipe in the case of the same mass flow. The results obtained show that bulk temperature decreases with x as heat is transferred from the inside flow to the external flow. Higher values of λ_o/λ_i and Ra/Pe increase the heat transfer rate, which results in more rapid decrease in the bulk temperature. Analyzing data shows that $T_w < T_{bl}$ and also that the difference between these values is greater for larger values of λ_o/λ_i. This occurs because the internal resistance becomes greater when the ratio λ_o/λ_i increases, which leads to an increase in the temperature difference. Comparison of the data for $\mathrm{Pr}_o = 0.7$ and $\mathrm{Pr}_o = 5$ reveals that the bulk temperature is affected slightly by the Prandtl number Pr_o.

Example 7.7: Transient Heat Transfer in a Pipe with Laminar Fully Developed Flow and an Outer Surface Heated by Constant Temperature T_o [28]

Under usual assumptions, the governing equations and boundary conditions are

$$\frac{\partial\theta}{\partial t} + \mathrm{Pe}(1-y^2)\frac{\partial\theta}{\partial x} = \frac{1}{y}\frac{\partial}{\partial y}\left(y\frac{\partial\theta}{\partial y}\right) + \frac{\partial^2\theta}{\partial x^2} \quad \frac{\alpha}{\alpha_w}\frac{\partial\theta}{\partial x} = \frac{1}{y}\frac{\partial}{\partial y}\left(y\frac{\partial\theta}{\partial y}\right) + \frac{\partial^2\theta}{\partial x^2}$$

$$\theta_w(x,R_o,t) = 1, \quad \frac{\partial\theta}{\partial y}(x,R_o,t) = 0 \tag{7.44}$$

The first boundary condition is for the heated part of the pipe, and the other is for both insulated parts of the pipe before and after the heated zone. The dimensionless variables are scaled similar to those in a previous example: R_i for x and y, R_i^2/α for t, $\theta = (T - T_e)/(T_o - T_e)$.

Both equations for fluid and walls are solved, and solutions are conjugated using the finite-difference scheme and recommendations of Patankar [2]. A detailed description of discretization and specifics of performing numerical computing and employing grids may be found in Reference [28]. The analysis of results yields the following conclusions:

1. The basic characteristics, the bulk temperature, the interfacial heat flux, and the Nusselt number are calculated by the following expressions:

$$\theta_{bl} = \int_0^1 (y-y^3)\theta\,dy \Big/ \int_0^1 (y-y^3)\,dy, \quad q_{wi} = -\frac{\partial\theta}{\partial y}\Big|_{y=1}, \quad \mathrm{Nu} = \frac{2q_{wi}}{\theta_{wi}-\theta_{bl}} \tag{7.45}$$

These characteristics depend on four dimensionless parameters: the ratio of wall-to-fluid conductivity λ_w/λ, the ratio of radii R_o/R_i, Peclet number $Pe = 2R_i u_e/\alpha$, and the ratio of wall-to-fluid diffusivity α_w/α.

During the initial transient time ($t < 0.016$), the interfacial heat flux increases stepwise from zero to the maximum value and then monotonically decreases to its steady-state value. Because at this period the heat is basically transported by radial conduction in the wall, the fluid temperature increases less than the interfacial temperature. Hence, a large wall–fluid temperature difference forms that results in a rapid increase in fluid temperature. For $t < 0.016$, q_{wi} is relatively uniform across the heated section, but near the entrance and the exit of the heated section, the interfacial heat flux peaks are observed. After reaching the maximum, q_{wi} decreases until it becomes steady-state conditions. Analysis shows that steady state is reached more quickly near the exit end than at other positions of the heated section. The negative values of q_{wi} appear in the downstream region close to the heated section and in the upstream region close to the heated section but for the smaller α_w/α.

2. A decrease in thermal resistance in the wall that corresponds to large values of λ_w/λ leads to higher interfacial temperatures. In the upstream region of the heated section, both the interfacial heat and preheating increase as the conductivity ratio grows.

3. Thinner wall thicknesses correspond to smaller resistances and energy storage of the walls; hence, the energy from the outside surface to the fluid is easier to transport. This causes a rapid response of the interfacial heat flux and a shorter time for reaching the steady state. The time to reach the steady state becomes shorter also by decreasing the conductivity ratio. For the small thickness after the early period, the convective effect becomes dominant, but for a thicker wall, the conduction remains dominant. Reverse heat transfer is observed earlier downstream of the heated section for a smaller value of radii ratio.

4. A lower thermal diffusivity ratio corresponds to a greater wall capacity and, hence, delays the increase of interfacial temperature, causing an increase of the time required to reach the steady state. In the early time period, the response of q_{wi} is faster for the larger diffusivity ratio. Later, the convective heat transfer becomes dominant, and its effect increases with decreasing α_w/α.

5. At larger Pe, the greater convection results in faster heat transportation from the wall to the fluid, and a decrease in Pe decreases the interfacial heat flux. According to the computation results, the preheating length in the upstream region decreases with an increase of the Peclet number. Conversely, the postheated length downstream of the directly heated portion is increased with Pe. The results show also that reverse heat transfer takes place earlier for larger values of Pe. The time required to reach the steady state is shorter for increasing Peclet numbers.

The same conjugate problem with different conditions at the outer surface of the pipe or channel is considered in three other papers. In Reference [29], the uniform heat flux is applied to the outer pipe surface instead of the uniform temperature applied in Reference [30]. In Reference [31], the case of stepwise temperature variation on the outer pipe surface is studied, and the case of stepwise ambient temperature change on the outer wall channel surface is considered. The governing equations, solution approach, and basic conclusions in these papers are almost the same as those found in Reference [28].

Example 7.8: Unsteady Heat Transfer in a Duct with Convection from the Ambient [32]

A fluid flowing inside the duct with a steady laminar, fully developed velocity profile is at initial temperature T_i when suddenly the outside of the duct walls are exposed to an ambient fluid at temperature T_0 and constant surface coefficient h_0. Under usual assumption, the governing equations for fluid and thin walls are

$$\frac{\partial\theta}{\partial t}+3\left(y-\frac{y^3}{2}\right)\frac{\partial\theta}{\partial x}=\frac{\partial^2\theta}{\partial y^2}, \quad \frac{h_0 R}{\lambda_0}(1-\theta)=-\frac{\partial\theta}{\partial y}+\frac{\Delta}{Lu}\frac{\partial\theta}{\partial t} \tag{7.46}$$

$$t=0, \; x=0, \; \theta=0, \; y=1, \; \frac{\partial\theta}{\partial y}=0 \tag{7.47}$$

Dimensionless variables are scaled by half channel height R for y and Δ, $R^2 u_m/\alpha$ for x, R^2/α for t. $\theta=(T-T_i)/(T_0-T_i)$, and $Lu=\rho c_p/\rho_w c_{pw}$ is the Luikov number (see Section 6.6.1).

The problem is solved by using the finite-difference method. Details of the numerical scheme and proof of consistency of the finite-difference energy balance for the wall and boundary conditions with the partial differential equation for the fluid may be found in Reference [32]. To control the accuracy of the numerical program, some comparisons with available analytical solutions are made.

The following results are obtained:

1. Comparison of the results obtained for the limiting case $Lu \to \infty$ and for a large but finite value of Luikov number shows that the agreement between these two results at fixed magnitude of Lu depends on the value of parameter $h_0 R/\lambda_0$, which determines the value of the minimum time when the approximation $Lu \to \infty$ is acceptable. As the parameter $h_0 R/\lambda_0$ increases, the values of the minimum time become smaller. Thus, for $h_0 R/\lambda_0 = 10$, the approximation is acceptable for $t > 0.2$, and for $h_0 R/\lambda_0 = 2$, it becomes acceptable only for times $t > 0.5$.
2. Both wall and fluid bulk temperatures increase monotonically to their state distribution in x for all values of Lu and $h_0 R/\lambda_0$. The greater the Luikov number is, the shorter is the time required to

reach the steady state. For a fixed value of Lu, this time is shorter for the larger values of parameter $h_0 R / \lambda_0$.

3. The behavior of the surface heat flux is strongly dependent on the value of Lu and is significantly different for Lu $\to \infty$ and for any other finite value of Lu. For Lu $\to \infty$, the surface heat flux increases stepwise to its maximum value right at $t = 0$ and then monotonically decreases to its steady-state distribution. On the other hand, for the finite values of Lu, heat flux is zero at the beginning and then exhibits quite complicated behavior. At some values of Lu, $h_0 R / \lambda_0$, and x, the heat flux increases from an initial value of zero to its maximum and then decreases below the steady-state value and finally approaches the state value from below. Such behavior occurs in the second time domain after the transient period. This effect was also observed in some earlier works, and results are compared with data obtained in Reference [32].

4. Higher values of parameter $h_0 R / \lambda_0$ lead to higher values of both wall and bulk temperatures for all Lu, t, and x. The heat flux is greater at higher values of this parameter and at lower values of time. For larger values of time, heat flux is greater for lower x but is lower at the larger x.

5. Comparison of data obtained for the conjugate model for Lu $\to \infty$ and a simple nonconjugate approach, when it is used as a heat transfer coefficient of an isothermal wall, shows that large errors could result from such a simple calculation. The analysis presented in Reference [32] indicates that the results of simple and numerical approaches would get closer as the parameter $h_0 R / \lambda_0$ or Luikov number become smaller, as time gets larger in the first time domain, or as x gets larger in the second time domain.

Example 7.9: Heat Transfer in a Pipe with Developing Hydrodynamically Steady and Thermally Unsteady Laminar Flows [33]

Under the usual assumption, the governing equations for the fluid and the wall and thermal boundary conditions are ($t = 0, x \geq 0, t > 0$):

$$\frac{\partial u}{\partial x} + \frac{1}{r}\frac{\partial rv}{\partial r} = 0 \quad u\frac{\partial u}{\partial x} + v\frac{\partial u}{\partial r} = -\frac{\partial p}{\partial x} + \frac{1}{r}\frac{\partial}{\partial r}\left(r\frac{\partial u}{\partial r}\right)$$

$$\frac{\partial \theta}{\partial t} + u\frac{\partial \theta}{\partial x} + v\frac{\partial \theta}{\partial r} = \frac{1}{r\,\mathrm{Pr}}\frac{\partial}{\partial r}\left(r\frac{\partial \theta}{\partial r}\right)$$

$$(7.48)$$

$$x = 0, \ \theta = \theta_s = 1, \qquad \frac{\partial \theta}{\partial t} = \frac{\lambda_w}{r\lambda\,\mathrm{Pr}}\frac{\partial}{\partial r}\left(r\frac{\partial \theta}{\partial r}\right)$$

$$(7.49)$$

$$r = r_o, \ \theta_s = 0, \ \theta = \frac{T - T_w}{T_i - T_w}$$

Others are the boundary and conjugate conditions for a wall–fluid interface, far away from body and symmetry conditions. The dimensionless variables are scaled by inner radius R for r, $R\,\mathrm{Re}/2$ for x, u_m for u, v/R for v, ρu_m^2 for p, R^2/v for t, $\lambda(T_i - T_w)/R$ for q_w.

The problem is solved by using the finite-difference method. Because the wall and fluid physical properties are constant, the velocity and temperature equations are uncoupled. The first is obtained velocity field which is used in solving the equations for the fluid temperature field. This temperature field is conjugated with the temperature field of the body to get the conjugate problem solution. To obtain finally the fully developed velocity profile and steady-state temperature condition, the following steps in computing are used $\Delta x = 0.0001, \Delta r = 0.005, \Delta t = 10^{-5}$. Other details of the numerical scheme are given in Reference [34].

The calculations are performed for $\mathrm{Pr} = 0.7$ and different values of $\lambda_w/\lambda, \alpha_w/\alpha$ and r_o and yields the following basic results:

1. The response of the hydrodynamically developing flow to the variation in the thermal conditions of the outer wall surface is faster than that in the case of the hydrodynamically fully developed flow. This is due to the fact that the amount of the flow in the thermal boundary layer for the last case is less than that for the former case. As a result, the temperature distribution in the fully developed flow is broader, and the heat transfer coefficient is lower. The deviation of these two cases increases with time.

2. Increasing the thermal conductivity ratio λ_w/λ leads to enhancing the wall heat flux because increasing the wall conductivity intensifies the fluid response to any variation in the thermal boundary condition.

3. The thermal entrance length required for reaching the fully developed temperature profile increases as the thermal conductivity ratio λ_w/λ or the thermal diffusivity ratio α_w/α increases or the ratio of outer to inner radii decreases.

OTHER WORKS: The reviews of heat transfer studies in pipes and channels including the solutions of conjugate problems may be found in Reference [20] for relatively early results, in Reference [31] for papers published until the end of the last century and in Reference [26] or [14] for more recent publications. Some conjugate heat transfer problems different from those discussed above may be mentioned. The conjugate heat transfer in compressible flows is investigated in Reference [16]. The simultaneously developing velocity and temperature fields in a rectangular channel are studied in Reference [35]; the conjugate heat transfer from radiating fluid in a rectangular channel is considered in Reference [36]; the conjugate heat transfer by natural convection in enclosures with openings and in a square cavity are studied in References [37] and [38], respectively; and the measurement results of water flow boiling conjugate heat transfer in a channel are presented in Reference [39].

7.6 Examples of Numerical Studies of the Conjugate Convective Heat Transfer in Flows around and inside Bodies

Example 7.10: Heat Transfer from a Rectangular Slab in Incompressible Flow [40]

Because the slab is of finite length and height, the problem is governed by Navier-Stokes, elliptic energy equations, and boundary conditions:

$$\nabla^2\psi = -\omega, \quad u\frac{\partial\omega}{\partial x} + v\frac{\partial\omega}{\partial y} = \frac{1}{Re}\nabla^2\omega, \tag{7.50}$$

$$u\frac{\partial\theta}{\partial x} + v\frac{\partial\theta}{\partial y} = \frac{1}{Re}\nabla^2\theta, \quad \nabla^2\theta_s = 0$$

$$y = 0, \quad |x| > \frac{1}{2}, \quad \frac{\partial\theta}{\partial y} = 0, \quad x = \pm\frac{1}{2}, \quad -H \le y \le 0, \quad \frac{\partial\theta_s}{\partial x} = 0 \; y = -H, \tag{7.51}$$

$$|x| \le \frac{1}{2}, \quad \theta_{bt} = 1, \quad y \to \infty, \quad |x| \to \infty, \quad \omega \to 0$$

Here ψ and ω are the stream function and vorticity. Boundary conditions stand: the first for uniform temperature (similar condition for uniform velocity $\partial^2\psi/\partial y^2 = 0$) before and behind the slab, the second for insulated side surfaces of the slab, the third for temperature θ_{bt} of the bottom slab surface, and the fourth for irrotational flow. Others are the usual boundary and conjugate conditions on the slab–fluid interface, and far away from the body for a fluid. The dimensionless variables are scaled by slab length L for x, y and slab height H, U_∞ for u, v, LU_∞ for $\psi, U_\infty/L$. for ω and $\theta = (T - T_\infty)/(T_{bt} - T_\infty)$.

The problem is solved by using the finite-difference method of Patankar [2] (Section 7.3). Because Equation (7.50) is elliptic, boundary conditions at infinity before and behind the slab should be specified. The estimations for ψ and ω are adopted from Reference [41], while for θ these are derived employing the balance between the heat lost at the slab surface and that transported downstream. The results are

$$\theta = \omega = 0, \psi \quad y + C_d[-1 + (1/\pi)]/\tan(y/x) \text{ for } x = -x_\infty, \; 0 \le y \le y_\infty \text{ and}$$

$$y = y_\infty, -x_\infty \le x \le 0 \; \theta \sim Cx^{-1/2}\exp(-Pey^2/4x), \psi \sim y + C_d[1/\pi]\tan(y/x)$$

$$-erf(Re^{1/2}\,y/2x^{1/2}), \omega \sim C_d[Re^{3/2}\,y/(\pi x)^{1/2}] \times \exp(-Rey^2/4x) \text{ for } x \tag{7.52}$$

$$= x_\infty, \; 0 \le y \le y_\infty \text{ and } y = y_\infty, \; 0 \le x \le x_\infty$$

where $2C_d$ and C are a drag coefficient and constant as defined below:

$$C_d = -\frac{1}{Re}\int_{-1/2}^{1/2}\omega(x,0)dx, \quad C = \frac{1}{(\pi Pe)^{1/2}}\int_{-1/2}^{1/2}\frac{\partial\theta}{\partial y}(x,0)dx \tag{7.53}$$

Computations are performed for $10^2 \leq \mathrm{Re} \leq 10^4$, $\mathrm{Pr} = 10^{-2}$, 10^2, $\lambda_w/\lambda = 1, 2, 5, 20$, and aspect ratio $H = 0.25$ and 1. The basic results are as follows:

1. The isotherms show that at lower $\mathrm{Pr} = 10^{-2}$ and lower ratio λ_w/λ, the temperature drop across the slab is greater than that in the case of a higher value of λ_w/λ. They indicate also that there is a kink at the boundary between fluid and slab which occurs due to a high slab conductivity. In the case of high $\mathrm{Pr} = 10^2$, the temperature drop across the slab exists for both considered values of $\lambda_w/\lambda = 1$ and 20.

2. The numerical solutions of the Navier-Stokes equation for low $\mathrm{Pr} = 10^{-2}$, $\mathrm{Re} = 5 \cdot 10^2$, 10^4, $H = 0.25, 1$ and all considered values of $\lambda_w/\lambda = 1, 2, 5, 20$ indicate that high local Nusselt number at the left-hand end of the boundary first decreases monotonically along most of the slab length and then increases considerably by reaching the other end of the slab. At the large $\mathrm{Pr} = 10^2$ and the same other parameters, the computations show that Nusselt number is approximately constant along the slab surface and increases going to the right-hand end of the slab as well.

3. It is clear from the results obtained for the low and large Prandtl numbers and the same combinations of other parameters that the increase in Peclet number $\mathrm{Pe} = \mathrm{RePr}$ or in aspect ratio H decreases the boundary temperature, while an increase in the conductivity ratio λ_w/λ leads to an increase of boundary temperature.

4. The dependences $\overline{\mathrm{Nu}}(\mathrm{Re})$ indicate that the average Nusselt number increases as usual with increasing Reynolds and Prandtl numbers and with increasing the aspect ratio and thermal conductivity ratio.

5. The numerical finite-difference results are compared for the case of $\mathrm{Re} \gg 1$ with data obtained by two analytical approximate methods. One of these approaches, an averaging method, gives good results for the mean Nusselt number and average boundary temperature at $\lambda_w/\lambda \leq 5$ and $\mathrm{Pr} \gg 1$. In particular, the comparison with numerical data shows that the averaged Nusselt and boundary temperature can be determined using two simple formulae depending on only two and one parameter, respectively:

$$\overline{\mathrm{Nu}} = \frac{(\lambda_w/\lambda H)B}{1+B}, \quad \overline{\theta}_w = \frac{1}{1+B}, \quad B = \frac{H\,\mathrm{Pr}^{1/3}\,\mathrm{Re}^{1/2}}{1.596(\lambda_w/\lambda)} \tag{7.54}$$

A second method is derived considering two-dimensional conduction in a slab and convection in an adjacent boundary layer. According to this method for $\mathrm{Pr} \geq 1$, the set of governing parameters is $(\mathrm{Re}^{1/2}\,\lambda/\lambda_w, \mathrm{Pr}, H)$, and for $\mathrm{Pr} \ll 1$, this set is $(\mathrm{Pe}^{1/2}\lambda/\lambda_w, H)$. However, applicability of the second method is restricted, as there are some difficulties in applying this method for $\mathrm{Pr} \gg 1$.

Example 7.11: Heat Transfer in an Inclined Open Shallow Cavity [42]

The model is presented as an incline by the angle φ two-dimensional opening reservoir with a thick facing wall heated by a constant heat flux q. The other perpendicular to the heated wall sides are adiabatic.

Usual assumptions including the Boussinesq approximation are used. The system of governing equations and boundary conditions are is follows:

$$\frac{\partial u}{\partial x}+\frac{\partial v}{\partial y}=0, \quad \frac{\partial u}{\partial t}+u\frac{\partial u}{\partial x}+v\frac{\partial u}{\partial y}=-\frac{\partial p}{\partial x}+\mathrm{Ra}\,\mathrm{Pr}\,\theta\cos\varphi+\mathrm{Pr}\,\nabla^2 u,$$

$$\frac{\partial\theta}{\partial t}+u\frac{\partial\theta}{\partial x}+v\frac{\partial\theta}{\partial y}=\nabla^2\theta$$

$$\frac{\partial v}{\partial t}+u\frac{\partial v}{\partial x}+v\frac{\partial v}{\partial y}=-\frac{\partial p}{\partial y}+\mathrm{Ra}\,\mathrm{Pr}\,\theta\sin\varphi+\mathrm{Pr}\,\nabla^2 v$$

$$p=0, \quad \frac{\partial v}{\partial x}=0, \quad \frac{\partial u}{\partial x}=-\frac{\partial v}{\partial y}, \quad \left(\frac{\partial\theta}{\partial x}\right)_{out}=\theta_{in}=0$$

(7.55)

These conditions should be satisfied at the opening walls. The other conditions are as follows: zero velocities on the walls and zero derivatives of the temperature on the adiabatic surfaces. The subscripts "out" and "in" denote outside and into the cavity. Here the dimensionless variables are scaled by L for x and y, L/α for u and v, $L^2/\rho\alpha^2$ for p, Lq/λ for T, α/L^2 for t, L and H are cavity width and height, and $\mathrm{Ra}=g\beta qL^4/\nu\alpha\lambda$.

The system of governing equations is solved using the SIMPLER algorithm developed in Reference [2] and briefly described in Section 7.3. The details of domain discretization may be found in Reference [42]. Special computations for extended domain with different parameters are performed to verify the computational technique for considering a restricted cavity. The calculations are performed for the Rayleigh number from 10^6 to 10^{12}, aspect ratio from 1 to 0.125 and angle φ from 90° (heated vertical wall) to 45° and $\mathrm{Pr}=0.72$. The basic results are formulated as follows:

1. The isotherms show that at low Rayleigh numbers of 10^6, there is a conduction dominated regime. As Ra is increased to 10^{10}, the flow becomes close to boundary layer type and at $\mathrm{Ra}=10^{12}$, the heat transfer and flow finally turn to boundary layer type, and the convection becomes the dominant regime.
2. The nondimensional temperature profiles on the wall and at the opening follow a nonlinear variation, a little lower near the bottom and higher at the top. The cold fluid entering the cavity is heated along the wall and cools it. As the heated fluid moves along the wall, the wall temperature increases. The temperature profile at the opening shows that the temperature is lower at the bottom portion and the hot fluid is at the remaining upper portion. The outgoing flow occupies only 30% of the opening.
3. Normalized Nusselt number and nondimensional volume flow rate in and out of the cavity $\dot V$ were calculated using the following formulae:

$$Nu = \frac{\overline{Nu}_{Ra}}{\overline{Nu}_{Ra=0}}, \quad \overline{Nu} = -\int_0^A \frac{\partial \theta}{\partial x} dy / \int_0^A (\theta_1 - \theta_2) dy,$$

$$\dot{V} = -\int_{x=1} u_{in} dy, \quad u_{in} = u_{x=1} \quad \text{if } u_{x=1} < 0, \quad u_{in} = 0 \quad \text{if } u_{x=1} \geq 0$$

(7.56)

where indices 1 and 2 indicate the section at the facing wall and opening, respectively. The results show that volumetric flow rate is an increasing function of Ra and aspect ratio $A = H/L$. At a given Ra, the shallower the cavity, the smaller is the volumetric flow rate. As the aspect ratio increases, the volumetric flow rate tends to an asymptote $A \to \infty$.

For shallower cavities (e.g., for a cavity with $A = 0.125$) the conduction regime dominates even at high Ra, but at Ra = 10^9, the transition to the convection regime occurs. Up to this value of Ra, the Nusselt number is an increasing function of A, but then a mixed pattern emerges that remains until Ra = $2 \cdot 10^{11}$. The reason for this is that heat flux is transferred partly by conduction through the wall and partly by convection through the opening. When the parameters vary, these parts of heat flux are varied in such a way that the function Nu = $f(A)$ in the range of Ra = $10^6 - 10^{12}$ becomes a type of mixed pattern. Detailed analysis of this phenomenon is given in Reference [42].

The volumetric flow rate and the Nusselt number are similarly affected by relative facing wall thickness Δ / L. Both $\dot{V}$ and Nu are the increasing functions of wall thickness at all Rayleigh numbers. These characteristics are also similarly affected by the conductivity ratio λ_w / λ. Both $\dot{V}$ and Nu decrease when the conductivity ratio grows. The reason for this effect is also a variation of the heat parts transferring through the wall and opening.

4. Isotherms and streamlines for different inclination angles of the range $\varphi = 90° - 45°$ show that as φ decreases, the velocity gradient increases at the upper adiabatic wall and the strength of circulation increases. Flow of the hot fluid from the opening becomes choked as the inclination decreases. As a result, the volumetric flow rate and Nusselt number increase as the angle φ decreases. At low Rayleigh numbers, the increase is small, but as Rayleigh numbers increase, the increase in flow rate becomes appreciable.

Example 7.12: Free Convection on Thin Vertical and Horizontal Plates [43]

The problem is governed by the following set of equations:

$$u \frac{\partial u}{\partial x} + v \frac{\partial u}{\partial y} = -\frac{1}{\rho} \frac{\partial p}{\partial x} + \frac{\partial^2 u}{\partial y^2} + g\beta(T - T_\infty) \sin \varphi,$$

$$-\frac{\partial p}{\partial y} + g\beta(T - T_\infty) \cos \varphi = 0, \quad u \frac{\partial T}{\partial x} + v \frac{\partial T}{\partial y} = \frac{\partial^2 T}{\partial y^2}$$

(7.57)

supplemented by the continuity differential equation and common
conjugate and boundary conditions on the fluid–plate interface and far
away from the plate. For a horizontal plate $\varphi = 0$, and for a vertical plate
$\varphi = \pi/2$ and pressure gradients in both coordinates are equal to zero.
Using the conjugate condition for a thin plate with linear temperature
distribution and scale analysis leads to the conjugate criterion in the
following form:

$$-\lambda\left(\frac{\partial T}{\partial y}\right)_{y=0} = \frac{\lambda_w(T_0 - T_w)}{\Delta}, \qquad \frac{T_0 - T_w}{T_w - T_\infty} \sim \frac{\lambda\Delta}{\lambda_w\delta} = \zeta \tag{7.58}$$

where T_0 is the temperature of the outer plate surface. Because the heat
transfer coefficient $h \sim \lambda/\delta$, it is seen from the last equation that this
criterion, like the others, in essence is the Biot number, as it should be
according to general theory (Section 6.1).

It is assumed that the conjugate heat transfer may be investigated by
employing a combination of two well-known limiting cases of constant
wall temperature and constant heat flux. Applying this idea and formulae for boundary layer thickness in the limiting cases of free convection
on a vertical plate $\delta_T \sim x(\sigma \mathrm{Ra}_T)^{-1/4}$ and $\delta_q \sim x(\sigma \mathrm{Ra}_q)^{-1/5}$ and $\sigma = \mathrm{Pr}/(1+\mathrm{Pr})$,
the thickness for the conjugate free convection on a vertical plate is
defined as follows:

$$\delta \sim \left|\delta_T^4 + \delta_q^4\right| \sim x/\gamma, \gamma = \left[(\sigma \mathrm{Ra}_T)^{-1} + (\sigma \mathrm{Ra}_q)^{-4/5}\right]^{-1/4},$$

$$\mathrm{Ra}_T = g\beta(T_q - T_\infty)x^3/\alpha\nu, \frac{\mathrm{Ra}_T}{\mathrm{Ra}_q} = \lambda\Delta/\lambda_w x \tag{7.59}$$

Introducing new dependent and independent variables

$$f = \psi/\alpha\gamma, \quad \theta = [(T - T_\infty)/(T_0 - T_\infty)]\xi^{-1}$$

$$\xi = \left[1 + \sigma \mathrm{Ra}_T/(\sigma \mathrm{Ra}_q)^{4/5}\right]^{-1}, \quad \eta = y/\delta = (y/x)\gamma \tag{7.60}$$

transforms the system of governing equations and boundary condition
into the form

$$\mathrm{Pr}\, f''' + \frac{16-\xi}{20} ff'' - \frac{6-\xi}{10} f'^2 + (1+\mathrm{Pr})\theta = \frac{1}{5}\xi(1-\xi)\left[f'\frac{\partial f'}{\partial \xi} - f''\frac{\partial f}{\partial \xi}\right],$$

$$f(\xi,0) = f'(\xi,0) = f(\xi,\infty) = 0 \tag{7.61}$$

$$\theta'' + \frac{16-\xi}{20} f\theta' - \frac{1-\xi}{5} f'\theta = \frac{1}{5}\xi(1-\xi)\left[f'\frac{\partial \theta}{\partial \xi} - \theta'\frac{\partial f}{\partial \xi}\right],$$

$$\xi\theta(\xi,0) - (1-\xi)^{5/4}\theta'(\xi,0) = 1, \theta(\xi,\infty) = 0$$

A similar consideration leads to analogous results for the horizontal plate:

$$\Pr f''' + \frac{10-\xi}{15} f'' - \frac{5-2\xi}{15} f'^2 + \frac{1}{15}(1+\Pr)[(5+\xi)\eta p' - (10-4\xi)p]$$

$$= \frac{1}{3}\xi(1-\xi)\left[f'\frac{\partial f'}{\partial \xi} - f''\frac{\partial f}{\partial \xi} + (1+\Pr)\frac{\partial p}{\partial \xi} \right], \quad \theta = p', \qquad (7.62)$$

$$\theta'' + \frac{10-\xi}{15} f\theta' - \frac{1-\xi}{3} f'\theta = \frac{1}{3}\xi(1-\xi)\left[f'\frac{\partial \theta}{\partial \xi} - \theta'\frac{\partial f}{\partial \xi} \right]$$

with almost the same variables and boundary conditions, additional dimensionless pressure scaled by $\rho a v \gamma^4 / \sigma x^2$, and the following changes:

$$\gamma = \left[(\sigma \mathrm{Ra}_T)^{-1} + (\sigma \mathrm{Ra}_q)^{-5/6} \right]^{-1/5}, \quad \xi = \left[1 + \sigma \mathrm{Ra}_T / (\sigma \mathrm{Ra}_q)^{5/6} \right]^{-1}, \qquad (7.63)$$

$$\xi\theta(\xi,0) - (1-\xi)^{6/5}\theta'(\xi,0) = 1, \quad p(\xi,\infty) = 0$$

The set of these equations are solved by Keller's finite-difference method, details of which may be found in References [43] and [44]. The following results are obtained:

1. The profiles of dimensionless velocities and interface temperatures at $\Pr = 0.7$ for both vertical and horizontal plates show that the maximum value of velocity decreases while the interface temperature increases from T_∞ to T_0 as the coordinate ξ increases from 0 to 1. It is also seen that both sets of profiles develop from the typical profiles of constant wall heat flux at $\xi = 0$ to these for an isothermal wall at $\xi = 1$. The calculations also indicate that the interface temperature and friction increase as the conjugation criterion ξ or local Biot number decreases.

2. The heat transfer rate q_w / q_{wq} decreases from one asymptotical value of constant wall heat flux to another asymptotical value of constant wall temperature as the conjugation criterion decreases from $\zeta \to \infty$ to $\zeta \to 0$. This occurs due to the growing thermal boundary layer thickness.

3. Comparison of the numerical data with well-known results for constant wall temperature or heat flux shows that the ranges of conjugation criteria in which the problem should be treated as conjugate depend on the Prandtl number. The analysis of such data presented in this paper reveals that this dependence is slight. Thus, these ranges vary from $\zeta = 0.087 - 25.5$ to $\zeta = 0.102 - 30.8$ for vertical plate and from $\zeta = 0.146 - 23.5$ to $\zeta = 0.159 - 30.8$ for horizontal plate for the entire diapason of $\Pr = 0.001 - \infty$. For other values of conjugation parameter ζ, the heat transfer problem can be solved using a simple approach with error less than 5%.

4. The local Nusselt number on the interface in the form Nu/γ decreases almost linearly with increasing variable ξ. This yields

simple correlations for vertical and horizontal plates. Knowing the local Nusselt number, one determines the local temperature on the interface. Then the local heat transfer rate can be obtained by $q_w = \text{Nu}(\lambda/x)(T_0 - T_\infty)\xi\theta$:

$$\text{Nu}/\gamma = \xi\left[\text{Nu}_T/(\sigma\text{Ra}_T)^{1/4}\right] + (1-\xi)\left[\text{Nu}_q/(\sigma\text{Ra}_q)^{1/5}\right]$$

$$= [1-\xi\theta]/(1-\xi)^{5/4}\theta$$

$$\text{Nu}/\gamma = \xi\left[\text{Nu}_T/(\sigma\text{Ra}_T)^{1/5}\right] + (1-\xi)\left[\text{Nu}_q/(\sigma\text{Ra}_q)^{1/6}\right] \tag{7.64}$$

$$= [1-\xi\theta]/(1-\xi)^{6/5}\theta$$

Example 7.13: Heat Transfer between an Elliptical Cylinder and a Laminar Flow [23]

This is an example of using the method of reducing a conjugate problem to an equivalent conduction problem in the case of flow with nonzero pressure gradient (Sections 6.2 and 6.3). As shown in Chapter 3 in the case of a nonzero pressure gradient, the role of the longitudinal coordinate is as the Görtler's variable so that in Duhamel's integral (Equation 1.38) the variable Φ should be substituted for x. Then, as well as with Equation (1.38), a conjugate problem is reduced to a conduction problem with an integral

$$q_w = h_*\left[\theta_w(0) + \int_0^\Phi \left[1-\left(\frac{\xi}{\Phi}\right)^{C_1}\right]^{-C_2} \frac{d\theta_w}{d\xi}\,d\xi\right] \tag{7.65}$$

as a boundary condition. Exponents C_1 and C_2 are given in Figure 3.4.

The analytical solution of this problem is considered in Example 6.7. The conduction problem for an elliptical cylinder with the uniform heat generation q_v is governed in elliptic coordinates (u,v) by equation and boundary conditions (Equation 6.34). The surface heat flux q_w incorporated in the first boundary conditions is determined by Equation (7.65). The numerical solution is performed by finite-difference method with grid nodes $u_{i,j} = (i-0.5)\delta_i$ and $v_{i,j} = (j-0.5)\delta_j$, where $\delta_i = (1/2M)\ln[(a+b)/(a-b)]$, $0 \le i \le M+1$, $\delta_j = \pi/N$, $0 \le j \le N+1$. The last three boundary conditions (Equation 6.32) are presented using node temperatures as $\theta_{i,0} = \theta_{i,1}, \theta_{0,j} = \theta_{1,j}, \theta_{i,N+1} = \theta_{i,N}$.

Let the vector $\boldsymbol{\theta}_i$ consist of temperatures $\theta_{i,1}, \theta_{i,2}, \theta_{i,3}, ..., \theta_{i,N-1}, \theta_{i,N}$ and accordingly the vector $\mathbf{F}_i$ is made up of $f_{i,1}\delta_i^2, f_{i,2}\delta_i^2 \cdot\cdot f_{i,N-1}\delta_i^2, f_{i,N}\delta_i^2$, where $f_{i,j} = -(q_v c^2/\lambda_w T_\infty)(sn^2 u_{i,j} + \sin^2 v_{i,j})$ is the right-hand part of Equation (6.34). The finite-difference solution of Equation (6.34) subjected to the last three boundary conditions may be presented in the vector-matrix form as $\boldsymbol{\theta}_{i+1} + A_i\boldsymbol{\theta}_i + \boldsymbol{\theta}_{i-1} = \mathbf{F}_i$, where $(i = 1, 2, ..., M)$ and a diagonal matrix A_i given in Reference [23] depend on the ratio $(\delta_i/\delta_j)^2$.

The first boundary condition (Equation 6.34) that contains the integral (Equation 7.65) is modified using Equations (6.37) for variables Φ and ε to obtain

$$-\frac{\partial \theta}{\partial u}\bigg|_{u=\frac{1}{2}\ln\frac{a+b}{a-b}} = \frac{\mathrm{Nu}_* \mathrm{Bi}}{\sqrt{\mathrm{Re}}}\sqrt{1-\frac{c^2}{a^2}\cos^2 v}$$

$$\times\left\{\int_0^v \left[1-\left(\frac{1-\cos\varepsilon}{1-\cos v}\right)^{C_1}\right]^{-C_2}\frac{d\theta_w}{d\varepsilon}d\varepsilon+\theta_w(0)\right\}, \quad \mathrm{Bi}=\frac{\lambda\sqrt{\mathrm{Re}}}{\lambda_w} \tag{7.66}$$

Because this condition should be satisfied between the temperature θ_{M+1} and θ_M, an arithmetic mean temperature is applied to approximate it (Equation 7.66). Assuming also that between nodes the temperature changes linearly giving the following form for condition (7.66):

$$\theta_{M,1}\left(\frac{\beta_{2,j}}{\delta_j}-1\right)\frac{\sigma_j}{2}+\theta_{M,2}(\beta_{3,j}-\beta_{2,j})\frac{\sigma_j}{2\delta_j}+\ldots+\theta_{M,j}\left(\frac{1}{\delta_i}-\frac{\beta_{j,j}\sigma_j}{2\delta_j}\right)$$

$$=\theta_{M+1,1}\left(1-\frac{\beta_{2,j}}{\delta_j}\right)\frac{\sigma_j}{2}+\theta_{M+1,2}(\beta_{2,j}-\beta_{3,j})\frac{\sigma_j}{2\delta_j}+\ldots+\theta_{M+!,j}\left(\frac{1}{\delta_i}+\frac{\beta_{j,j}\sigma_j}{2\delta_j}\right) \tag{7.67}$$

$$\sigma=\frac{\mathrm{Nu}_*(v)\mathrm{Bi}}{\sqrt{\mathrm{Re}}}\sqrt{1-\frac{c^2}{a^2}\cos^2 v}\quad \beta_{mn}=\int_m^n\left\{1-\left\{\frac{1-\cos\varepsilon}{1-\cos[(j-0.5)\delta_j]}\right\}^{C_1}\right\}^{-C_2}d\varepsilon \tag{7.68}$$

Here $\beta_{2,j},\beta_{3,j},\beta_{j,j}$ are integrals with limits: from $m=0.5\delta_j,1.5\delta_j,\ (j-0.5)\delta_j$ to $n=1.5\delta_j,\ 2.5\delta_j,\ 1+(j-0.5)\delta_j$, respectively. The last equation is valid at each $j>1$. At $j=1$ due to relation $\theta_{i,0}=\theta_{i,1}$, a simpler relation $\theta_{M,1}[(1/\delta_i)-(\sigma/2)]=\theta_{M+1,1}[(1/\delta_i)-(\sigma/2)]$ is valid.

The last three equations may also be written in the vector-matrix form. Then, the numerical solution of the problem in question is reduced to the following system of vector equations containing three matrices: A_i, R_M and R_{M+1}.

$$\theta_{i+1}+A_i\theta_i+\theta_{i-1}=\mathbf{F}_i \quad \theta_0=\theta_1 \quad R_M\theta_M=R_{M+1}\theta_{M+1}$$

These matrices as well as additional details of derivation of some equations are given by Davydenko [23]. The vector-matrix system has been solved by TDMA (*progonka* in Russian terminology). The calculations have been performed for $a/b=2,4,8$, Bi $=0.1,1,10$ and Pr $=0.1,1$. The numerical results differ about 1% from those obtained analytically and discussed in Example 6.7.

Example 7.14: Heat Transfer from a Sphere Heated by an X-Ray or Laser in External Flow at a Low Reynolds Number [45]

A sphere cooled by a uniform stream of air is heated by X-ray. Two cases are considered: the first when the sphere absorbs the full energy of the oncoming X-ray beam and the other when the energy of the X-ray beam is absorbed only near the sphere's surface. The first process simulates heating a particle by intense X-ray, which is used in crystallographic applications. The other case is similar to laser heating applied in microfabrication and micromachining.

The three-dimensional domain and full system for the steady state of continuity, Navier-Stokes and energy equations for fluid and Poisson's equation with heat source $S = (I/L_{att})\exp(-L/L_{att})$ for a sphere are used. Here I is an incident intensity of the beam, L is the depth traveled by the beam through the sphere, and L_{att} is the beam attenuation length depending on the beam energy level (see formula for S). This system of equation is solved using boundary and conjugate conditions for the sphere surface and for far away from the sphere for a fluid.

A special parameter $L_{en} = D/L_{att}$, where D is a sphere diameter, is used to determine the beam penetration into the sphere. At high values of $L_{en} \approx 100$, the X-ray energy is absorbed only near the sphere surface. As this parameter decreases, the penetration depth increases, and at $L_{en} = 4.6$, it becomes equal to the sphere diameter. In the case of the higher values of L_{en}, only a part of beam energy is absorbed by the sphere. At $L_{en} = 1$, heat absorption along the centerline decays exponentially, and at $L_{en} = 0.1$, the heat distribution inside the sphere becomes almost uniform.

The size of the beam is another important characteristic. In the case of a full beam, its diameter is equal to the diameter of the sphere. For the focused regime, the beam diameter is kept at 0.1D. Nevertheless, by increasing the incident intensity in the focused regime, total energy in both cases is the same.

The solution is performed iteratively by using the finite-volume element method. The SIMPLE algorithm (Section 7.3) is employed to solve the continuity equation determining the pressure and velocity components. The numerical results are obtained for sphere diameter 2mm, Re = 100, Pr = 0.73, $\lambda_w = 430, 0.6, 0.00882$ W/m K, $L_{en} = 0.1, 1, 100$, and $I = 1.92 \times 10^4$ W/m² (full regime), 1.92×10^6 W/m² (focused regime). The above indicated values of thermal conductivity stand for a typical biocrystal (0.6) and two limited cases such as metal and gas. The basic result and conclusions are as follows:

1. The data for $\lambda_w = 0.6$ and different L_{en} show that the maximum temperature increases considerably and shifts to the front as the parameter L_{en} increases in both full and focused cases. This occurs due to an increase in absorbed energy. These effects are greater for the focused regime, because in this case, about the same amount of energy is absorbed by a much smaller sphere region. As a result, at $L_{en} = 100$, the maximum temperature in this regime is the highest, which decreases with decreasing L_{en} and growing the beam size from focused to full. The temperature variation along

the centerline also depends on L_{en}. The curves $T(x)$ change from sharply falling at $L_{en} = 100$ to almost horizontal at $L_{en} = 0.1$.

2. The same behavior is observed for other values of conductivities 430 and 0.00882 in which cases the temperature increases with L_{en} increasing as well. An exception is the case of full beam at $\lambda_w = 0.00882$ and $L_{en} = 100$, in which the temperature is lower than at $L_{en} = 1$. The reason for this is the high heat transfer rate at the surface and low thermal conductivity of the sphere which result in a significant amount of heat being removed by convection rather than being absorbed by the sphere. The curves $T_{max} - T_\infty = f(\lambda_w)$ indicate that the highest temperature difference $T_{max} - T_\infty = 4,650°C$ takes place for the focused beam, the least sphere conductivity $\lambda_w = 0.00882$, and the maximum $L_{en} = 100$. At the same time, the lowest difference at the least $L_{en} = 0.1$ and the highest conductivity $\lambda_w = 430$ is only about 5°C for both full and focused regimes.

3. The results for the surface sphere temperature in the form $T_{sf} - T_\infty = f(\varphi)$, where φ defines the angular position starting from the stagnation point, show that the variations of this difference in both full and focused cases are different. Although for the highest conductivity this temperature difference is almost constant along the surface at all considered values of L_{en} for both regimes, the temperature changes for two other values of conductivity are more significant in the full beam case. In the focused regime at $\lambda_w = 0.6$ and $\lambda_w = 0.00882$, there are small temperature variations at the front and back of the sphere which increase as the L_{en} increases and the thermal conductivity decreases. As a result, at $L_{en} = 100$, the sphere front is essentially hotter than the other part of the surface with rapidly decreasing temperature toward the back. In contrast, for the full beam case at $L_{en} = 0.1$ and $L_{en} = 1$ and the same two conductivities, the gradual increase to the back minimum temperature difference is observed at the front. The reverse kind of temperature profile is observed at $L_{en} = 100$, where due primarily to deposition of energy, the highest temperature is at the front of the sphere.

4. For the better comparison with obtained result, the local Nusselt number is compared with the uniform sphere temperature. As indicated by the data, for the case of the highest conductivity $\lambda_w = 430$, Nu in both regimes is almost identical with the other reported results for all values L_{en}. This is expected because in this case the real surface temperature is almost uniform. As is well known, in such a situation the Nusselt number is higher in the front half of the sphere surface. For the focused regime at $\lambda_w = 0.6$ and $\lambda_w = 0.00882$, Nu is always higher at the front region as compared with the case of $\lambda_w = 430$ for all values of L_{en}. For the full beam regime, the results are similar, except for the case of $L_{en} = 100$ and $\lambda_w = 430$ and 0.6 where the local Nusselt number is high in the front surface part and decreases along the other portion of the surface. At $\lambda_w = 0.00882$ in this case, Nu reduces sharply after $\varphi > 60°$ and becomes constant thereafter. Thus, in contrast to the focused

regime for the full beam, Nu at the front is nearly the same for all values of other parameters, which tells us that in the case of full beam, the internal heat distribution does not significantly affect the convection at the sphere front.

5. The data for the average Nusselt number indicate that as L_{en} increases, average $\bar{N}u$ decreases for a fixed thermal conductivity. For both regimes, the maximum of an average Nusselt number is at $\lambda_w = 0.6$ and $L_{en} = 0.1$ and at $L_{en} = 100$ it is the lowest.

6. This analysis results in at least two important conclusions applicable for large times when the steady state is achieved. For the laser heating that corresponds to $L_{en} = 100$ and higher, there is a steady peak temperature and total heat transfer rate, except for the sphere with high thermal conductivity. The other result corresponds to the low value as $L_{en} = 0.1$ and less when the heating is performed by the deep-penetrating x-ray. In this case, the maximum internal temperature and heat transfer rate for both focused and full beam regimes can be estimated with reasonable accuracy by using the average Nusselt number, which is not much different as in the usual case of the isothermal temperature, at least for the spheres with thermal conductivity 0.6 and higher.

Example 7.15: Unsteady Radiative-Convective Heat Transfer in a Laminar Compressible Boundary Layer on a Thin Plate [46]

It is considered a gray, absorbing, and scattering airflow on a thin plate with an insulated outer surface and back end. The governing system consists of an averaging across the thickness energy equation for the plate and the steady-state energy equations (quasi-steady approach) for the fluid:

$$\frac{\partial T_w}{\partial t} = \frac{\partial^2 T_w}{\partial x^2} - \frac{1}{\lambda_w \Delta}\left[-\lambda\frac{\partial T}{\partial y} + E\right]_{y=0},$$

$$\rho c_p\left(u\frac{\partial T}{\partial x} + v\frac{\partial T}{\partial y}\right) = \frac{\partial}{\partial y}\left(\lambda\frac{\partial T}{\partial y}\right) - \frac{\partial E}{\partial y} + \mu\left(\frac{\partial u}{\partial y}\right)^2. \tag{7.69}$$

This system is supplemented by no-slip on the plate and by the asymptote far away from the plate conditions. The conjugate conditions and radiation flux are defined as

$$-\lambda_w\frac{\partial T_s}{\partial y}\bigg|_{y=0} = \lambda\frac{\partial T}{\partial y}\bigg|_{y=0} + E \quad E = 2\pi\int_{-1}^{1} I(y,\gamma)\gamma d\gamma \tag{7.70}$$

Using the dimensionless variables $\theta = T/T_\infty$, $\xi = x/L$, the dimensionless time scaled by L^2/α_w (the Fo number) and Blasius velocity profile $f(\eta)$, the equations for the fluid, the plate, and boundary condition are transformed to the forms

$$\frac{\partial^2\theta}{\partial\eta^2} + \frac{f}{2}\frac{\partial\theta}{\partial\eta} - \Pr\xi\frac{\partial f}{\partial\eta}\frac{\partial\theta}{\partial\xi} - \frac{Sk}{Re}\tau_L\xi(1-\omega)(\theta^4-\phi_I) + \Pr Ec\left(\frac{\partial^2 f}{\partial\eta^2}\right)^2 = 0,$$

$$\frac{\partial\theta_w}{\partial t} = \frac{\partial^2\theta_w}{\partial\xi^2} - \frac{\lambda L}{\lambda_w\Delta}SkQ_w$$

$$\phi_I(\tau) = 2\pi\int_{-1}^{1}\frac{I(\tau,\gamma)}{4\sigma T_\infty^4}d\gamma \quad t=0, \ \xi=\xi_0 \ \theta_w=\theta_{w0}, \ \xi=\xi_L \ \partial\theta/\partial\xi=0$$

$$\phi(\tau) = \frac{E(\tau)}{4\sigma T_\infty^4} = 2\pi\int_{0}^{1}\frac{I(\tau,\gamma)}{4\sigma T_\infty^4}\gamma d\gamma$$

$$\eta = \left(\frac{\rho_\infty U_\infty}{\mu_\infty x}\right)^{1/2}\int_{0}^{y}\frac{\rho}{\rho_\infty}d\hat{y}, \ Sk=\frac{4\sigma T_\infty^4 L}{\lambda_\infty}, \ Ec=\frac{U_\infty^2}{c_p T_\infty},$$

$$Q_w = \frac{Re^{1/2}}{Sk\xi^{1/2}}\frac{\partial\theta}{\partial\eta}\bigg|_{\eta=0} + \phi(0)$$

(7.71)

Here, η is a Dorodnizin's-type variable (Section 3.7); Ec, Sk are Eckert and Starks numbers; Q_w is the total heat flux at the wall; $\phi(\tau)$ and $\phi_I(\tau)$ are the dimensionless radiation flux and radiant energy density; $\tau = \tau_L(\xi/Re)^{1/2}\eta$ and $\tau_L = bL$ are the optical depth of boundary layer in section ξ_L and characteristic optical thickness, where b is the extinction coefficient; ω is the single scattering albedo; I is the integral radiation intensity; σ is the Stefan-Boltzmann constant; $\theta_0(\eta)$ is the self-similar temperature profile for the radiation free heat transfer case; ξ_0 and ξ_L are dimensionless coordinates of the origin and the end of the heated plate section (without a small part at $\xi=0$ which is assumed to be kept at ambient temperature to avoid the singularity).

Radiation flux $\phi(\tau)$ and radiant energy density $\phi_I(\tau)$ are determined by employing the equation of radiative transfer written for emitting, absorbing, and scattering medium. The modified mean flux method [47] is applied to reduce the integro–differential radiative transfer equation to a system of two ordinary nonlinear differential equations with corresponding boundary conditions:

$$\frac{d(\phi^+-\phi^-)}{d\tau} + (1-\omega)(m^+\phi^+ - m^-\phi^-) = (1-\omega)\theta^4,$$

$$\frac{d(m^+\delta^+\phi^+ - m^-\delta^-\phi^-)}{d\tau} + (1-\omega\hat{\zeta})(\phi^+-\phi^-) = 0$$

(7.72)

$$\tau=0, \ \phi^+ = \varepsilon\theta_w^4/4\theta \ +r\phi^-. \ \tau=\tau_\infty, \ \phi^- = \theta_\infty^4/4$$

$$m(\tau) = \int_{0}^{1}I(\tau,\gamma)d\gamma\bigg/\int_{0}^{1}I(\tau,\gamma)\gamma d\gamma, \ \delta(\tau) = \int_{0}^{1}I(\tau,\gamma)\gamma^2 d\gamma\bigg/\int_{0}^{1}I(\tau,\gamma)d\gamma$$

$$\hat{\zeta} = \frac{1}{2}\int_{-1}^{1}z(\zeta)\zeta d\zeta$$

Here $\tau_{\infty} = \tau_L (\xi / \text{Re})^{1/2} \eta_{\infty}$, η_{∞} is the value of the external boundary layer edge, r and $\varepsilon = 1 - r$ are the reflectivity and the emissivity of the plate surface, $\hat{\zeta}$ is the mean cosine of the scattering angle, $z(\zeta)$ is the scattering indicatrix, and ζ is the cosine of the angle between incident and scattering beams. Expression (7.71) for ϕ, m and δ contained in Equation (7.72) are denoted as ϕ^+, m^+, and δ^+, and similar expressions for ϕ^-, m^-, and δ^- differ from those only by integral limits that are (–1 and 0) instead of (0 and 1). The radiation flux and radiant energy density contained in Equation (7.71) are determined as $\phi = \phi^+ - \phi^-$ and $\phi_I = m^+\phi^+ - m^-\phi^-$, where m and δ are the transfer coefficients.

The system of Equations (7.71) and (7.72) is solved iteratively using the finite-difference approach. The energy equation (Equation 7.71) for fluid is solved jointly with system (Equation 7.72) using an assumed value of θ_w. Then the total heat flux Q_w is calculated to solve the energy Equation (7.71) for a plate and to obtain the new value of θ_w. Usually three to five iterations are required to achieve the converging results. The coordinate ξ_0 where the self-similar temperature profile can be used is determined from the condition of the optically thin boundary layer $\tau_L (\xi_0 / \text{Re})^{1/2} \leq 0.01$. Details of the numerical scheme and the authors' mean flux method are given in Reference [46] and more completely in original articles [47,48].

The calculations are performed for plate of length and thickness $L = 1m$ and $\Delta = 001m$. The conjugate parameter $K = \lambda_{\infty} L / \lambda_w \Delta$ is used which is varied from 0.01 to 10.

The basic results are as follows:

1. In the case of transparent medium ($\tau_L = 0$), the spatial-temporal plate temperature distribution depends considerably on the conjugate parameter. The greater K, the faster the steady state is achieved. Thermal radiation yields a higher level of temperature than that in the nonradiated case, and as a result, the steady-state situation is achieved faster. The conjugate effects are more noticeable at small values of the conjugation parameter. The results for different values of Reynolds and Starks numbers compare the intensity of the convective and radiative heating. In particular, it follows from these data that when $\text{Re}^{1/2} / \text{Sk} < 0.1$, the convection heating component may be neglected.

2. The results for absorbing, emitting, and scattering medium for three values of optical thickness $\tau_L = 1, 10$ and 100 indicate that the time required for achieving the steady state increases as the optical thickness grows. The reason for this is the radiation flux attenuation in the boundary layer. The effect of scattering is also more considerable at the small values of K. In the case of anisotropic scattering and a black wall, the maximum increase in plate temperature is noticed, and the minimum is obtained for a reflective surface and an isotropic scattering.

3. The viscous dissipation significantly affects the temperature distribution on a plate and influences the behavior of the radiation component in the total heat flux. In the case of Ec = 0 when

the viscous dissipation is ignored, the radiation flux monotonically decreases along the plate. In contrast, at the significant dissipation (Ec > 1), the behavior of the radiation flux is complicated. When the Eckert numbers are appreciable, there are two cases with the maximum in radiation flux distribution along the plate — namely, at moderate thickness ($\tau_\infty \approx 1$) and thick ($\tau_\infty > 4$) optical regions. These features are most remarkable in these specific regions of optical thickness under no monotonous temperature field. On the other hand, at the small optical thickness, the local radiation flux is almost constant. The described behavior at $\tau_\infty \approx 1$ should be most significant in the absorbing media ($\omega \approx 0$), while in the scattering media ($\omega \to 1$), with usual smoothing radiation flux distribution this phenomenon seems to be only moderate. Another complex form of flux distribution on the plate occurs because in thick optical medium the resulting radiation flux releases on the plate, while in the transparent medium it heats the flow.

4. The data of this study indicate that the conjugate approach is required to be used in the radiative-convective heat transfer investigation, especially in the cases of complex temperature distribution.

Example 7.16: Turbulent Natural Convection with Radiation in a Rectangular Enclosure

Interaction between turbulent natural convection and wall radiation in a rectangular enclosure filled with air is studied [49]. An enclosure is heated from below and the floor is maintained at constant temperature T_0. Air is considered as a transparent incompressible fluid. The walls, which are assumed to be thin with diffuse gray surfaces, are exposed to ambient temperature T_∞ with heat transfer coefficient h_∞. The Reynolds averaging Navier-Stokes (RANS) equations, analogously averaged energy equation for fluid and energy balance for the thin walls along with the $k - \varepsilon$ (turbulent energy-dissipation) model and Boussinesq hypothesis make up the governing system:

$$\frac{\partial \rho u}{\partial x} + \frac{\partial \rho v}{\partial y} = 0, \quad \frac{\partial \rho uu}{\partial x} + \frac{\partial \rho vu}{\partial y} = -\frac{\partial p}{\partial x} + \frac{\partial}{\partial x}\left(2\mu_{eff}\frac{\partial u}{\partial x}\right) + \frac{\partial}{\partial y}\left[\mu_{eff}\left(\frac{\partial u}{\partial y} + \frac{\partial v}{\partial x}\right)\right],$$

$$\mu_{eff} = \mu + \mu_{tb}$$

$$\frac{\partial \rho uv}{\partial x} + \frac{\partial \rho vv}{\partial y} = -\frac{\partial p}{\partial y} + \frac{\partial}{\partial x}\left[\mu_{eff}\left(\frac{\partial v}{\partial x} + \frac{\partial u}{\partial y}\right)\right] + \frac{\partial}{\partial y}\left(2\mu_{eff}\frac{\partial u}{\partial y}\right) + \rho\beta g(T - T_\infty),$$

$$\mu_{tb} = 0.09\frac{\rho k^2}{\varepsilon} \tag{7.73}$$

$$\frac{\partial \rho u c_p T}{\partial x} + \frac{\partial \rho v c_p T}{\partial y} = \frac{\partial}{\partial x}\left(\lambda_{eff}\frac{\partial T}{\partial x}\right) + \frac{\partial}{\partial y}\left(\lambda_{eff}\frac{\partial u}{\partial y}\right),$$

$$P_k = \mu_{tb}\left[2\left(\frac{\partial u}{\partial x}\right)^2 + 2\left(\frac{\partial v}{\partial y}\right)^2 + \left(\frac{\partial u}{\partial y} + \frac{\partial v}{\partial x}\right)^2\right]$$

$$\frac{\partial \rho u k}{\partial x} + \frac{\partial \rho v k}{\partial y} = \frac{\partial}{\partial x}\left[\left(\mu + \frac{\mu_{tb}}{\sigma_k}\right)\frac{\partial k}{\partial x}\right] + \frac{\partial}{\partial y}\left[\left(\mu + \frac{\mu_{tb}}{\sigma_k}\right)\frac{\partial k}{\partial y}\right] + P_k + G_k - \rho\varepsilon,$$

$$\lambda_{eff} = \lambda + \frac{\mu_{tb}c_p}{\sigma_T}$$

$$\frac{\partial \rho u \varepsilon}{\partial x} + \frac{\partial \rho v \varepsilon}{\partial y} = \frac{\partial}{\partial x}\left[\left(\mu + \frac{\mu_{tb}}{\sigma_k}\right)\frac{\partial \varepsilon}{\partial x}\right] + \frac{\partial}{\partial y}\left[\left(\mu + \frac{\mu_{tb}}{\sigma_k}\right)\frac{\partial \varepsilon}{\partial y}\right]$$
$$+ \left[1.44\left(P_k + G_k \tanh\frac{v}{u}\right) - 1.92\varepsilon\right]\frac{\varepsilon}{k} \tag{7.74}$$

$$\lambda_w \Delta \frac{\partial^2 T_w}{\partial s^2} + q_1 + q_2 + \varepsilon_{e1}G_1 + \varepsilon_{e1}G_2 = (\varepsilon_{e1} + \varepsilon_{e2})\sigma T_{w1}^4$$

$$G_k = \frac{\mu_{tb}g\beta}{\sigma_T}\frac{\partial T}{\partial y}, \quad G_i = \sum_{j=1}^{N} B_j F_{i-j} \quad \sum\left[\frac{\delta_{ij}(1-\varepsilon_{ei})F_{i-j}}{\varepsilon_{ei}}\right]B_j = \sigma T_i^4$$

$$B_i = \varepsilon_{ei}\sigma T_{wi}^4 + (1-\varepsilon_{ei})G_i$$

Here $\sigma_T = \sigma_k = 1$ and $\sigma_\varepsilon = 1.3$ are Prandtl numbers for temperature and for $k-\varepsilon$ parameters, respectively. The energy balance (Equation 7.74) determines the wall temperature taking into account radiation, natural convection, and conduction. In this equation, s, ε_e and G are the length measured along the wall, the hemispherical emissivity, and the irradiation falling on the surface. Both convective heat fluxes, the external from ambient and the internal from the fluid, are $q_1 = h_\infty(T_w - T_\infty)$ and $q_2 = \lambda(T - T_w)/\Delta n$, respectively, where Δn is the normal distance between the wall and the first fluid node. The irradiation G_i is related to radiosity B_i, where the subscript i is the index of the segments forming the enclosure, F_{i-j} is the shape factor from segment i to segment j, N is the number of radiating segments, and δ_{ij} is the Kronecker delta.

The system of Equations (7.73) and (7.74) under usual boundary and conjugate conditions at the walls is solved numerically using the SIMPLE algorithm. The nonuniform staggered grid with very fine spacing at the walls is used to resolve the sharp gradients, including those even in sublayers. The radiosities of the wall surfaces are presented as a function of the wall temperature, the emissivity, and the shape factors. View factors are estimated using the crossed string method [50]. The convergence of iterations is controlled by computing the normalized sum of the errors of all grid points and all variables. It is necessary to have this sum be less than 10^{-3}. Other details of discretization as well as of validation of the grid size and of the comparison with available known data to control the accuracy of the software can be found in Reference [49].

The basic results and conclusions are as follows:

1. In the case of pure convection, the calculations are performed for four aspect ratio 0.5, 1, 1.5, 2 and Rayleigh numbers from 10^8 to 10^{12}. For aspect ratio $H/W = 0.5$ and Ra $= 10^{12}$, the streamlines

and isotherms show a double-cell flow pattern similar to Rayleigh-Benard convection with the plume. When the aspect ratio increases, a primary counterclockwise vortex cell with a secondary one at the top forms. It is found that the location of these vortex cells becomes interchanged when an initial disturbance is given to the flow. The distribution of the convective Nusselt number at the bottom wall indicates that for turbulent natural convection in enclosures, the flow field is not symmetric, and the flow pattern may change from monocellular to bicellular, which is similar to that observed in laminar natural convection. At small values of Ra, the isotherms show that the temperature field in an enclosure is relatively uniform. As the Rayleigh number increases, isotherms concentrate at the walls, while at the central part, a nearly isothermal bulk field is established. The correlation for average convection Nusselt number is obtained as $Nu_C = 0.152(H/W)^{0.27} Ra^{0.34}$, where Ra is defined on the width enclosure. This correlation is valid for $05 \leq H/W \leq 2$ and $10^8 \leq Ra \leq 10^{12}$.

2. In general, the wall temperature depends on a balance between convection, radiation, conduction in the walls, and convection from external surfaces. To study the interaction of natural convection with radiation, the effects of wall conduction and emissivity are neglected. The external convection leads to an increase in the temperature of the cold walls, and as a result, the natural convection velocity is reduced. The surface radiation in both regimes at $Ra = 10^8$ and 10^{12} increases the wall temperature farther and affects farther the natural convection. Comparison of the convective, radiative, and overall Nusselt numbers for pure natural convection with the case when the external convection is involved indicates that the latter Nusselt numbers are only about 60% of the former data. This is the result of the reduction of convective flow due to decreasing the temperature difference between the hot and cold walls. The contribution of radiation is similar to that of pure natural convection, and the total contribution of both of them is very close to that of natural convection affected by external convection.

3. The effect of wall emissivity from $\varepsilon = 0$ to $\varepsilon = 0.9$ was analyzed for the same range of Rayleigh numbers, for unity aspect ratio, and with external convection. The data for convective, radiative, and total Nusselt numbers show that emissivity reduces the convective Nusselt number. The larger the emissivity, the greater is the reduction of the convective Nusselt number and the higher is the radiative Nusselt number. For high emissivity $\varepsilon = 0.9$, both convective and radiative Nusselt numbers became almost equal.

4. The effect of external convection is studied for the heat transfer coefficient from 5 to 40 W/m² K, which corresponds to values obtained for the wind velocities existing around real enclosures. As shown by the results, the convective, radiative, and total Nusselt numbers increase when the heat transfer coefficient grows. This yields in decreasing the cold wall temperatures due to enhancing the driving temperature difference for natural convection. It is found that

'   the increase in convective heat transfer is considerable, as it is compared to that for the radiative part of total heat transfer.

5. The obtained data emphasize the necessity of the conjugate approach for accurate prediction of heat transfer intensity in enclosures. Although the considered enclosure is simpler than real ones, the fundamental mechanisms governing the heat transfer and flow in natural convection under the gravity in the presence of radiation are studied in detail. It is hoped that these results would improve the capabilities of the nuclear subsystem and building design to help save energy and improve thermal comfort.

Example 7.17: Unsteady Heat Transfer from a Translating Fluid Sphere at Moderate Reynolds Numbers [51]

A fluid sphere is falling at constant velocity U_∞ in another immiscible infinite fluid. The radius of the sphere is a and the initial temperature is T_0. At the moment $t = 0$, the surrounding temperature undergoes a step change from T_0 to T_∞. The transient heat transfer between the fluid sphere and its ambient fluid is studied.

The basic assumptions are as follows: (a) The sphere size and shape are unchanged, and both flows outside and inside are fully developed and steady; (b) the flow and heat transfer are uncoupled because the physical properties are assumed to be constant and dissipation and buoyancy effects are neglected; (c) there are no surface active agents and no oscillation and rotation of the sphere; (d) the Reynolds numbers are moderate, no higher than 50.

This problem is not of the boundary layer type because the Reynolds number is low in comparison with those in boundary layer problems when the inertia terms are important, but one of the viscous terms is omitted. In contrast, at low (Re << 1) and moderate Reynolds numbers, the viscous terms play the dominant role, and inertia terms can be omitted for the case of low Reynolds number or can be taken partly into account. Thus, the equation of motion in terms of stream function in spherical coordinates for the problem in question has the following form:

$$E^4 \psi = \frac{\mathrm{Re}\sin\varphi}{2}\left[\frac{\partial\psi}{\partial r}\frac{\partial}{\partial\varphi}\left(\frac{E^2\psi}{r^2\sin^2\varphi}\right) - \frac{\partial\psi}{\partial\varphi}\frac{\partial}{\partial r}\left(\frac{E^2\psi}{r^2\sin^2\varphi}\right)\right],$$

$$(7.75)$$

$$E^2 = \frac{\partial^2}{\partial r^2} + \frac{\sin\varphi}{r^2}\frac{\partial}{\partial\varphi}\left(\frac{1}{\sin\varphi}\frac{\partial\psi}{\partial\varphi}\right)$$

Here $E^4\psi$ is the harmonic function that consists of the pressure and the viscous terms and results in equation $E^4\psi = 0$ for the low Reynolds numbers. Equation (7.75) is valid for both flows inside and outside of the sphere. The velocities, components, and boundary conditions are

$$u = \frac{1}{r^2\sin\varphi}\frac{\partial\psi}{\partial\varphi}, \quad v = \frac{1}{r^2\sin\varphi}\frac{\partial\psi}{\partial r}, \quad a)\frac{\partial\psi}{\partial\varphi} = 0,$$

$$b)\frac{\partial\psi_1}{\partial r} = \frac{\partial\psi_2}{\partial r}, \quad \tau_{r\varphi_1} = \tau_{r\varphi_2}, \quad c)\,\psi_2 = \frac{r^2\sin^2\varphi}{2}$$

$$(7.76)$$

These conditions should be satisfied: (a) at the axis symmetry ($\varphi = 0, \pi$), and $\psi_1 = 0$, (b) at the interface ($r = 1$) and $\psi_1 = \psi_2 = 0$, and (c) far away from interface ($r \to \infty$) The variables are scaled by $U_\infty a$ for ψ and a for r, $\mathrm{Re}_1 = 2\rho_1 U_\infty a/\mu_1$, $\mathrm{Re}_2 = 2\rho_\infty U_\infty a/\mu_\infty$, indices 1 and 2 denote the sphere and ambient, respectively.

The series-truncation method [52] is used to solve Equation (7.75). The basic idea of this method is to define the stream function and then velocity components as a series of Legendre polynomials P_n with radial functions $F_n(r)$:

$$\psi = \sum_{n=1}^{\infty} F_n(r) \int_{\cos\varphi}^{1} P_n(t)dt \quad u = \sum_{n=1}^{\infty} \frac{F_n(r)}{r^2} P_n(\cos\varphi) \quad v = -\sum_{n=1}^{\infty} \frac{F'_n}{r} \frac{P'_n(\cos\varphi)}{n(n+1)} \tag{7.77}$$

Employing these relations transforms Equation (7.75) into an infinite series of ordinary nonlinear differential equations that are truncated. The remaining ordinary equations are solved by finite-differential method. Details for $0.5 < \mathrm{Re}_2 < 50$ are presented in Reference [53].

The energy equation and initial and boundary conditions for a fluid sphere are

$$\frac{\alpha_2}{\alpha_1}\left[\frac{\partial\theta}{\partial t_2} + \frac{\mathrm{Pe}_2}{2}\left(u\frac{\partial\theta}{\partial r} + \frac{v}{r}\frac{\partial\theta}{\partial\varphi}\right)\right] = \frac{\partial^2\theta}{\partial r^2} + \frac{2}{r}\frac{\partial\theta}{\partial r} + \frac{\cos\varphi}{r^2}\frac{\partial\theta}{\partial\varphi} + \frac{1}{r^2}\frac{\partial^2\theta}{\partial\varphi^2},$$

$$\theta = \frac{T - T_\infty}{T_{1,0} - T_\infty}, \quad \mathrm{Pe}_2 = \frac{2aU_\infty}{\alpha_2} \tag{7.78}$$

$$t_2 = 0, \; 0 < r < 1, \; \theta = 1, \; 1 < r < \infty, \; \theta = 0 \;\; \varphi = 0, \; \pi, \;\; \frac{\partial\theta}{\partial\varphi} = 0,$$

$$r = 1, \; (\lambda_1/\lambda_2)\left(\frac{\partial\theta}{\partial r}\right)_1 = \left(\frac{\partial\theta}{\partial r}\right)_2, \; r = 0, \; \theta = \text{finite}$$

Here t is dimensionless time (Fo) scaled by a^2/α_2, in contrast to time in Equation (7.77).

Because the temperature is not defined at the droplet center, a new variable should be introduced $\phi = r\theta$ that is $\phi = 0$ at $r = 0$. At the same time, a new variable $\eta = 1/r$. is introduced for the ambient region to provide a high density of nodes near the interface. The energy equations for internal and external problems and the initial conditions, which are different for grid points on the interface and for all other points, take the form

$$\frac{\alpha_2}{\alpha_1}\left[\frac{\partial\phi}{\partial t_2} + \frac{\mathrm{Pe}_2}{2}\left(u\frac{\partial\phi}{\partial r} + \frac{\phi}{r}\right)\right] = \frac{\partial^2\phi}{\partial r^2} + \frac{\cos\varphi}{r^2}\frac{\partial\phi}{\partial\varphi} + \frac{1}{r^2}\frac{\partial^2\phi}{\partial\varphi^2}$$

$$t = 0, \; \eta = r = 1, \; \theta_1 = \theta_2 = \mathrm{Lu}(1 + \Delta r) + \mathrm{Lu} \tag{7.79}$$

$$\frac{\partial\theta}{\partial t_2} + \frac{\mathrm{Pe}_2}{2}\left(-\eta^2 u\frac{\partial\theta}{\partial r} + \eta v\frac{\partial\theta}{\partial\varphi}\right) = \eta^2\left(\eta^2\frac{\partial^2\theta}{\partial\eta^2} + \cos\varphi\frac{\partial\theta}{\partial\varphi} + \frac{\partial^2\theta}{\partial\varphi^2}\right)$$

$$t = 0, \; 0 < r < 1, \; \theta = 1, \; 0 < \eta < 1, \; \theta = 0$$

$$r = 0, \ \phi = 0, \ \eta = r = 1, \quad \frac{\lambda_1}{\lambda_2}\left(\frac{\partial \phi}{\partial r} - \frac{\phi}{r}\right) + \frac{\partial \theta}{\partial \eta} = 0,$$

$$\phi = 0, \pi, \ \frac{\partial \phi}{\partial \varphi} = \frac{\partial \theta}{\partial \varphi} = 0, \ \eta = 0, \ \theta = 0 \tag{7.80}$$

Here the Luikov number is a ratio of volumetric heat capacities $Lu = \rho_1 c_1 / \rho_2 c_2$ (Section 6.6.1). The two last energy equations are solved by alternating direction implicit (ADI) finite-difference method [54,55]. This method is based on solving an elliptical equation with the additional first derivative of time as a parabolic equation. The solution of the steady-state problem is achieved when the time derivative becomes zero.

The basic numerical results are obtained for the range of Reynolds number Re_2 from 0 to 50 for both viscosity ratio μ_1/μ_2 and thermal conductivity ratio λ_1/λ_2 from 0.333 to 3 and for the thermal diffusivity ratio $\alpha_1/\alpha_2 = 1$ and for $Pe_2 = 300$. The following provides more information:

1. The dependencies of Nusselt number and dimensionless time for different parameters show that as the Reynolds number increases, the intensity of heat transfer grows considerably. This increase is a result of strengthened circulation in flows both inside and outside of the sphere. When the thermal conductivity ratio increases, the Nusselt number increases as well, and simultaneously, the fluctuating amplitudes decrease. Both increasing the heat transfer intensity and decreasing the fluctuating amplitudes inside the droplet occur due to increased heat transfer from the surface to the center of the fluid sphere. These results are obtained for small and for large viscosity ratios $\mu_1/\mu_2 = 0.333$ and 3. The increase of viscosity ratio leads to the lower Nusselt numbers because a higher viscosity causes weaker internal circulation and yields a longer cycle of oscillation.

2. In the limiting case of a solid sphere ($\mu_1 \rightarrow \infty$), the velocity at the surface is always zero and there is no internal circulation. As a result, the dependence of heat transfer on Reynolds number is weaker than that in the case of a fluid sphere. For another limiting case of a gas bubble ($\mu_1 \approx 0$), the internal circulation is always the maximum for the given parameters, and the steady-state Nusselt number is almost independent of the Reynolds number. The data of this study show that formula $1/Nu = [(\lambda_2/\lambda_1)(1/Nu_{in}) + (1/Nu_{ext})]$ obtained in References [53] and [55] reasonably predicts the asymptotic steady-state Nusselt number for conjugate heat transfer from a solid sphere at moderate Reynolds number and for the case of $\alpha_1 = \alpha_2$.

Two other similar studies of unsteady heat transfer from solid and fluid spheres are performed for low Reynolds number. The case of identical thermal properties in flows both inside and outside of a sphere is investigated in Reference [55]. The effect of variable ratios of volumetric heat capacities but at equal thermal diffusivities is studied in Reference [53].

OTHER WORKS A comprehensive review of studies of radiative-convective heat transfer from early works in the 1960s to current results, but basically without conduction effects, is given in Reference [56]. The review of early works that considered the coupling of free convection with conduction on the plate until the 1990s may be found in Reference [43]. Analysis of results of interaction between a natural convection, radiation, and conduction in enclosures obtained in the last 20 to 25 years is presented in Reference [49]. A survey of forced heat transfer from flat plate beginning with early studies and proceeding to the end of the last century is given in Reference [40]. Some examples of other topics studied as conjugate problems that differ from those discussed in this section may be found in the literature [57–62]. In these papers are investigated the following problems: the rising of inert bubbles [57], the heat transfer between jet and slab [58], the dimensionless variables analysis of heat transfer from a flat plate [59], the unsteady heat transfer from blunt bodies in supersonic flow [60], the semi-analytical analysis of heat transfer from rods by Galerkin's method [61], and heat transfer from a circular cylinder at low Reynolds numbers [62].

References

1. Panov, D., 1951. *Handbook of Numerical Treatment of Partial Differential Equations*, 5th ed. (in Russian). Isdatel'stro ANSSR, Moscow.
2. Patankar, S. V., 1980. *Numerical Heat Transfer and Fluid Flow.* Taylor & Francis, Boca Raton, FL.
3. Brebbia, C. A., and Walker, S., 1980. *Boundary Element Technique in Engineering.* Butterworths, London.
4. Brebbia, C. A., and Dominguez, J., 2001. *Boundary Elements. An Introductory Course*, 2nd ed. WIT Press, Boston.
5. Acharya, S., Baliga, B. R., Karki, K., Murthy, J. Y., Prakash, C., and Vanka, S. P., 2007. Pressure-based finite-volume methods in computational fluid dynamic, *ASME J. Heat Transfer* 129: 407–424.
6. Patankar, S. V., and Spalding, D. B., 1972. A calculation procedure for heat, mass and momentum transfer in three-dimensional parabolic flows, *Int. J. Heat Mass Transfer* 15: 1787–1806.
7. Garetto, I. S., Gosman, A. D., Patankar, S. V., and Spalding, D. B., 1972. Two calculation procedures for steady, three-dimensional flows with recirculation, *Proc. 3rd Int. Conf. Num. Methods Fluid Dyn.*, Paris II: 60–68.
8. Patankar, S. V., 1975. Numerical prediction in three-dimensional flows. In *Studies in Convection: Theory Measurement and Application*, edited by B. E. Launder, vol. 1, pp. 1–9. Academic, New York.
9. Van Doormaal, J. P., and Raithby, G. D., 1984. Enhancements of the SIMPLE method for predicting incompressible fluid flows, *Numer. Heat Transfer* 7: 147–163.
10. Moukalled, F., and Acharya, S., 1989. Improvements to incompressible flow calculation on non-staggered curvilinear grids, *Numer. Heat Transfer, Part B* 15: 131–152.

11. Issa, R. I., Gosman, A. D., and Watkinc, A. P., 1986. Computation of compressible and incompressible recirculating flows by a non-iterative implicit scheme, *J. Comput. Phys.* 62: 66–82.

12. Baliga, B. R., and Patankar, S. V., 1983. A control-volume finite-element method for two-dimensional incompressible fluid flow and heat transfer. *Numer. Heat Transfer* 6: 245–261.

13. Kansa, E. J., and Hon, Y. C., 2000. Circumventing the ill-conditioning problem with multiquadratic radial basis functions: Applications to elliptic partial differential equations, *Comput. & Math. with Appl.* 30: 123–137.

14. Divo, E., and Kassab, A. J., 2007. An efficient localized radial basis function meshless method for fluid flow and conjugate heat transfer, *ASME J. Heat Transfer* 129: 124–136.

15. Patankar, S. V., 1978. A numerical method for conduction in composite materials: Flow in irregular geometries and conjugate heat transfer, *Proc. 6th Int. Heat Transfer Conf.*, Toronto, 3: 297.

16. Divo, E., Steinthorsson, E., Kassab, A. J., and Biaecki, R., 2002. An iterative BEM/FVM protocol for steady-state multi-dimensional conjugate heat transfer in compressible flows, *Engn. Analys. Bound. Elem.* 26: 447–454.

17. Kassab, A., Divo, E., Heidmann, J., Steinthorsson, E., and Rodriguez, F., 2003. BEM/FVM conjugate heat transfer analysis of a three-dimensional film cooled turbine blades, *Int. J. Num. Meth. Heat Fluid Flow* 13: 581–610.

18. He, M., Bishop, P. J., Kassab, A. J., and Minardi, A., 1995. A coupled FDM/BEM solution for the conjugate heat transfer vertical pipe-internal forced convection and external natural convection, *ASME J. Heat Transfer* 102: 402–407.

19. Sohal, M. S., and Howell, J. R., 1973. Determination of plate temperature in case of combined conduction, convection and radiation heat exchange, *Int. J. Heat Mass Transfer* 16: 2055–2066.

20. Barozzi, G. S., and Pagliarini, G., 1985. A method to solve conjugate heat transfer problems: The case of fully-developed laminar flow in a pipe, *ASME J. Heat Transfer* 107: 77–83.

21. Dorfman, A. S., 1982. *Heat Transfer in Flow around Nonisothermal Bodies* (in Russian). Mashinostroenie, Moscow.

22. Dorfman, A. S., 1985. A new type of boundary condition in convective heat transfer problems, *Int. J. Heat Mass Transfer* 28: 1197–2003.

23. Davydenko, B. V., 1984. Finite difference solution of conjugate heat transfer problems by reducing them to a equivalent heat conduction problem (in Russian), *Promyshlennaia Teplotekhnika* 6: 55–59.

24. Chiu, W. K. S., Richards, C. J., and Jaluria, Y., 2001. Experimental and numerical study of conjugate heat transfer in a horizontal channel heated from below, *ASME J. Heat Transfer* 123: 688–697.

25. Shah, R. K., and London, A. L., 1978. *Laminar Flow Forced Convection in Ducts.* Academic Press, New York.

26. Choi, C. Y., 2006. A boundary element solution approach for conjugate heat transfer problem in thermally developing region of a thick walled pipe, *J. Mech. Sci. Techn.* 20: 2230–2241.

27. Sparrow, E. M., and Faghri, M., 1980. Fluid-to-fluid conjugate heat transfer for a vertical pipe-internal forced convection and external natural convection, *ASME J. Heat Transfer* 102: 402–407.

28. Lee, K. T., and Yan, W. M., 1993. Transient conjugated forced convection heat transfer with fully developed laminar flow in pipes, *Numer. Heat Transfer* 23: 341–359.

29. Lin, T. F., and Kuo, J. C., 1988. Transient conjugated heat transfer in fully developed laminar pipe flows, *Int. J. Heat Mass Transfer* 31: 1093-102.

30. Yan, W. M., Tsay, Y. L., and Lin, T. F., 1989. Transient conjugated heat transfer in laminar pipe flows, *Int. J. Heat Mass Transfer* 32: 775–777.

31. Yan, W. M., 1993. Transient conjugated heat transfer in channel flows with convection from ambient, *Int. J. Heat Mass Transfer* 36: 1295–1301.

32. Sucec, J., 1987. Unsteady conjugated forced convective heat transfer in a duct with convection from the ambient, *Int. J. Heat Mass Transfer* 30: 1963–1970.

33. Al-Nimr, M. A., and Hader, M. A., 1994. Transient conjugated heat transfer in developing laminar pipe flow, *ASME J. Heat Transfer* 116: 234–236.

34. Hader, M. A., 1992. *Unsteady Conjugated Heat Transfer in Concentric Annuli*. M.Sc. thesis, Jordan University of Science and Technology, Irbid, Jordan.

35. Al-Bakhit, H., and Fakheri, A., 2006. Numerical simulation of heat transfer in simultaneously developing flows in parallel rectangular duct, *Appl. Therm. Engn.* 26: 596–603.

36. Mohammad, K., 1987. *Conjugated Heat Transfer from a Radiating Fluid in a Rectangular Channel*. Ph.D. thesis, Akron University, Ohio.

37. Yamane, T., and Bilgen, E., 2004. Conjugate heat transfer in enclosures with openings for ventilation, *Int. J. Heat Mass Transfer* 40: 401–411.

38. Wansophark, N., Malatip, A., and Dechauphai, P., 2005. Streamline upwind finite element method for conjugate heat transfer problems, *Acta Mechanica. Sinica*, 21: 436–443.

39. Boyd, R. D., and Zhang, H., 2006. Conjugate heat transfer measurement with single-phase and water flow boiling in a single-side heated monoblock flow channel, *Int. J. Heat Mass Transfer* 49: 1320–1328.

40. Vynnycky, M., Kimura, S., Kaneva, K., and Pop, I., 1998. Forced convection heat transfer from a flat plate: The conjugate problem, *Int. J. Heat Mass Transfer* 41: 45–59.

41. Robertson, G. E., Seinfeld, J. H., and Leal, L. G., 1973. Combined forced and free convection flow past a horizontal flat plate, *AIChE J.* 19: 998–1008.

42. Polat, O., and Bilgen, E., 2003. Conjugate heat transfer in inclined open shallow cavities, *Int. J. Heat Mass Transfer* 46: 1563–1573.

43. Yu, W. S., and Lin, H. T., 1993. Conjugate problems of conduction and free convection on vertical and horizontal flat plates, *Int. J. Heat Mass Transfer* 36: 1303–1313.

44. Cebeci, T., and Bradshaw, P., 1984. *Physical and Computational Aspects of Convective Heat Transfer*. Springer, New York.

45. Mhaisekar, A., Kazmierczak, M. J., and Banerjee, R. K., 2005. Steady conjugate heat transfer from x-ray or laser-heated sphere in external flow at low Reynolds number, *Numer. Heat Transfer, Part A* 47: 849–874.

46. Rubtsov, N. A., and Timofeev, A. M., 1990. Unsteady conjugate problem of radiaive-convective heat transfer in a laminar boundary layer on a thin plate, *Numer. Heat Transfer* 17: 127–142.

47. Rubtsov, N. A., Timofeev, A. M., and Ponomarev, N. N., 1987. On behavior of transfer coefficients in direct differential methods of theory of radiative heat transfer in scattering media, *Izv. SO SSSR Ser. Tekhn. Nauk.* 18: 3–8.

48. Rubtsov, A. M., and Ponomarev, N. N., 1984. Heat transfer in laminar boundary layer of absorbing, emitting and scattering media on a permeable plate, *Izv. SO SSSR Ser. Tekhn. Nauk.* 10: 65–73.
49. Sharma, A. K., Velusamy, K., Balaji, C., and Venkateshan, S. P., 2007. Conjugate turbulent natural convection with surface radiation in air filled rectangular enclosure, *Int. J. Heat Mass Transfer* 50: 625–639.
50. Siegel, R., and Howell, J. R., 1973. *Thermal Radiation Heat Transfer.* McGraw-Hill, New York.
51. Oliver, D. L. R., and Chung, J. N., 1990. Unsteady conjugate heat transfer from a translating fluid sphere at moderate Reynolds numbers, *Int. J. Heat Mass Transfer* 33: 401–408.
52. Van Dyke, M. D., 1965. *A Method of Series Truncation Applied to Some Problem in Fluid Mechanics.* Stanford University, Report SUDAER 247.
53. Oliver, D. L. R., and Chung, J. N., 1987. Flow about a fluid sphere at low to moderate Reynolds number, *J. Fluid Mech.* 177: 1–18.
54. Richtmyer, R. D., and Morton, K. W., 1967. *Difference Methods for Initial Value Problems.* Interscience, New York.
55. Abramzon, B. M., and Borde, I., 1980. Conjugate unsteady heat transfer from a droplet in creeping flow, *A.I.Ch.E. J.* 26: 536–544.
56. Wecel, G., 2006. BEM/FVM solution of the conjugate radiative and convective heat transfer problem, *Arch. Comput. Mech. Engn.* 13: 171–248.
57. Lai, H., Zhang, H., and Yan, Y., 2004. Numerical study of heat and mass transfer in rising inert bubbles using a conjugate flow model, *Numer. Heat Transfer, Part A* 46: 79–98.
58. Kanna, P. R., and Das, M. K., 2007. Conjugate heat transfer study of a two-dimensional laminar incompressible wall jet over a backward-facing step, *ASME J. Heat Transfer* 129: 220–231.
59. Chida, K., 2000. Surface temperature of a plate of finite thickness under conjugate laminar forced convection heat transfer condition, *Int. J. Heat Mass Transfer* 43: 639–642.
60. Reviznikov, D. L., 1995. Coefficients of nonisothermality in the problem of unsteady-state conjugate heat transfer on the surface on the blunt bodies, *High Temperature* 33: 259–264.
61. Hovart, A., Mavko, B., and Catton, I. 2003. Application of Galerkin method to conjugate heat transfer calculation, *Num. Heat Transfer*, Part B, 44: 509–531.
62. Sunden, B., 1979. A coupled conduction-convection problem at low Reynolds number flow. In *Numerical Methods in Thermal Problems, Proceedings of the Ist International Conference*, Swansea, Wales, July 2–6, pp. 412–422.

Part III

Applications

8

Thermal Treatment of Materials

Improving the effectiveness of such manufacturing processes as forming films, rolling of metals, glass production, continuous casting, extrusion, and other types of a thermal treatment of continuously moving materials stimulates investigations of heat transfer in flow on a tape or a thread moving out from a slot and pulling through a coolant. As shown in the first fundamental studies in the 1960s [1,2], the boundary layer forming in this case differs from that developed on a moving or streamlined plate that up to the 1960s was well investigated. The difference between these two types of boundary layer is discussed in Section 3.8 where an exact solution for heat transfer from a continuously moving surface with an arbitrary temperature distribution is derived.

After publication of the above-mentioned first papers, numerous investigations of flow and heat transfer on moving plates and rods including conjugate formulations are performed, and now there is an expanded literature. A comprehensive survey of results in this area was given by Jaluria in 1992 [3] who considers three types of approaches: the problems with a given heat transfer coefficient, the problems studying heat transfer from isothermal or uniform heat flux moving surfaces, and conjugate heat transfer problems.

In this chapter, some examples of conjugate heat transfer simulating different manufacturing processes of thermal treatment of moving materials are presented.

8.1 Moving Materials Undergoing Thermal Processing

Example 8.1: Heat Transfer between a Moving Continuous Plate and Surrounding Medium

An infinite flat plate or circular cylinder is moving out of a slot with constant velocity U_w into a viscous medium at temperature T_0 and cools pulling through it (Figure 3.7). The governing equation [4] is as follows:

$$\frac{\partial u}{\partial x} + \frac{\partial v}{\partial y} = 0, \quad u\frac{\partial u}{\partial x} + v\frac{\partial u}{\partial y} = \nu\frac{\partial^2 u}{\partial y^2}, \quad u\frac{\partial T}{\partial x} + v\frac{\partial T}{\partial y} = \alpha\frac{\partial^2 T}{\partial y^2}, \quad U_w\frac{\partial T}{\partial x} = \alpha_w\frac{\partial^2 T}{\partial y^2}$$

(8.1)

$$y = 0, \ \Delta, u = U_w, \ v = 0, \ T = T_w(x), \ q = q_w(x); \qquad y \to \infty, u \to 0, T \to T_\infty$$

289

Introducing dimensionless variables transforms these equations:

$$\frac{d^3 f}{d\eta^3} + \frac{f}{2}\frac{d^2 f}{d\eta^2} = 0, \quad \frac{\partial^2 \theta}{\partial \eta^2} + \Pr\frac{f}{2}\frac{\partial \theta}{\partial \eta} = \Pr\frac{df}{d\eta}\xi\frac{\partial \theta}{\partial \xi}, \quad b\frac{\partial \theta}{\partial \xi} = \frac{\partial^2 \theta}{\partial \zeta^2} \tag{8.2}$$

$$\eta = 0, f = 0, \frac{df}{d\eta} = 1, \eta \rightarrow \infty, \frac{df}{d\eta} = 0; \quad \xi = 0, \theta_w = 1, \quad \theta = \theta_0;$$

$$\eta = \xi = 0, \quad \theta = \theta_w, \quad -\frac{\partial \theta}{\partial \eta} = \frac{(\Pr\xi)^{1/2}}{b}\frac{\partial \theta}{\partial \zeta} \tag{8.3}$$

$$f = \frac{\psi}{(vxU_w)^{1/2}}, \quad \xi = \frac{x\lambda \mathrm{Lu}}{U_w\Delta^2(\rho c_p)_w}, \quad \zeta = \frac{y}{\Delta}, \quad \theta = T - T_\infty, \quad \eta = y\sqrt{U_w/vx}$$

The numerical solution of an energy equation at constant plate temperature is used as the initial temperature profile for fluid. The first equation was solved by Sakiadis [1]. Another equation was solved using a similar technique. Discretization and other details may be found in Reference [4]. The problem for a cylinder is solved by applying the approximate integral method which is not presented here.

The experiments are performed by pulling stainless steel and plastic endless belts through the air and water. The calculation data of the plate temperature distribution for different Prandtl numbers agree with the experimental results for air ($\Pr = 0.7$) and show that the conjugate formulation of the problem using the conjugation parameter $b = (\lambda/\lambda_w)\mathrm{Lu}$ governed the heat transfer process.

Example 8.2: Heat Transfer between a Moving Surface and Surrounding Medium (Method of Reduction to a Conduction Problem) [5]

The same problem of heat transfer between a moving surface and surrounding medium is solved using a method of reduction of the conjugate problem to an equivalent heat conduction problem [5].

The governing equations in the dimension variables are the same as in Example 8.1. However, in this case, the general exact expressions for heat flux on an arbitrary nonisothermal moving plate in differential (Equation 3.32) and integral (Equation 3.40) forms are used with the isothermal coefficient of heat transfer determined by Equation (3.93).

$$q_w = h_* \left[\theta_w(x) + \sum_{k=1}^{\infty} g_k x^k \frac{d^k \theta_w}{dx^k} \right],$$

$$q_w = h_* \left\{ \int_0^x \left[1 - \left(\frac{\xi}{x}\right)^{C_1} \right]^{-C_2} \frac{d\theta_w}{d\xi} d\xi + \theta_w(0) \right\}, \quad h_* = g_0\sqrt{U_w/vx} \tag{8.4}$$

Coefficients g_0 and g_k are given in Figures 3.8 and 3.9. The correspond-ing exponents C_1 and C_2 are plotted in Figure 3.10. These quantities are plotted as the functions of Pr and velocity ratio $\phi = U\infty/U_w$, where U_∞ is the free stream velocity.

It is shown [4,6] that the conjugate solution of the problem under con-sideration should be solved in two parts. The first part from $x = 0$ to $x = x_{cr}$ (Figure 3.7) is a conjugate problem of heat transfer between a mov-ing semi-infinite slab and the surrounding medium when the thermal boundary layer is formed analogous to that on the plate in an external medium. For this case, the equation for plate (Equation 8.1) in variables x/L and $\varphi = y\sqrt{U_w/2vx}$ coincides with Equation (3.1) without the right-hand part for the case $Pr = v/\alpha_w \to 0$ and, consequently, has the solution with coefficients g_k (Equation 3.25):

$$2x\frac{\partial\theta}{\partial x}+\varphi\frac{\partial\theta}{\partial\varphi}=\frac{1}{Pr}\frac{\partial^2\theta}{\partial\varphi^2}, \quad q_w=\frac{\lambda_w\sqrt{Re}}{\theta_w}\sqrt{\frac{Pr}{x\pi}}\left(T_w-T_0+\sum_{k=1}^{\infty}g_kx^k\frac{d^kT_w}{dx^k}\right),$$

$$g_{k\infty}\frac{(-1)^{k+1}}{k!(2k+1)} \tag{8.5}$$

Substituting heat fluxes from the plate and from fluid sides in conjugate condition yields the expression whence it follows that the plate tempera-ture at the die is constant:

$$K(T_w-T_\infty)+(T_w-T_0)+\sum_{k=1}^{\infty}(Kg_k+g_{k\infty})x^k\frac{\partial^kT_w}{\partial x^k}=0,$$

$$\frac{\partial^kT_w}{\partial x^k}=0, \quad \frac{T_w-T_\infty}{T_0-T_\infty}=\frac{1}{1+K}, \tag{8.6}$$

where $K = g_k\,Pr^{1/2}\sqrt{\pi/b}$ and $b = c_p\rho\lambda/(c_p\rho\lambda)_w = (\lambda/\lambda_w)Lu$ is the same conjugation parameter as in the previous study. Thus, the temperature profiles in the fluid and in the solid are similar. The constancy of the surface temperature is a consequence of the proportional growth of the thermal boundary layer in the moving solid and in viscous flow entrained by it. The jump of the surface temperature at the point $x = 0$ is attributable to a general property of the equations of a parabolic type describing transport processes involving the propagation of the distur-bances with infinite speed.

For the second part of the plate from $x > x_{cr}$, where the boundary lay-ers on the inner surfaces of the solid interact, the conjugate problem of heat transfer between the moving continuous flat plate and surround-ing medium is solved numerically by reduction to an equivalent con-duction problem outlined in Section 6.3. Equation (8.1) for the plate and general boundary condition (Equation 6.9) are transformed to variables $\tilde{x} = x\alpha_w/\Delta^2U_w$ and $\tilde{y} = y/\Delta$:

$$\frac{\partial \theta}{\partial \tilde{x}} = \frac{\partial^2 \theta}{\partial \tilde{y}^2}, \quad q_w = h_* \left[\theta_w(\tilde{x}) + g_1 \tilde{x} \frac{d\theta_w}{d\tilde{x}} + \varepsilon(\tilde{x}) \right],$$

$$\varepsilon(x) = \frac{q_w^{\mathrm{int}}}{h_*} + \theta_w(0) - \theta_w(\tilde{x}) - g_1 \tilde{x} \frac{d\theta_w}{d\tilde{x}} \tag{8.7}$$

where q_w^{int} is the integral expression for heat flux (Equation 3.40). This equation with boundary condition applied at $\tilde{y} = 1$ and $\tilde{y} = 0$ is solved numerically using an implicit differential scheme inscribed on a four-point stencil. In the first iteration, the correction $\varepsilon(\tilde{x})$ is taken to be zero. After the first iteration, this correction is calculated and the computation is repeated. This process is continued until the wall temperature is practically the same in two successive iterations. The solution is obtained in the third iteration. More details are given in Reference [5].

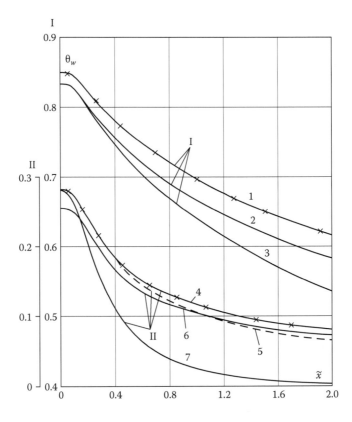

FIGURE 8.1
Temperature of the plate surface in symmetrical flow. (I) $b = 0.042$, $Pr = 5.5$; (II) $b = 8.51$, $Pr = 6.1$; (1) $\phi = 0$; (2) $\phi = 0.8$; (3) $\phi = 0$, $g_1 = 0$, $\varepsilon = 0$; (4) $\phi = 0$; (5) $\phi = 0$, $\varepsilon = 0$; (6) $\phi = 0.8$; (7) $\phi = 0$, $\varepsilon = 0$, $g_1 = 0$.

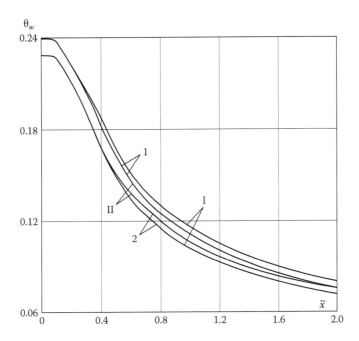

FIGURE 8.2
Temperature of the plate surface: symmetrical (I) and asymmetrical (II). (1) $\phi = 0$; (2) $\phi = 0.8$.

The results for various Pr, ϕ and b are plotted in Figures 8.1 and 8.2 for symmetrical and asymmetrical flow around the plate. It is evident from Equation (8.7) that for $\varepsilon = 0$ and $g_1 = 0$, the usual conduction problem with boundary condition of the third kind is obtained. The result of its solution is represented by curve 7 in Figure 8.1. The dashed curve in this figure corresponds to the first iteration when only the first derivative is taken into account and $\varepsilon = 0$. The crosses on the curves giving the temperature distribution indicate the experimental data from Reference [4]. It is seen that the results obtained after correction (curves 1 and 4) agree with experimental data, and the significant deviation of the temperature distribution calculated using the boundary condition of the third kind implies that the problem of heat transfer between a moving continuous plate and the surrounding medium must be solved strictly as a conjugate problem.

Example 8.3: Heat Transfer from Continuously Moving Horizontal and Vertical Flat Plates Including the Effects of a Transient Response, Thermal Buoyancy, and Effects near the Slot [7]

Considered here is the infinite plate or sheet moving with constant velocity U_w and uniform temperature T_0 as in previous examples. However, it is shown by the authors in References [8,9] that elliptic equations should be considered to study nonboundary layer effects near the slot. Thus, the governing equations are as follows ($\mathbf{v} = iu + jv$ is velocity vector):

$$\nabla \cdot \mathbf{v} = 0, \quad \frac{\partial \mathbf{v}}{\partial t} + \mathbf{v} \cdot \nabla \mathbf{v} = -\frac{\nabla p}{\rho} + \nu \nabla^2 \mathbf{v} - g\beta(T - T_\infty), \quad \frac{\partial T}{\partial t} + \mathbf{v} \cdot \nabla T \quad (8.8)$$

$$= \alpha \nabla^2 T \quad (\rho c_p)_w \frac{\partial T}{\partial t} + \rho c_p \frac{\partial T}{\partial x} = \lambda_w \nabla_s^2 T$$

The boundary conditions are as usual no-slip at the moving plate and on the walls at the slot, zero velocity far away from the plate, conjugate conditions on the interface, and the symmetry condition about the x axis. The last condition is not valid if the buoyancy effect is taken into account, except for the cases of vertical plate or when the surface with the x axis is insulated. The condition $-x_{in} < x < 0, T = T_0$ is used as the initial condition, which implies that at time $t = 0$, the long plate with beginning length x_{in} upstream of the slot at temperature T_0 starts moving with velocity U_w, and simultaneously, the heat is also turned on. This condition represents an extrusion die, furnace outlet, and so forth.

The governing equations for a plate moving vertically upward are transformed in stream function-vortex ($\psi - \omega$) form:

$$\nabla^2 \psi = -\omega, \quad \frac{\partial \omega}{\partial t} + \mathbf{v} \cdot \nabla \cdot \omega = \frac{\nabla^2 \omega}{Re} - \frac{Gr}{Re^2} \frac{\partial \theta}{\partial y}, \quad \frac{\partial \theta}{\partial t} + \mathbf{v} \cdot \nabla \theta = \frac{\nabla^2 \theta}{Re\,Pr},$$

$$\frac{\partial \theta}{\partial t} + \frac{\partial \theta}{\partial x} = \frac{\nabla^2 \theta}{Pe}, \quad Gr = \frac{g\beta(T_0 - T_\infty)\Delta^3}{8\nu^2} \quad (8.9)$$

Here all variables are dimensionless and are scaled by $\Delta/2$ for x and y, U_w for u and v, $\Delta/2U_w$ for t and $2U_w/\Delta$ for ψ and ω. $Re = U_w \Delta/2\nu$, $Pe = U_w \Delta/2\alpha_w$. For the vertical plate moving down, the buoyancy term is reversed, For the horizontal plate, the pressure gradient normal to the plate arises due to buoyancy, and therefore, the sign at the last term in the second equation should be changed to plus. The plate is divided into two parts. For the first part, starting from the slot the elliptic equations are solved because here the nonboundary layer effects are important. Downstream this solution is joined with the solution of boundary layer equations which is valid for the second part of the plate. The equality of velocity components and temperature in both parts are used as conditions for joining.

In the first part, the elliptic equations are solved using the alternate direction implicit (ADI) method (see Example 7.17). The Poisson's equation for stream function determining the temperature in a fluid is solved using an overrelaxation method with factor 1.75, and in this case the convergence is obtained after a few iterations. The procedure of iterations and other details of numerical solution are described comprehensively in Reference [7].

The basic numerical results obtained for different Pe, Gr, Pr, λ/λ_w and Re are as follows:

1. Streamlines and isotherms for an aluminum plate moving vertically upward show that flow starts close to the moving plate. Then a recirculating flow appears near the slot and moves away downstream. At first, heat is conducted along the plate and afterwards is conducted to ambient fluid. Finally, the transient process approaches the steady state. When the plate starts heating after the full flow is developed, no recirculating flow arises, and the entrained fluid comes from ambient fluid. Although this transient process is different, the steady-state velocity and temperature distributions are the same.

2. The buoyancy affects the corresponding transient velocity and temperature distribution only in time after beginning the transient process, and the velocity maxima occurs during this process but not at the steady state. The Nusselt number decreases with the distance and after approaching some minimum value increases. The minimum value of Nu occurs in the recirculation zone mentioned in (Number 1).

3. As the same aluminum plate moves horizontally, the steady state is reached later than in the case of a vertically moving plate. The reason for this is the different directions of buoyancy forces, which aid the flow in the case of the vertical plate but are normal to flow direction for the horizontal plate. The temperatures of a horizontal plate are higher than those for a vertical plate. This occurs due to higher velocities near the vertical plate which results in a higher amount of heat removal from the plate.

4. In the transient process, the maximum velocity u/U_w monotonically increases near the slot. As the distance from the slot increases, the overshoot phenomenon arises, and due to buoyancy forces increasing, it becomes larger with distance.

5. Although in the steady state the flow patterns are similar for different values of parameters, the flow in the case of air (Pr = 0.7) is stronger than that in the case of water (Pr = 7) because the buoyancy effects are larger in the air. The temperature on the aluminum plate is almost uniform which results due to high thermal conductivity. Near the slot the strong cooling is observed. This effect is not always desirable because it may result in residual thermal stresses. For the case of air and Teflon with low thermal conductivity, the plate is not cooling much.

6. As the Grashof number increases, the flow becomes stronger, and the heat transfer rate from the plate to the fluid as well as the Nusselt number grows. The plate velocity affects the induced flow and finally changes the plate temperatures and Nusselt number. Therefore, when the plate velocity is higher, these quantities are also higher. However, a certain distance is observed where despite the higher Nu in the case of higher plate speed, the distance required to cool the plate to a given level would be greater.

7. The data obtained for different ratio (λ/λ_w) show that the heat transfer coefficient near the slot is basically determined by plate thermal conductivity so that the aluminum plate has the highest Nu in this

region. Comparison of these data with results obtained for stationary semi-infinite isothermal or with uniform heat flux plate indicates that the temperature distributions in the transverse direction in both cases are similar. At the same time, the velocity distributions are very different, and the Nusselt number is much higher for the moving plate.

8.2 Simulation of Industrial Processes

Some examples of simulation of complicated practical processes given here are performed by Jaluria and coauthors. It is very difficult to simulate such complex processes in their entirety. To simplify this problem, the real computation domain is divided into several subdomains with relatively simple processes that could be easier to model. Solutions obtained for the subdomains are then coupled using proper boundary conditions to get the model of the whole process. Such problems are also of the conjugate type even if the effect of wall conduction is not taken into account as in the next example.

Example 8.4: Simulation of Process in a Twin-Screw Extruder [10]

The actual complicated flow domain as a circular region is divided into two subdomains named translation and intermeshing regions. The translation domain is modeled similar to a single-screw extruder as a channel with cross-section $H/W \ll 1$, where H and W are height and width of the channel. For steady developing, the creeping approximation is used for two-dimensional flow at a low Reynolds number. In contrast to the boundary layer approximation, in this case the inertia terms are neglected, while the viscose terms are important (see Example 7.17).

The momentum conservation equations are considered in the $y - z$ cross-section, where z is directed along the screw helix, y is normal to this direction, and x is normal to the $y - z$ cross-section. The heat convection equation should be written by taking into account the temperature dependence of physical properties and the dissipation effect, because the treatment materials are strongly viscous. Thus, the governing system for the translating domain is (for more details see [11])

$$\frac{\partial p}{\partial x} = \frac{\partial}{\partial y}\left(\mu \frac{\partial u}{\partial y}\right), \quad \frac{\partial p}{\partial z} = \frac{\partial}{\partial y}\left(\mu \frac{\partial w}{\partial y}\right), \quad \frac{\partial p}{\partial y} = 0,$$

$$\rho c_p w \frac{\partial T}{\partial z} = \frac{\partial}{\partial y}\left(\lambda \frac{\partial T}{\partial y}\right) + \mu\left(\frac{\partial u}{\partial y}\right)^2 + \mu\left(\frac{\partial w}{\partial y}\right)^2 \tag{8.10}$$

The rheology power law for non-Newtonian fluid is used to describe the dependence $\mu(T,\dot{\gamma})$, where $\dot{\gamma}$ is the strain rate and m is the temperature coefficient:

$$\mu = \mu_0 \left(\frac{\dot{\gamma}}{\dot{\gamma}_0}\right)^{n-1} \exp\left[-m(T-T_0)\right], \quad \dot{\gamma} = \left[\left(\frac{\partial u}{\partial y}\right)^2 + \left(\frac{\partial w}{\partial y}\right)^2\right] \tag{8.11}$$

These governing equations for adiabatic screw under usual no-slip boundary conditions and specified barrel temperature are solved using the finite difference method. In contrast to similar solutions in Reference [11], here the results are presented in dimensional variables.

The flow in the intermeshing domain is also simplified using steady creeping flow equations for the case of a screw channel completely filled with fluid. Because the axial length of most extruders is much larger than other sizes, the velocity gradients in the z direction are taken as much smaller than those in the x and y directions. The governing equations are as follows (see details in [12]):

$$\frac{\partial u}{\partial x} + \frac{\partial v}{\partial y} = 0 \quad \frac{\partial p}{\partial x} = \frac{\partial \tau_{xx}}{\partial x} + \frac{\partial \tau_{xy}}{\partial y}, \quad \frac{\partial p}{\partial y} = \frac{\partial \tau_{xy}}{\partial x} + \frac{\partial \tau_{yy}}{\partial y},$$

$$\frac{\partial p}{\partial z} = \frac{\partial \tau_{xz}}{\partial x} + \frac{\partial \tau_{yz}}{\partial y}, \quad \tau_{ij} = \mu(u_{i,j} + u_{j,i}) \tag{8.12}$$

where Einstein's notations are used and the velocity vector is $\mathbf{V} = u\mathbf{i} + v\mathbf{j} + w\mathbf{k}$. The same rheology power law (Equation 8.11) for non-Newtonian fluid is used, but the expression for strain rate as well as the energy equation in this case are more complicated due to dissipation function S:

$$\frac{\partial}{\partial x}\left(\lambda\frac{\partial T}{\partial x}\right) + \frac{\partial}{\partial y}\left(\lambda\frac{\partial T}{\partial y}\right) + \rho c_p\left(u\frac{\partial T}{\partial x} + v\frac{\partial T}{\partial y} + w\frac{\partial T}{\partial z}\right) + \dot{Q} = 0$$

$$\tag{8.13}$$

$$S = 2\left(\frac{\partial u}{\partial x}\right)^2 + \left(\frac{\partial u}{\partial y} + \frac{\partial v}{\partial x}\right)^2 + 2\left(\frac{\partial v}{\partial y}\right)^2 + \left(\frac{\partial w}{\partial x} + \frac{\partial w}{\partial y}\right)^2, \quad \dot{Q} = \mu S, \quad \dot{\gamma} = S^{1/2}$$

In this case, the velocity component w is directed along the extruder axis. As in the previous domain no-slip boundary conditions, assumptions of adiabatic screw and specified barrel temperature are applied. The second and third equations in Equation (8.12) — in the x and y directions — are solved by finite element method, and the last one in the z direction is solved using the Galerkin method. The axial velocity component w is perpendicular to the x–y plane where the u, v velocity components are located. These components are coupled through the law for the non-Newtonian fluids, but they are uncoupled in the case of the Newtonian fluids. Therefore, in the case of non-Newtonian fluid, the equations for velocity components, pressure, and temperature should be solved iteratively.

It is found numerically by varying the location of the interface between two separated domains that the intermeshing domain should occupy two-thirds of the total circular region. There are two techniques of modeling the flow in a screw extruder by the screw moving as it is in the actual circumstances and by moving the barrel. Because the flow in the translating domain is simulated by barrel formulation and the flow in the intermeshing domain is modeled by screw formulation, the change from one type of modeling to another is made at the interface. To match the profiles of temperature and velocities components, the iterative procedure is used. Starting with a guessed profile at the inlet, the iterations are continued until the profile at the outlet becomes close to the one obtained at the outlet of another domain. More details on the coupling procedure may be found in Reference [12].

The calculations are performed for a concrete screw with diameter 23.3 mm, channel depth 4.77 mm, and barrel diameter 30.84 mm. Other characteristics can be found in Reference [10] and more details are provided below:

1. The results for the translating region show that the flow temperature increases above the barrel temperature, which is 275°C. The reason for this increasing is the viscous dissipation. As a result of such a temperature difference, the heat goes from the fluid to the barrel. The temperature variations slightly affect the velocity field that is defined mainly by flow rate. Knowing the velocity and temperature distributions, one can calculate other characteristics of the screw, like stress or strain distributions.

2. The flow and pressure are screw-symmetric in the intermeshing domain. At the center of this region, a significant pressure change can be seen. The linear velocity profile at the outlet yields a nonlinear pressure change along the annulus. The temperature at the root of the screw is lower than that at the barrel, and the temperature rise in this domain is small in comparison with that for the translating domain. A portion of total flow in the screw channel goes into the other channel, while the remaining flow is retained in the same channel. The ratio of flow portion leaving the channel to the total flow is defined as the division ratio. This ratio decreases as the index n in the power law (Equation 8.11) increases, but it increases when the throughput increases or the depth of the screw channel decreases.

3. The complete flow is obtained by coupling both models of translating and intermeshing regions. The velocity, temperature, shear, and pressure fields indicate that the region where both flows interact is relatively small. The pressure patterns show the different regions with low, high, and uniform pressure. The pressure rise in the intermeshing region is small as it is compared with that in the

translating region, and the same conclusion is valid for the bulk temperature. At the smaller die openings when the throughputs are smaller also, the bulk temperature is higher, because in this case the velocity profiles are much steeper, giving higher viscous dissipation.

Example 8.5: Polymer Melt Flow through Extrusion Dies [13]

A flow of non-Newtonian polymeric fluid with significant viscous dissipation and temperature-dependent viscosity is considered. The considered fluid flows through a circular pipe with a contraction angle of 45°. Although such a problem simulates a wide variety of applications of polymer and food melts flowing through the extrusion dies, very few studies were performed especially in conjugate formulation. One of the reasons for this is the complication of such a problem that results not only due to the above-mentioned features, but also because the involved materials are chemically reactive and the viscosity is a function of the temperature and moisture content.

The entire system of governing equations in cylindrical two-dimensional coordinates with temperature-dependent viscosity is complicated:

$$\frac{\partial u}{\partial x} + \frac{1}{r}\frac{\partial rv}{\partial r} = 0, \quad u\frac{\partial u}{\partial x} + v\frac{\partial u}{\partial r}$$

$$= -\frac{\partial p}{\partial x} + \frac{1}{Re}\left\{\left[\frac{1}{r}\frac{\partial}{\partial r}\left(\mu r\frac{\partial u}{\partial r}\right) + \frac{\partial}{\partial x}\left(\mu\frac{\partial u}{\partial x}\right)\right] + \left[\frac{\partial \mu}{\partial r}\frac{\partial v}{\partial x} + \frac{\partial \mu}{\partial x}\frac{\partial u}{\partial x}\right]\right\}$$

$$u\frac{\partial v}{\partial x} + v\frac{\partial v}{\partial r} = -\frac{\partial p}{\partial r}$$

$$+ \frac{1}{Re}\left\{\left[\frac{1}{r}\frac{\partial}{\partial r}\left(\mu r\frac{\partial v}{\partial r}\right) + \frac{\partial}{\partial x}\left(\mu\frac{\partial v}{\partial x}\right)\right] + \left[\frac{\partial \mu}{\partial r}\frac{\partial v}{\partial r} + \frac{\partial \mu}{\partial x}\frac{\partial u}{\partial r} + \mu\frac{v}{r^2}\right]\right\}$$

$$\tag{8.14}$$

$$u\frac{\partial \theta}{\partial x} + v\frac{\partial \theta}{\partial r} = \frac{1}{Pe}\left[\frac{\partial}{\partial x}\left(\lambda\frac{\partial \theta}{\partial x}\right) + \frac{1}{r}\frac{\partial}{\partial r}\left(\lambda r\frac{\partial \theta}{\partial r}\right)\right] + \frac{Ec}{Re}\mu S$$

$$S = 2\left[\left(\frac{\partial u}{\partial x}\right)^2 + \left(\frac{v}{r}\right)^2 + \left(\frac{\partial v}{\partial r}\right)^2\right] + \left(\frac{\partial u}{\partial r} + \frac{\partial v}{\partial x}\right)^2$$

All variables are dimensionless and are scaled by the radius of the channel R for x and r, average velocity U for u and v, ρU^2 for p, reference values μ_0 and λ_0 for μ and λ. The rheology power law (Equation 8.11) is used for polymer viscosity with strain rate $\dot{\gamma} = S^{1/2}$. For food materials, such as cornmeal and other reactive polymers, a more complex law is used:

$$\mu = \mu_0 \left(\frac{\dot{\gamma}}{\dot{\gamma}_0}\right)^{n-1} \exp\left[-m(T - T_0)\right]\exp(-m_cC) \text{ or}$$

$$\mu = 0.49\dot{\gamma}^{-0.32} \exp(3969/T)\exp(-0.03C) \qquad\qquad (8.15)$$

where C is the concentration of moisture on dry-weight in percent. The second expression is used for simulation of the flow of cornmeal and for comparison with corresponding experimental data [14].

The boundary conditions are as follows: (1) at the inlet for flow: the fully developed velocity and uniform temperature profiles; (2) At the outlet for flow: developed condition; (3) at inlet and outlet for walls: adiabatic condition; and (4) at the outer surface: isothermal conditions or convection from ambient with given heat transfer coefficient:

1) $x = 0, 0 \leq r \leq 1, u = \dfrac{3n+1}{n+1}[1 - r^{(n+1)/n}], v = 0, \theta = \theta_{in}$,

2) $x = L, 0 \leq r \leq 1, \dfrac{\partial^2 u}{\partial x^2} = \dfrac{\partial v}{\partial x} = \dfrac{\partial^2 \theta}{\partial x^2} = 0$

$$\qquad\qquad\qquad\qquad\qquad\qquad\qquad\qquad\qquad (8.16)$$

3) $x = 0, x = L, 1 \leq r \leq R_0, \dfrac{\partial \theta}{\partial x} = 0$,

4) $0 \leq x \leq L, r = R_0, \theta = \theta_0$ or $\text{Bi}(\theta_0 - \theta_{am}) = -\dfrac{\partial \theta}{\partial r}\bigg|_0, \text{Bi} = \dfrac{hR}{\lambda_w}$

where the subscripts *in*, 0 and *am* refer to inlet, outer surface (also reference value), and ambient, respectively.

The solution method is similar to the SIMPLER method, and zero velocity in walls is achieved by setting viscosity to a large value such as 10^{20} (Section 7.3). The constricted part of the pipe is approximated using rectangular steps. Discretization and other details of the numerical scheme may be found in Reference [15].

The results are presented in four separate parts:

1. *Unconstricted tubes* — The flow is studied in the case with specified temperature θ_0 at the outer surface. The bulk temperature decreases along with decreasing θ_0. This occurs because the heat from the wall decreases and viscosity becomes higher which results in a more rapid drop of pressure. The effect of the outer surface temperature increases as the conductivity ratio λ_w/λ grows, because the thermal resistance is smaller when the conductivity is greater. Thus, as λ_w/λ increases, the fluid and internal surface temperatures as well as the Nusselt number become larger. At high Peclet numbers, the convective heat transfer dominates and, as a result, the role of the conductivity ratio decreases. Therefore, the effect of

small polymer conductivity is relatively small. As expected, the data show that the temperatures inside the tube and Nusselt number are smaller for greater wall thicknesses.

2. *Constricted tubes with specified temperature at the outer surface* — Due to restructured flow at the inlet to the constricted region, the largest variation of the fluid temperature is observed near the internal surface. The fluid temperature increases downstream so that at the exit the fluid temperature achieved the highest value. This occurs in cases of both cold and hot walls because the largest viscous dissipation is here. Although the pressure drops much faster in the constricted part, decreasing pressure in the unconstricted part of the tube is also significant. There are peaks of heat fluxes at the inlet to the constricted part where the temperature gradients are large. Downstream at the exit, the fluid loses heat to the wall due to high temperature that arises from the dissipation.

3. *Constricted tubes with specified Biot number at the outer surface* — Although the results show that for high (100) Biot numbers the temperature gradients are greater than those for low numbers (10), in both cases the maximum fluid temperature is observed at the exit close to the internal surface due to the effect of dissipation as indicated in other situations considered above. The larger Biot number causes larger heat losses from fluid that has a higher inlet temperature than that of ambient. Due to that, the temperature of the internal surface is lower for the higher Bi. This leads to a higher fluid velocity at the centerline in comparison with the fluid velocity at the internal surface where the temperature is lower and hence the viscosity is larger. Despite the fluid temperature being lower at the internal surface, at higher Bi the bulk temperature downstream at the centerline is higher when the Biot number is higher. The reason for this is the high effect of dissipation that affects the flow far from walls much stronger than it does the outer boundary condition.

4. *Food extrusion* — The calculation is made for an actual food extruder as a long, straight tube with die of three narrow steps. The food is cornmeal, the ambient temperature is 20°C, and the heat transfer coefficient at the outer surface is 10 W/m² K as these quantities were in an experiment. The comparison of data shows agreement. Using results obtained for different values of parameters, the correlation is found between pressure drop in MPa and the mass flow rate $\dot{m}$ in kg/h, moisture content m_c in percent, and temperature

$$\Delta p = KT^a m^b m_c^c, \text{ where}$$

$$K = \exp(55.129), a = -8.723, b = 0.432, c = -1.0276 \qquad (8.17)$$

Detailed calculations performed for data $m_c = 21\%$, $T_{in} = 382K$, $\dot{m} = 51.8kg/h$ show an almost isothermal temperature field near the centerline and the largest temperature gradients near the boundary. Due to high viscosity, no recirculation arises. The largest pressure drop takes place at the narrow exit region. The heat transfer coefficient strongly affects the pressure and temperature distribution. As the mass flow rate increases, the required pressure drop grows, while the bulk temperature decrease becomes smaller.

The data presented in this study indicate that wall conduction significantly influences the flow and heat transfer; hence, the process in an extruder should be investigated using conjugate formulation of the problem.

Example 8.6: Thermal Transport of Non-Newtonian Fluid in the Channel of an Extruder

To simplify the problem, the following assumptions are used [16]: (1) The coordinate system for two-dimensional axisymmetric approximation is fixed to the stationary screw, while the barrel is rotating; (2) The screw channel is shallow and the flight effects are negligible. Such a formulation is similar to the flow between infinite parallel plates; (3) The creeping flow approximation is used so the inertia and gravity terms are negligible; and (4) Fluid recirculation is avoided by using axial formulation [17]. The validation of those assumptions is presented in the papers [17,18].

The governing equations consist of the momentum and continuity equations in tangential (x) and axial (z) directions, an energy equation for fluid, and the same equation for the barrel and the screw:

$$\frac{\partial p}{\partial x} = \frac{\partial}{\partial y}\left(\mu \frac{\partial u}{\partial y}\right), \quad \frac{\partial p}{\partial z} = \frac{\partial}{\partial y}\left(\mu \frac{\partial w}{\partial y}\right), \quad \int_{R_s}^{R_b} u\,dy = \frac{\dot{m}\cos\phi}{\rho V_b WH}, \quad \int_{R_s}^{R_b} w\,dy = \frac{\dot{m}\sin\phi}{\rho V_b WH},$$

$$G = \frac{\mu_0 V_b^2}{\lambda_0 (T_{in} - T_0)} \tag{8.18}$$

$$Pe\,w\frac{\partial\theta}{\partial z} = \frac{\partial}{\partial y}\left(\lambda \frac{\partial\theta}{\partial y}\right) + G\mu\left[\left(\frac{\partial u}{\partial y}\right)^2 + \left(\frac{\partial w}{\partial y}\right)^2\right], \quad \frac{\partial^2\theta}{\partial y^2} + \frac{1}{y}\frac{\partial\theta}{\partial y} + \frac{\partial^2\theta}{\partial z^2} = 0$$

Here y is coordinate distance perpendicular to the screw and barrel surfaces; G is the Griffith number; $\dot{m}$ is the mass flow rate; $V_b = 2\pi R_b N/60$ is the tangential velocity of the barrel; H and W are height and width of the channel; N is screw speed, rpm; ϕ is the helix angle of the screw; $Pe = V_b H/\alpha$ is the Peclet number; and the subscripts s, b, in, and 0 refer to screw, barrel, inlet condition, and reference value or outer barrel surface, respectively.

All variables are dimensionless and are scaled by H for x, y and z, V_b for u and w, $H/\mu_0 V_b$ for p, μ_0 and λ_0 for μ and λ. The boundary conditions are as follows:

For the screw:

a. Symmetry at the centerline
b. At the outer surface
c. Adiabatic conditions at the beginning and the end

For the barrel:

a. At the internal surface
b. At the outer surface with Bi $= h\Delta/\lambda_b$
c. Adiabatic conditions at the beginning and the end

$$1) \; a) \; y = 0, \; \frac{\partial \theta}{\partial y} = 0, \; b) \; y = R_s, \; u = w = 0, \; \theta = \theta_s, \; \lambda \frac{\partial \theta}{\partial y} = \lambda_s \left(\frac{\partial \theta}{\partial y} \right)_s,$$

$$c) \; z = 0, \; z = L, \; \frac{\partial \theta}{\partial z} = 0$$

$$2) \; a) \; y = R_b, \; u = V_b, \; w = 0, \; \theta = \theta_b, \; \lambda \frac{\partial \theta}{\partial y} = \lambda_b \left(\frac{\partial \theta}{\partial y} \right)_b,$$

$$b) \; y = R_0, \; \text{Bi}\left(\theta_b - \theta_0\right) = -\frac{\partial \theta}{\partial y}, \; c)z = 0, \; z = L, \; \frac{\partial \theta}{\partial z} = 0$$

(8.19)

The problem is solved iteratively using separate solutions for screw, barrel, and for fluid in a channel, because the governing equations for fluid and solids are different. As usual, the procedure starts at guessing the temperature distributions in the screw and barrel, and the temperature field in the channel is obtained. Then employing the conjugate condition, the boundary conditions for the next calculation are obtained. The iterations are continued until the convergence is achieved.

The calculations are performed for a single-screw extruder with characteristics determined experimentally. As in the previous study, cornmeal food with the rheology power law (Equation 8.15) are used. The experimental setup consists of four jackets where six thermocouples near the barrel internal surface are installed. The controlled numerical results for three different operations show agreement with measured data:

1. The isotherms obtained for the screw channel, barrel, and screw body show that the fluid temperature near the barrel in the third heating section is higher than that near the screw. The fluid, barrel, and screw temperatures slightly increase near the end of the second jacket. Because the temperature difference between the second and third jackets is high, the largest temperature gradients in the barrel are observed here. The heat conduction in the barrel and in the screw occurs in the negative axial direction because the temperature gradient in the z direction is positive. The internal

and external temperatures are almost identical to those in the first and in part of the second jackets. In the third jacket, the external barrel temperature is about 5% to 10% higher than the internal temperature. The heat conduction from the third section to the second leads to an increase in the barrel temperature despite maintaining a constant oil temperature in the second section.

2. The radial temperature profiles indicate that the screw temperature increases monotonically with z. The screw temperatures grow in the axial direction and are much lower than the barrel temperatures.

3. The axial conduction significantly influences the axial heat flux distribution at the internal barrel surface. Due to axial conduction, a considerable amount of heat from the third heating jacket penetrates upstream and downstream through the barrel wall in the sections where for the zero axial conduction case the heat flux is zero. The variation of the barrel internal surface temperature shows that the wall is preheated upstream of the heating section and then the temperature continues to rise to a maximum value. Barrel preheating by fluid conduction is much smaller than that by wall axial conduction. This is expected because the barrel thermal conductivity is much higher. These results demonstrated the inadequacy of ignoring the barrel axial conduction.

4. The bulk temperature is greater than zero upstream of the heated jacket. This occurs due to the barrel and fluid conductivity. Internal surface temperature increases until it reaches the maximum at the third jacket and then decreases in the adiabatic sections because the fluid temperature is lower than the barrel temperature. Two cases of different boundary conditions $Bi_1/Bi_2/Bi_3 = 0.90/0.90/0.30$ and $0.225/0.225/0.075$ are considered. The temperatures of the internal barrel surface for the first case are higher than those in the second case in both portions before and after the heating section, but the temperatures in the heating section are lower in the first case. This yields the higher bulk fluid temperature in the first case.

5. The heat flux on the outer screw surface is not negligible in comparison with that at the internal barrel surface. The fluid temperature is higher downstream. Therefore, downstream the fluid heats the screw, while upstream the heat goes in the opposite direction. The conduction in the screw significantly affects the fluid temperature near the outer screw surface. In the case without conduction, the fluid temperature at the screw is the lowest, and when the heat conduction in the screw is taken into account, the screw temperature becomes higher and the screw heats the fluid. The effect of the conjugate heat transfer in the screw leads to slightly higher internal barrel surface temperatures.

6. The numerical results show the considerable difference between data obtained with and without conjugate effects. According to the data obtained in the conjugate problem, the material in the channel starts melting at an earlier location.

Example 8.7: Heat Transfer in an Optical Fiber Coating Process [19]

A study of the coating process is very important because improper coating, in particular, nonuniform or missing coating, leads to microbending and finally affects the transmission loss. In the typical optical fiber system, a bare fiber is drawn from a furnace where it reaches over 1,600°C. After cooling to a proper wetting temperature, fibers move through a coating applicator with a coating fluid, with a laser micrometer to control the diameter, and finally get through an ultraviolet (UV) curing oven to a take-up a spool.

In this study, an axisymmetric two-dimensional process is considered in cylindrical coordinates with the radial distance r measured from the center of the fiber and the axial distance z measured upward from a die exit. The coating material is a UV-curable acrylate with viscosity that highly depends on the temperature. This yields the coupled momentum and energy equations with nonlinear diffusion terms.

The system of governing equations is used in the form of a general variable [20]:

$$\frac{1}{r}\frac{\partial(\rho r u f)}{\partial r} + \frac{\partial(\rho v f)}{\partial z} = \frac{1}{r}\frac{\partial}{\partial r}\left(r\mu\frac{\partial f}{\partial r}\right) + \frac{\partial}{\partial z}\left(\mu\frac{\partial f}{\partial z}\right) + I_f$$

$$f = \begin{pmatrix} 1 \\ u \\ v \end{pmatrix} \quad I_f = \begin{pmatrix} 0 \\ -\dfrac{\partial p}{\partial r} + \dfrac{1}{r}\dfrac{\partial}{\partial r}\left(r\mu\dfrac{\partial u}{\partial r}\right) + \dfrac{\partial}{\partial z}\left(\mu\dfrac{\partial v}{\partial r}\right) - 2\mu\dfrac{u}{r} \\ -\dfrac{\partial p}{\partial z} + \dfrac{1}{r}\dfrac{\partial}{\partial r}\left(r\mu\dfrac{\partial u}{\partial z}\right) + \dfrac{\partial}{\partial z}\left(\mu\dfrac{\partial v}{\partial z}\right) + \rho g_z \end{pmatrix} \tag{8.20}$$

$$\rho_w c_w v_w \frac{\partial T}{\partial z} = \frac{2\lambda_{in}}{r_w}\frac{\partial T}{\partial r}$$

$$\frac{1}{r}\frac{\partial(r\rho c u T)}{\partial r} + \frac{\partial(\rho c v T)}{\partial z} = \frac{1}{r}\frac{\partial}{\partial r}\left(\lambda r\frac{\partial T}{\partial r}\right) + \frac{\partial}{\partial z}\left(\lambda\frac{\partial T}{\partial z}\right) + \mu S$$

$$S = 2\left[\left(\frac{\partial u}{\partial r}\right)^2 + \left(\frac{u}{r}\right)^2 + \left(\frac{\partial v}{\partial z}\right)^2\right] + \left(\frac{\partial u}{\partial z} + \frac{\partial v}{\partial r}\right)^2$$

Here the first equation is valid for continuity (the first line in the second equation) and momentum equations in both directions for fluid. The energy equation for fiber (the third part of Equation 8.20) is simplified, taking into account that (1) the Biot number Bi= $2hr_w / \lambda_w$ is small due to very small fiber diameter $2r_w = 125\mu m$, and hence, the temperature can be radially averaged; (2) the axial conduction may be neglected because the energy carried from the fiber by convection is much higher than axial conduction; and (3) the buoyancy effects are neglected because $Gr/Re^2 \ll 1$. In this equation λ_{in} is harmonic thermal conductivity at the fiber–fluid interface [20].

The boundary conditions are as follows:

1. Shear free and adiabatic conditions at the top surface in the pressured applicator.
2. At the fiber surface a no-slip radial condition, a given axial speed, and conjugate thermal conditions.
3. No-slip condition at all walls, isothermal condition at die wall, an adiabatic condition or a given heat transfer coefficient at the applicator wall.
4. Uniform speed and temperature 298.15 K of coating material at the feed inlet.
5. At the die exit, a fully developed flow and thermal conditions and the specified meniscus.

Details of free surface modeling can be found in Reference [21].

The primitive variables (i.e., velocity, pressure, and temperature) after Landau transformation $\xi = (r - r_i)/(r_0 - r_i)$ and $\eta = z/L$, where r_i is the inner radius [22], are used to solve the governing equations. Highly clustered grids are employed in the regions with large velocity and temperature gradients. The second-order upwind scheme is applied. An algorithm like SIMPLE [20] (see also Section 7.3) for pressure is repeated until convergence is achieved. For validation, the final coating thicknesses at variable fiber speed in laminar flow in a circular duct are calculated to compare with known data.

The basic results are as follows:

1. The moving fiber creates the thermal field in which the cooler fluid removes the energy from the fiber. Nevertheless, the fiber temperature increases with speed. The reason for this is that higher temperatures of the fluid and the wall occur due to increased dissipation effects with the growing speed. As the moving fiber first meets the fluid, it loses energy to the cooler fluid due to high Nusselt numbers. As the fluid heats up, the temperature gradient becomes smaller. When the speed increases, the Nusselt number becomes negative, and the fluid heats the fiber. It follows from this fact that the smallest gap between the die wall and the fiber is responsible for the increase in the Nusselt number and the fall of the fiber temperature near the die exit.
2. The viscous dissipation increases considerably as the speed grows. Because of this, the fiber temperature is higher than that at the die entrance. The temperature gradient at the fiber is very sharp and the temperature increase is highest in the die. This very large temperature due to dissipation at high speed should be avoided because the polymer may start to cross-link and degrade. As the fiber moves away from the dynamic contact point to the die

exit, the fiber temperature decreases due to the removal of heat at the fiber surface. However, when the fiber speed increases, the temperature also increases, and the Nusselt number reaches an asymptotic value.

3. The die exit diameter is one of the critical variables determining the quality of the coating process. The numerical data show that the coating thickness increases linearly with the die exit diameter at the different fixed speed values. The ratio of thermal conductivities of the fiber and coating fluid affects the process significantly. At the fixed material properties, the increase in coating fluid conductivity leads to an increase in coating thickness. This effect becomes unnoticeable when the fiber speed grows. If the fluid thermal conductivity is high, the Nusselt number variation is strong as well. This and increasing the coating thickness with the growing fluid conductivity is due to enhanced thermal diffusion in the coating fluid.

4. The effect of entrance temperature is high. As entrance temperature increases, the coating thickness decreases linearly. The fiber speed affects this dependence only slightly. At the lower entrance temperature, viscous heating becomes a major factor. The condition on the outer wall of the applicator does not significantly affect the resulting coating thickness so that practically, the conditions at the applicator wall can be flexible.

5. To understand the overall coating process, this work should be used along with the other results such as isothermal flow and the exit meniscus modeling, which are important for improving the quality of the coating, its uniformity, and the production rate.

8.3 Drying of Continuous Moving Materials

Example 8.8: Heat and Mass Transfer in Drying and Moistening in the Initial Period

The present work is concerned with the conjugate problem of heat and mass exchange between a heat transfer agent and a continuous material pulled through it [23,25]. The model is schematically presented in Figure 3.7. Here and later, the subscripts 0, 1, 2, and 3 refer to heat transfer agent (air), vapor, liquid, and dry material, respectively. The material of thickness Δ is pulled at the speed U_x. The thermophysical properties of material and air are known, including the relative concentration of vapor in air $P_{10,\infty}$ as well as the initial temperature $T(0)$ and moisture content $M(0)$ of the material.

Although the problem is unsteady in the frame moving with the material, it is a steady-state problem in a fixed frame and is governed by velocity, energy, and diffusion boundary layer equations and corresponding boundary conditions:

$$\frac{\partial u}{\partial x} + \frac{\partial v}{\partial y} = 0 \quad u\frac{\partial u}{\partial x} + v\frac{\partial u}{\partial y} = \nu\frac{\partial^2 u}{\partial y^2}, \quad u\frac{\partial T}{\partial x} + v\frac{\partial T}{\partial y} = \alpha\frac{\partial^2 T}{\partial y^2},$$

$$u\frac{\partial \rho_{10}}{\partial x} + v\frac{\partial \rho_{10}}{\partial y} = D\frac{\partial^2 \rho_{10}}{\partial y^2} \qquad (8.21)$$

$$y = \Delta/2, \quad u = U_x, \quad v = U_y, \quad \rho_{10} = \rho_{10,w}^+(x), \quad T = T_w^+(x),$$

$$y \to \infty, \quad u = v \to 0, \quad \rho_{10} \to \rho_{10,\infty}, \quad T \to T_\infty$$

Here D is a vapor diffusion coefficient, superscripts + and − denote the values calculated at the interface from the agent side ($y > \Delta/2$) and from the material side ($y < \Delta/2$). The transverse velocity on the surface is determined as follows [24]:

$$U_y = -\frac{D}{1-\rho_{10,w}}\frac{\partial \rho_{10}}{\partial y}\bigg|_w \qquad (8.22)$$

The heat and mass transfer in a capillary-porous body is defined by the following equations [24]:

$$c_M \rho_3 \frac{\partial T}{\partial t} = \frac{\partial}{\partial x}\left(\lambda_M \frac{\partial T}{\partial x}\right) + \frac{\partial}{\partial y}\left(\lambda_M \frac{\partial T}{\partial y}\right) + \rho_3 \varepsilon \Lambda \frac{\partial M}{\partial t} \qquad (8.23)$$

$$\rho_3 \frac{\partial M}{\partial t} = \frac{\partial}{\partial x}\left[\rho_3 \alpha_M\left(\frac{\partial M}{\partial x} + \gamma\frac{\partial T}{\partial x}\right)\right] + \frac{\partial}{\partial y}\left[\rho_3 \alpha_M\left(\frac{\partial M}{\partial y} + \gamma\frac{\partial T}{\partial y}\right)\right] \qquad (8.24)$$

Here subscript M refers to moist material, γ is the relative thermal diffusion coefficient, ε is a phase change coefficient defining the part of a vapor in the moisture content, and Λ is the heat of evaporation.

Estimation of the order of magnitude of terms in these equations shows that the terms determining the heat and mass transfer in the x direction are relatively small and may be neglected [25]. Then the previous equations take the form

$$c_M \rho_3 \frac{\partial T}{\partial t} = \frac{\partial}{\partial y}\left(\lambda_M \frac{\partial T}{\partial y}\right) + \rho_3 \varepsilon \Lambda \frac{\partial M}{\partial t} \quad \rho_3 \frac{\partial M}{\partial t} = \frac{\partial}{\partial y}\left[\rho_3 \alpha_M\left(\frac{\partial M}{\partial y} + \gamma\frac{\partial T}{\partial y}\right)\right] \qquad (8.25)$$

The initial and symmetry conditions should be satisfied by solving these equations:

$$x = 0, T = T(0), M = M(0), \qquad y = 0, \partial T/\partial y = 0, \partial M/\partial y = 0 \qquad (8.26)$$

Conjugate conditions required to satisfy the relations between the quantities on the interface are calculated from agent and from material sides. Three of such conditions are the equalities of temperatures, vapor densities, and mass fluxes $I(x)$ from both sides:

$$T_w^+(x) = T_w^-(x), \qquad \rho_{10,w}^+(x) = \rho_{10,w}^-(x) \qquad I_w^+(x) = I_w^-(x) \tag{8.27}$$

The fourth condition is the heat balance on the material surface: the difference between heat incoming from coolant and absorbing by material is utilized for evaporation:

$$-q_w^+ + q_w^- = (1 - \varepsilon_w)I_w^- \Lambda \tag{8.28}$$

The temperatures are determined from the third part of Equation (8.21) using Equation (8.25). The vapor density at the material surface from the coolant side $\rho_{1,w}^+$ is defined using the relation for relative vapor concentration and considering the vapor and air as ideal gases:

$$\rho_{10,w}^+ = \frac{\rho_{1,w}^+}{\rho_{1,w}^+ + \rho_{0,w}} \qquad \rho_{0,w} = \frac{p_\infty - p_{1,w}^+}{R_0 T_w} \qquad p_{1,w}^+ = \rho_{1,w}^+ R_1 T_w$$

$$\rho_{1,w}^+ = \frac{p_\infty}{R_0 T_0[(1/\rho_{10,w}^+) - 1 + R_1/R_0]} \tag{8.29}$$

To determine the vapor density at the surface from the body side, the equation of the desorption isotherms should be used $\rho_{1,w}^-/\rho_{st}(T_w) = \phi(T_w, M_w)$, where $\rho_{st}(T_w)$ is the saturated vapor density. Substituting $\rho_{1,w}^+$ and $\rho_{1,w}^-$ from these relations into the second part of Equation (8.27) gives the conjugate condition that relates the temperature T_w and moisture content M_w at the material surface to the relative concentration of vapor $\rho_{10,w}$:

$$\frac{1}{\phi(T_w, M_w)} = \frac{p_{st}(T_w)}{p_\infty}\left[1 + \frac{R_0}{R_1}\left(\frac{1}{\rho_{10,w}} - 1\right)\right] \tag{8.30}$$

The heat flux at the surface from the coolant side is defined as a difference between the incoming coolant heat and the heat being carried away by transverse vapor flux. The vapor flux is found as the sum of diffusion and convective fluxes using Equation (8.22) for U_y:

$$q_w^+ = -\lambda \left.\frac{\partial T}{\partial y}\right|_w^+ - I_w^+ i_{1,w}, \quad I_w^+ = -\rho_\infty D \left.\frac{\partial \rho_{10}}{\partial y}\right|_w + \rho_{1,w}^+ U_y = \frac{\rho_\infty D}{1 - \rho_{10,w}} \left.\frac{\partial \rho_{10}}{\partial y}\right|_w \tag{8.31}$$

where i is enthalpy. Similarly defined is the heat flux at the surface from a body side that is found as a sum of the diffusion and thermal diffusion fluxes:

$$q_w^- = -\lambda_M \frac{\partial T}{\partial y}\bigg|_w^- - I_w^- i_{1,w}^-, \quad I_w^- = -\rho_3 \alpha_M \frac{\partial M}{\partial y}\bigg|_w - \rho_3 \alpha_M \gamma \frac{\partial T}{\partial y}\bigg|_w \tag{8.32}$$

Substituting I_w^+ and I_w^- into the third part of Equation (8.27) and q_w^- and q_w^+ into Equation (8.28) yields two other conjugate conditions:

$$\frac{\rho_\infty D}{1-\rho_{10,w}} \frac{\partial \rho_{10}}{\partial y}\bigg|_w = \rho_3 \alpha_M \left[\frac{\partial M}{\partial y}\bigg|_w + \gamma \frac{\partial T}{\partial y}\bigg|_w^- \right]$$

$$\lambda \frac{\partial T}{\partial y}\bigg|_w^+ - \lambda_M \frac{\partial T}{\partial y}\bigg|_w^- = -(1-\varepsilon_w)\Lambda\alpha_M\rho_3 \left[\frac{\partial M}{\partial y}\bigg|_w + \gamma \frac{\partial T}{\partial y}\bigg|_w \right]$$

$$\tag{8.33}$$

Thus, the conjugate problem under consideration is reduced to solving the system (Equation 8.21) together with Equation (8.25) under boundary conditions (Equation 8.21), initial and symmetry conditions (Equation 8.26), and conjugate conditions (Equations 8.30 and 8.33). The system (Equation 8.21) is solved using the general boundary condition in differential (Equation 3.32) and integral forms (Equation 3.40):

$$q_w = h_* \left[T_w - T_\infty + \sum_{k=1}^{\infty} g_k x^k \frac{d^k T_w}{dx^k} \right],$$

$$q_w = h_* \left\{ T_w(0) - T_\infty + \int_0^x \left[1 - \left(\frac{\xi}{x} \right)^{C_1} \right]^{-C_2} \frac{dT_w}{d\xi} d\xi \right\} \tag{8.34}$$

where the temperatures are used instead of $\theta_w = T_w - T_\infty$.

Because the diffusion equation is similar to that of energy, analogous relations give dependence between a mass flux j_w and the relative concentration of vapor $\rho_{10,w}$:

$$j_w = h_{m*} \left[\rho_{10,w} - \rho_{10,\infty} + \sum_{k=1}^{\infty} g_k x^k \frac{d^k \rho_{10,w}}{dx^k} \right]_w,$$

$$j_w = h_{m*} \left\{ \rho_{10,w}(0) - \rho_\infty + \int_0^x \left[1 - \left(\frac{\xi}{x} \right)^{C_1} \right]^{-C_2} \frac{d\rho_{10,w}}{d\xi} d\xi \right\} \tag{8.35}$$

At constant temperature and laminar flow, the heat transfer coefficient can be determined using Equation (3.93). The mass transfer coefficient at the constant concentration is determined using the Lewis number Le which is practically 1 for gases:

$$h_{m,*} = (h_*/c_p)\,\mathrm{Le}^{1/2} \qquad (8.36)$$

Coefficients g_k and exponents C in Equation (8.34) for laminar flow depending on the Prandtl number are given in Section 3.8. For air (Pr = 0.72), some first coefficients and the exponents are as follows: $g_0 = 0.351$, $g_1 = 1.35$, $g_2 = -0.18$, $g_3 = 0.03$, $C_1 = 1.2$, $C_2 = 0.57$. Accordingly, these coefficients and exponents in Equation (8.35) are dependent on the Schmidt number, Sc. For air, Sc = Pr, and hence, the same coefficients and exponents may be used for the equations in Equation (8.35).

For the initial period under consideration, when the moisture content of the material exceeds the maximum sorptive moisture content $M_{m,s}$, the partial pressure of the vapor above the surface equals the saturation pressure. In view of this, $\phi = 1$. When $M > M_{m,s}$, the phase change coefficient $\varepsilon = 0$. In this case, Equation (8.23), Equation (8.30), and the second part of Equation (8.33) simplify and take the form

$$\frac{c_M}{c_3}\frac{\partial\theta}{\partial\mathrm{Fo}} = \frac{\partial}{\partial\eta}\left(\frac{\lambda_M}{\lambda_3}\frac{\partial\theta}{\partial\eta}\right). \qquad \frac{\partial\vartheta}{\partial\mathrm{Fo}} = \frac{\partial}{\partial\eta}\left[\frac{\alpha_M}{\alpha_3}\left(\frac{\partial\vartheta}{\partial\eta} + \gamma\,\frac{T(0)-T\infty}{M(0)-M_{m,s}}\frac{\partial\theta}{\partial\eta}\right)\right] \qquad (8.37)$$

Here the dimensionless variables and corresponding boundary conditions are as follows:

$$\mathrm{Fo} = \frac{x\lambda_3}{c_3\rho_3 U_x\Delta^2} = \frac{t\alpha_3}{\Delta^2},\ \eta = \frac{y}{\Delta},\ \theta = \frac{T-T\infty}{T(0)-T\infty},\ \vartheta = \frac{M-M_{m,s}}{M(0)-M_{m,s}}$$

$$(8.38)$$

$$\theta(0) = 1,\ \vartheta(0) = 1,\ \frac{\partial\theta}{\partial\eta} = 0,\ \frac{\partial\vartheta}{\partial\eta} = 0$$

The conjugate conditions are simplified as well. The expression on the right-hand side of the second part of Equation (8.33) is replaced by the term on the left-hand side of the first part of Equation (8.33). Taking into account that $\varepsilon = 0$, the derivative $\partial T/\partial y\big|_w$ is found from the first part of Equation (8.33) and then the derivative $\partial M/\partial y\big|_w$ is obtained from the second part of Equation (8.33). The expressions $\lambda\,\partial T/\partial y\big|_w^+$ and $\rho_\infty D\partial\rho_{10}/\partial y\big|_w$ that determine the convective heat and diffusion mass fluxes approaching the surface on the side of the coolant, are substituted by integral forms of Equations (8.34) and (8.35) using Equation (8.36) and dimensionless variables in Equation (8.38). Moreover, taking into account that for $M > M_{m,s}$, the vapor density is equal to the saturation density and the derivative $d\rho_{10,w}/d\xi$ is presented in the form of the product $(d\rho_{10,w}/dT_w)(dT_w/d\xi)$. These transformations yield the relations

$$-\frac{\partial\theta}{\partial\eta}\bigg|_w = \frac{g_0(\mathrm{Pr})\lambda_3}{\lambda_M}\left(\frac{\lambda\mathrm{Lu}}{\lambda_3\,\mathrm{Pr}\,\mathrm{Fo}}\right)^{1/2}\left\{1+\int\limits_0^{\mathrm{Fo}}\left[1-\left(\frac{\widehat{\mathrm{Fo}}}{\mathrm{Fo}}\right)^{C_1}\right]^{-C_2}\frac{d\theta_w}{d\widehat{\mathrm{Fo}}}d\widehat{\mathrm{Fo}}\right.$$

$$+\Lambda\left\{(1-\rho_{10,w})\left[c_{p1}\rho_{10,w}+c_{p0}(1-\rho_{10,w})\right]\right\}^{-1}$$

$$\left\{\frac{\rho_{10,w}(0)-\rho_{10,\infty}}{T(0)-T_\infty}+\int\limits_0^{\mathrm{Fo}}\left[1-\left(\frac{\widehat{\mathrm{Fo}}}{\mathrm{Fo}}\right)^{C_1}\right]^{-C_2}\frac{d\rho_{10,w}}{dT_w}\frac{d\theta_w}{d\widehat{\mathrm{Fo}}}d\widehat{\mathrm{Fo}}\right\} \qquad (8.39)$$

$$\frac{\partial\vartheta}{\partial\eta}\bigg|_w = \frac{g_0(\mathrm{Pr})c_3}{(\alpha_M/\alpha_3)(1-\rho_{10,w})[c_{p1}\rho_{10,w}+c_{p0}(1-\rho_{10,w})]}\left(\frac{\lambda\mathrm{Lu}}{\lambda_3\,\mathrm{Pr}\,\mathrm{Fo}}\right)^{1/2}$$

$$\times\left\{\frac{\rho_{10,w}(0)-\rho_{10,\infty}}{M(0)-M_{m,s}}+\frac{T(0)-T_\infty}{M(0)-M_{m,s}}\right.$$

$$\times\int\limits_0^{\mathrm{Fo}}\left[1-\left(\frac{\widehat{\mathrm{Fo}}}{\mathrm{Fo}}\right)^{C_1}\right]^{-C_2}\frac{d\rho_{10,w}}{dT_w}\frac{d\theta_w}{d\widehat{\mathrm{Fo}}}d\widehat{\mathrm{Fo}}+\gamma\frac{T(0)-T_\infty}{M(0)-M_{m,s}}\frac{\partial\theta}{\partial\eta}\bigg|_w \qquad (8.40)$$

Here the physical properties of the moist air are calculated additively and $\mathrm{Lu}=c_p\rho/(c\rho)_3$ is the Luikov number (Section 6.6.1). Equation (8.30), for the period being considered, for which $\phi=1$, is employed to determine the relative vapor concentration:

$$\rho_{10,w}=\left\{1+\frac{R_1}{R_0}\left[\frac{p_\infty}{p_{st}(T_w)}-1\right]\right\}^{-} \qquad (8.41)$$

Analyzing the system of equations as well as the initial, boundary, and conjugate conditions, stated in dimensionless variables, enables the author to make the following conclusions, which are valid for the first period of drying:

1. The considered conjugate problem for the fixed material and heat-transfer agent is governed by four parameters $\rho_{10,\infty}, T_\infty, T(0)$ and $M(0)$.
2. When the dependence of the material properties on temperature and moisture content is neglected and calculations are confined to the mean values of thermophysical coefficients, the number of parameters that governed the problem at the fixed material and heat-transfer agent is reduced to two:

$$\frac{T(0)-T_{\infty}}{M(0)-M_{m,s}}, \frac{\rho_{10,w}-\rho_{10,\infty}}{M(0)-M_{m,s}}. \tag{8.42}$$

3. The body thickness is incorporated only in the Fourier number. Therefore, the time needed for the material to reach the definite state characterized by certain values of temperature and moisture content and the drying duration are proportional to material thickness squared.

4. The velocity at which the material is pulled through the coolant does not affect the drying rate and time. It only determines the distance from the die over which the material reaches a certain state. The explanation is that the heat and mass fluxes in the laminar flow regime are inversely proportional to the ratio x/U_x (i.e., to the time of the material pulling, as follows from Equations 8.34, 8.35, and 8.36). It can be readily demonstrated that in the turbulent flow regime, the drying duration depends on the pulling velocity, because the coefficients h_* and $h_{m,*}$ are proportional to $U_x^{0.6}$.

Equation (8.37) under initial and boundary conditions (Equation 8.38) and conjugate conditions (Equations 8.39, 8.40, and 8.41) are solved numerically by the differential technique using the tridiagonal matrix algorithm and implicit difference scheme. The first part of Equation (8.37) is solved with the first condition (Equation 8.39), and thereafter, the second part of Equation (8.37) with the second condition (Equation 8.39) that contained the already known derivative $d\theta/d\eta|_{w}$ calculated when solving the first part of Equation (8.37). To calculate the coefficients that depend on the functions sought, iterations are carried out.

Calculations are performed for the following conditions: $T_{\infty}=90°C$, $\rho_{10,\infty}=0.125$ and $M(0)=0.25$. Two cases are considered: with the initial temperature of the material equal to $T(0)=70$ and 50°C. The first of them is higher, and the second is lower than the dew point temperature corresponding to the assigned relative vapor concentration in the heat transfer agent $\rho_{10,\infty}=0.125$. Therefore, in the first case, drying proceeds from the very beginning, whereas in the second case, the material is first moistened, and drying begins after a designated time interval.

The calculations are made for the thermophysical properties characteristic of the paper-type material: $\rho_3=800kg/m^3$, $c_3=1500J/kg\ K$, $c_M=c_3+c_{H_2O}M$ the maximum sorptive moisture content at the dew point temperature $M_{m,s}=0.19$ [26], the thermal conductivity of the moist material is taken to be constant $\lambda_M=0.4W/m\ K$. The Fourier number and other governing parameters are determined using λ_M instead of λ_3, which is more accurate, as it follows from the equations and conjugate conditions. The conjugate parameter and dimensionless coefficient of moisture diffusion calculated using this value are $c_p\rho\lambda/(c\rho)_3\lambda_M=(\lambda/\lambda_M)Lu=6\cdot10^{-5}$ and $\alpha_M(c\rho)_3/\lambda_M=0.125$. In accordance with the foregoing, the thickness and the pulling velocity of the material are not specified.

Predictions show that, in the conditions under consideration, the temperature and moisture content vary little across the material, and their values on the surface do not actually differ from those mean integrals

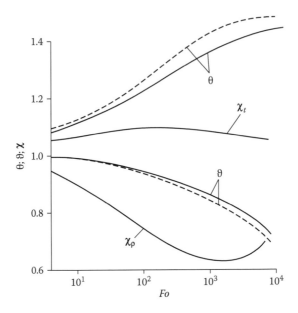

FIGURE 8.3
Variations in time of the dimensionless parameters during the drying process. Dashed curves: calculation with boundary condition of the third kind.

over the thickness. Figures 8.3 and 8.4 illustrate time dependence of the dimensionless temperatures and moisture content on the material surface. Figure 8.3 pertains to the case with the temperature higher than the dew point, and the Figure 8.4 pertains to the case with this temperature being lower than that of the dew point. The coefficients of nonisothermicity and nonisobaricity:

$$\chi_t = \frac{h}{h_*} \qquad \chi_p = \frac{h_m}{h_{m*}} \tag{8.43}$$

show how much the heat and mass transfer coefficients obtained in the conjugate problem differ from those obtained under the assumption of constant temperature and concentration heads. The analysis allows one to draw the following conclusions:

1. The rate of heat and mass transfer predicted with allowance for the interaction between the material and coolant is lower than that resulting from the calculation conducted for the third kind of boundary conditions using the heat and mass transfer coefficients for constant temperature and concentration heads. In the case of drying, the material moisture content is higher, and in the case of moistening lower than the corresponding values predicted by the third kind of boundary conditions.

2. The reason for this is considered in detail in Chapter 5, where it is shown that in general, when the head grows, either in the direction

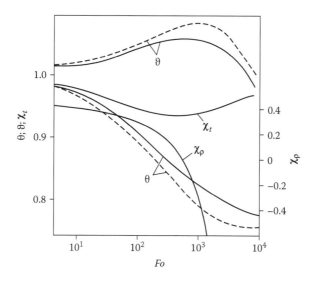

FIGURE 8.4
Variations in time of the dimensionless parameters during the moistening process. Dashed curves: calculation with boundary condition of the third kind.

of heat transfer agent flow or in time, the heat and mass transfer coefficients appear to be higher, while in the reverse case they turn out to be lower than the corresponding coefficients obtained for the constant heads. In the considered drying and moistening processes, the concentration heads diminish. The temperature head increases in the drying process and decreases during the moistening process. In conformity with this, the mass transfer coefficients in both cases are smaller than $h_{m*}(\chi_p < 1)$. The heat transfer coefficient is smaller than h_* for the moistening ($\chi_t < 1$) and larger than h_* for the drying ($\chi_t > 1$). Despite in the considered drying process $h > h_*$, the resulting heat transfer rate decreases. This occurs due to the fact that the mass transfer coefficient decreases much more than the heat transfer coefficient increases (Figure 8.3). Because the processes of the convective drying and moistening proceed under the conditions of falling concentration head, the established decrease in the rate will take place in any process of this type. Naturally, the quantitative results and the degree of this decrease will be determined by the specific conditions, but qualitatively, the results will be the same.

3. The analogy between heat and mass coefficients, frequently employed in predictions, is not observed. This is because for such an analogy, the coincidence is required not only of differential equations describing the heat and mass transfer, but also of the appropriate boundary conditions determined by the distribution of the temperature and concentration heads along the surface or in time. It is seen that these distributions substantially differ. Correspondingly, the heat and mass transfer coefficients

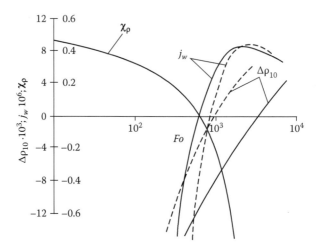

FIGURE 8.5
Variations of the mass flux and concentration head during the mass inversion. Dashed curves: calculation with boundary condition of the third kind.

also considerably differ, quantitatively and qualitatively. The heat transfer coefficients differ little from the isothermal coefficients h_*, so that the nonisothermicity coefficients χ_t are close to unity, the mass transfer coefficients differ drastically from the isobaric one h_{m*}, especially in the moistening process (Figure 8.4). Here the coefficient of partial nonisobaricity χ_p reaches 0.6 in one case (Figure 8.3), while in the other it even reduces to zero and becomes negative.

4. In the course of moistening, there occurs the mass flow inversion — the phenomenon similar to the well-known heat flow inversion (Section 5.3). It follows from the law of proportionality between the flux and the head that the moistening process should persist until the material temperature becomes equal to the dew point temperature, and the partial vapor pressure above the surface and far from it become identical. In fact, the mass flux reduces to zero earlier than does the concentration head. This is illustrated in Figure 8.5 which presents variations in the mass flux j_w and the concentration head $\Delta\rho_{10}$ obtained in the conjugate and nonconjugate problems.

It is seen that the mass flux reduces to zero much earlier than the concentration head does. In the nonconjugated problem (dashed curves), both points coincide. In the conjugate problem, at the point where the mass flux reduces to zero, the concentration head is finite and, therefore, both the mass transfer and nonisobaricity coefficients become zero. After this point, the mass flux changes its sign, and the drying process begins even though the direction of the concentration head remains the same ($\rho_{10,w} < \rho_{10,\infty}$). Thus, both the mass transfer and nonisobaricity coefficients are negative in this section. Such a pattern remains up to the point at which the concentration head vanishes. Because the mass flux is finite, both the mass transfer and the nonisobaricity coefficients at this point tend to infinity and virtually lose meaning.

Physically, the inversion phenomenon is explained by the inertial properties of the coolant, due to which the change in concentration near the wall is manifested much earlier in its immediate vicinity than far from the wall. More detailed explanation of the inverse phenomenon is given in Section 5.3.

The inversion processes cannot be investigated and calculated by traditional methods, because the boundary condition of the third kind essentially contradicts the inverse phenomenon.

Example 8.9: Heat and Mass Transfer of Drying in the Second Period [25]

In this study, the same problem of the drying process is considered for the second period, when the moisture content of the material M_w is less than the maximum sorptive moisture content $M_{m,s}$. The body is assumed to be thin, and the parameters of the material are averaged across the thickness. In that case, the problem is reduced to a system of two ordinary equations:

$$[1+(c_2/c_3)M_w]\frac{d\theta}{dz}+1+\int_0^z\left[1-\left(\frac{\xi}{z}\right)^{C_1}\right]^{-C_2}\frac{d\theta}{d\xi}d\xi$$

$$+\frac{\Lambda\left\{\rho_{10,w}(0)-\rho_{10,\infty}+\int_0^z\left[1-\left(\frac{\xi}{z}\right)^{C_1}\right]^{-C_2}\frac{d\rho_{10,w}}{d\xi}d\xi\right\}}{[T_\infty-T(0)][c_{p1}\rho_{10,w}+c_{p0}(1-\rho_{10,w})](1-\rho_{10,w})}=0$$

$$\frac{\partial M_w}{\partial\phi}\bigg|_w\left[\frac{\phi_w^2 p_{st}(T_w)R_0}{\rho_{10,w}^2 p_\infty R_1}\frac{d\rho_{10,w}}{dz}-\frac{\phi_w}{p_{st}(T_w)}\frac{\partial p_{st}}{\partial T}\bigg|_w\frac{dT_w}{dz}\right]+\frac{\partial M_w}{\partial T}\bigg|_w\frac{\partial T_w}{\partial z}$$ (8.44)

$$+\frac{c_3\left\{\rho_{10,w}(0)-\rho_{10,\infty}+\int_0^z\left[1-\left(\frac{\xi}{z}\right)^{C_1}\right]^{-C_2}\frac{d\rho_{10,w}}{d\xi}d\xi\right\}}{[c_{p1}\rho_{10,w}+c_{p0}(1-\rho_{10,w})](1-\rho_{10,w})}=0$$

These equations are written for dimensionless temperature but for dimension moisture content. The dimensionless longitudinal coordinate and dimensionless temperature are defined as

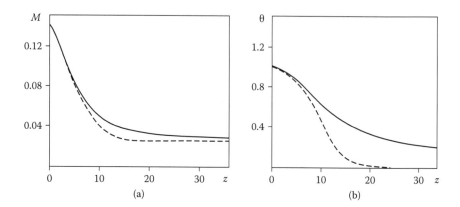

FIGURE 8.6
Dependence of moisture content (a) and dimensionless temperature (b) on dimensionless length of drying material. Dashed curves: calculation with boundary condition of the third kind.

$$z = \frac{2}{U_x \rho_3 c_3 \Delta} \int_0^x h_* d\zeta \qquad \theta = \frac{T_\infty - T_w}{T_\infty - T(0)} \qquad (8.45)$$

Calculations are performed for the same paper-type material as in the previous example for the initial period. The results presented in Figure 8.6 show that in the second drying period, the rate of heat and mass transfer predicted with the boundary condition of the third kind and heat and mass transfer coefficients for constant temperature and concentration heads are much lower than those resulting from the conjugate problem solution.

References

1. Sakiadis, B. C., 1961. Boundary layer behavior on a continuous solid surface, *A. I. Ch. E. J.* 7: 26–28, 221–225, Parts 1 and 2.
2. Tsou, F. K., Sparrow, E. M., and Goldstein, R. J., 1967. Flow and heat transfer in the boundary layer on a continuous moving surface, *Int. J. Heat Mass Transfer* 10: 219–235.
3. Jaluria, Y., 1992. Transport from continuously moving materials undergoing thermal processing. In *Annual Review of Heat and Mass Transfer* 4, Chapter 4, edited by C. L. Tien, Hemisphere Corp., Taylor & Francis, Boca Raton, FL.
4. Chida, K., and Katto, Y., 1976. Conjugate heat transfer of continuously moving surfaces, *Int. J. Heat Mass Transfer* 19: 461–470.
5. Dorfman, A. S., Grechannyy, O. A., and Novikov, V. G., 1981. Conjugate heat transfer problem for a moving continuous plate, in a fluid flow, *High Temperature* 19: 706–714.
6. Grechanniy, O. A., 1977. Conjugate heat transfer in the boundary layer of a continuously moving and cooling plate, Abstract no. 876. *Teplofiz. Vys.Tempr.* 15: 877.

7. Kang, B. H., Jaluria, Y., and Karwe, M. V., 1991. Numerical simulation of conjugate transport from a continuous moving plate in materials, *Numer. Heat Transfer* 19: 151–176.

8. Karwe, M. V., and Jaluria, Y., 1988. Fluid flow and mixed convection transport from a plate in rolling and extrusion processes, *ASME J. Heat Mass Transfer* 110: 655–661.

9. Karwe, M. V., and Jaluria, Y., 1988. Numerical simulation of thermal transport associated with a continuously moving flat sheet in rolling or extrusion, *Proc. ASME Natl. Heat Transfer Conf.* 3: 37–45.

10. Sastrohartono, T., Jaluria, Y., and Karwe, M. V., 1994. Numerical coupling of multiple-region simulations to study transport in a twin-screw extruder. *Numer. Heat Transfer Part A* 25: 541–557.

11. Gopalakrishna, S., Jaluria, Y., and Karwe, M. V., 1992. Heat and mass transfer in a single-screw extruder for non-Newtonian materials, *Int. J. Heat Mass Transfer* 35: 221–237.

12. Sastrohartono, T., 1992. *Finite element analysis of transport processes in single and twin screw extruders.* Ph.D. thesis, Rutgers University, New Brunswwick, NJ.

13. Lin, P., and Jaluria, Y., 1997. Conjugate transport in polymer melt flow through extrusion dies, *Polymer Engn. Sci.* 37: 1582–1596.

14. Fletcher, S. I., Master, T. J., Smith, A. C., and Richmon, P., 1985. Rheology and extrusion of maize grits, *Chem. Eng. Commun.* 32: 239–262.

15. Lin, P., and Jaluria, Y., 1996. Numerical approach to model heat transfer in polymer melts flowing in constricted channels, *Numer. Heat Transfer* 30: 103–123.

16. Lin, P., and Jaluria, Y., 1998. Conjugate thermal transport in the channel of an extruder for non-Newtonian fluids, *Int. J. Heat Mass Transfer* 41: 3239–3253.

17. Karwe, M. V., and Jaluria, Y., 1990. Numerical simulation of fluid flow and heat transfer in a single screw extruder for non-Newtonian fluids, *Numer. Heat Transfer Part A* 17: 167–190.

18. Chiruvella, R. V., Jaluria, Y., Esseghir, M., and Sernas, V., 1996. Extrusion of non-Newtonian fluids in a single-screw extruder with pressure back flow, *Polymer Eng. Sci.* 36: 358–367.

19. Yoo, S. Y., and Jaluria, Y., 2007. Conjugate heat transfer in an optical fiber process, *Numer. Heat Transfer Part A* 51: 109–127.

20. Patankar, S. V., 1980. *Numerical Heat Transfer and Fluid Flow.* Taylor & Francis, Boca Raton, FL.

21. Yoo, S. Y., and Jaluria, Y., 2008. Numerical simulation of the meniscus in nonisothermal free surface flow at the exit of a coating die, *Numer. Heat Transfer Part A* 53: 111–131.

22. Landau, H. G., 1950. Heat conduction in a melting solid, *Appl. Math. Q.* 8: 81–94.

23. Dolinskiy, A. A., Dorfman, A. S., and Davydenko, B. V., 1991. Conjugate heat and mass transfer in continuous processes of convective drying, *Int. J. Heat Mass Transfer* 34: 2883–2889.

24. Luikov, A. V., 1968. *Drying Theory* (in Russian). Izd. Energiya, Moscow.

25. Grechannyy, O. A., Dolinskiy, A. A., and Dorfman, A. S., 1987. Conjugate heat and mass transfer in continuous processes of the convective drying of thin bodies (in Russian), *Prom. Teplotekhn* 9 (4): 27–37.

26. Nikitina, L. M., 1964. *Tables of the Mass Transfer Coefficients of Moist Material* (in Russian). Izd. Nauka I Tekhnika, Minsk.

9

Technological Processes

9.1 Multiphase and Phase-Change Processes

Example 9.1: Mass Transfer in a Wetted-Wall Absorber [1]

The model consists of a tube with gas flow and the walls filled by flowing liquid. Such a model simulates many chemical, metallurgical, and other systems involving the absorption process. To simplify the problem, it is assumed that the gas flow rate is high, so that the influence of solute on the flow and mass transfer is negligible, the velocity profiles in both phases are fully developed, and the liquid film thickness is small versus the tube radius. In such a case, the problem is governed by two diffusion equations:

$$u\frac{\partial c}{\partial z} = D_L\frac{\partial^2 c}{\partial r^2} + k(c - c_{eq}), R \le r \le R_e, \qquad U\frac{\partial C}{\partial z} = D_G\frac{1}{r}\frac{\partial}{\partial r}\left(r\frac{\partial c}{\partial r}\right), 0 \le r \le R \qquad (9.1)$$

The lower letters and capitals as well as subscripts L and G are used for liquid and gas parameters, respectively; c and C are concentrations; R and R_e are internal and external radii; k is the first-order reaction constant; and c_{eq} is the equilibrium concentration.

In the case of $\Delta/R_e \ll 1$, the velocity distributions may be presented as

$$u = a - \kappa\frac{\mu_G}{\mu_L}\frac{y}{2} - u_c\frac{y^2}{R_e^2}, \qquad U = a - \kappa\phi_G\frac{y}{2} - u_c\frac{y^2}{R_e^2},$$

$$u_c = \frac{R_e^2}{4}\left[\frac{\kappa}{(2-\kappa)}\left(\phi_L - \phi_G\frac{\mu_G}{\mu_L}\right)\right] \qquad U_c = \frac{\kappa^2\phi_G R_e^2}{4} \qquad (9.2)$$

$$\phi = \frac{1}{\mu}\left(\rho g - \frac{\Delta p}{L}\right) \qquad a = \frac{R_e^2}{4}\left[\phi_L(1-\kappa^2) - \frac{\kappa(\kappa-1)(\kappa-3)}{(2-\kappa)}\left(\phi_L - \frac{\mu_G\phi_G}{\mu_L}\right)\right]$$

Here, $y = r - R$ is the transverse coordinate such that the origin ($y = 0$) is at the gas–liquid interface, $k = 1 - (\Delta/R_e)$, and u_c and U_c are characteristic velocities.

The initial, boundary (no flux), and conjugate conditions for Equation (9.1) are

$$c(y,0) = c_0, \quad C(y,0) = C_0, \quad \frac{\partial c}{\partial y}(\Delta, z) = \frac{\partial C}{\partial y}(-R, z) = 0 \tag{9.3}$$

$$C(0.z) = K_{eq}c(0,z), \quad D_G \frac{\partial C}{\partial y}(0,z) = D_L \frac{\partial c}{\partial y}(0,z) \tag{9.4}$$

The Green's functions associated with Equation (9.1) have the following forms:

$$G_L(\eta_L, \zeta_L - \zeta_L', \eta_L') = \sum_{n=1}^{\infty} \frac{Y_{nL}(\eta_L')Y_{nL}(\eta_L)}{\|Y_{nL}\|^2} \exp[-\gamma_{nL}^2(\zeta_L - \zeta_L')] \tag{9.5}$$

$$G_G(\eta_G, \zeta_G - \zeta_G', \eta_G') = \sum_{n=1}^{\infty} \frac{Y_{nG}(\eta_G')Y_{nG}(\eta_G)}{\|Y_{nG}\|^2} \exp[-\gamma_{nG}^2(\zeta_G - \zeta_G')] \tag{9.6}$$

$$\eta_L = \frac{y}{\Delta}, \quad \zeta_L = \frac{z}{\Delta Pe_L}, \quad \eta_G = \frac{\kappa R_e + y}{\kappa R_e}, \quad \zeta_G = \frac{z}{\kappa R_e Pe_G}, \quad Pe_L = \frac{\Delta u_c}{D_L}, \quad Pe_G = \frac{\kappa R_e U_c}{D_G}$$

The eigenfunctions should satisfy the following Sturm-Liouville equations:

$$\frac{d^2 Y_{nL}}{d\eta_L^2} + [\gamma_{nL}^2 w(\eta_L) - \alpha^2]Y_{nL} = 0, \quad Y_{nL}(0) = \frac{dY_{nL}}{d\eta_L}(1) = 0, \quad w(\eta_L) = a + b\eta_L - \eta_L^2 \tag{9.7}$$

$$\frac{1}{\eta_G} \frac{d}{d\eta_G}\left(\eta_G \frac{dY_{nG}}{d\eta_G}\right) + \gamma_{nG}^2(B - \eta_G^2)Y_{nG} = 0, \quad \frac{dY_{nG}}{d\eta_G}(0) = Y_{nG}(1) = 0 \tag{9.8}$$

It is shown in Reference [2] that these eigenfunctions equations can be solved in terms of the confluent hypergeometric function $F(A,B,X)$:

$$Y_{nL} = \exp(-X/2)[d_n F(A_0, B_0, X) + X^{1/2} F(A_1, B_1, X)], \quad X = \gamma_{nL}(\eta_L - b/2)^2 \tag{9.9}$$

$$Y_{nG} = \exp(-\gamma_{nG}\eta_G^2/2)F\left(\frac{2 - \gamma_{nG}\hat{B}}{4}, 1, \gamma_{nG}\eta_G^2\right) \tag{9.10}$$

$$\hat{B} = \frac{R_e^2}{4U_c}[\phi_L + \kappa^2(\phi_G - \phi_L) + \phi_L(1 - \kappa^2)] - \frac{a}{U_c}, \quad b = -\frac{\kappa R_e^2 \phi_L}{2u_c} \frac{\mu_G}{\mu_L} \tag{9.11}$$

The constants A, B, d_n and equation determining eigenvalues γ_n are given in Reference [1]. Applying these results yields solutions of Equation (9.1) for concentrations

$$c(\eta_L, \zeta_L) = c_0 \int_0^1 w(\eta')G(\eta, \zeta - 0, \eta')d\eta' + \int_0^\zeta \int_0^1 c_w(\zeta')w(\eta')$$

$$\times \frac{\partial}{\partial \zeta'} G(\eta, \zeta - \zeta', \eta')d\eta'd\zeta' - \sigma^2 \int_0^\zeta \int_0^1 c_w(\zeta')G(\eta, \zeta - \zeta', \eta')d\eta'd\zeta' \quad (9.12)$$

$$+ \sigma^2 c_{eq} \int_0^\zeta \int_0^1 G(\eta, \zeta - \zeta', \eta')d\eta'd\zeta'$$

$$C(\eta_G, \zeta_G) = C_0 \int_0^1 (\hat{B} - \eta'^2)\eta'G(\eta, \zeta - 0, \eta')d\eta' + \int_0^\zeta \int_0^1 C_w(\zeta')(\hat{B} - \eta'$$

$$\frac{\partial}{\partial \zeta'} G(\eta, \zeta - \zeta', \eta')d\eta'd\zeta'$$

Here $\sigma = (kR_e^2/D_L)^{1/2}$. It is evident that all variables in the first and the second equations should have omitted subscription L and G, respectively.

These final equations contain unknown concentrations on the interface c_w and C_w. The first conjugate condition (Equation 9.3) is used for eliminating the unknown C_w by expressing it in terms of c_w. To obtain the remaining unknown concentration c_w, the second conjugate condition (Equation 9.3) is employed. Substituting the heat fluxes in this condition leads to an integral equation that is solved by the Laplace transform. The detailed calculation and the result for unknown interfacial concentration expressed in series of polynomials $p(Z_k)$ and $q(Z_k)$ are given in Reference [1]:

$$c_w(\zeta_L) = \sum_{k=1}^\infty \frac{p(Z_k)}{q'(Z_k)} \exp(Z_k \zeta_L), \quad Z_k < 0 \quad (9.13)$$

Analyses show that interfacial concentration c_w is determined by three parameters $\sigma, \varepsilon = D_L kR/D_G \Delta$ and $\omega = \zeta_G/\zeta_L$ as well as by the initial concentrations and absorption equilibrium constant k. In the case of no chemical reaction, one has $\sigma = 0$. If the parameter ε is small, the gas phase resistance is negligible, and for large ε, the gas phase basically determines the system behavior.

Numerical results for carbon dioxide absorption in water are obtained:

1. Ten eigenvalues for each phase are calculated and listed. It is found that further eigenvalues did not affect the results. Series (9.13) rapidly converges so that only three or four terms are needed to get the result. However, near $\zeta = 0$, this series as well as the eigenvalues expansions converge slowly as it usually does for Graetz solutions.

2. Variations of mass fluxes of both phases defined by Sherwood numbers

$$\text{Sh}_L = \frac{1}{c_w - \bar{c}} \frac{\partial c}{\partial \eta_L}(0) \qquad \text{Sh}_G = \frac{1}{C_w - \bar{C}} \frac{\partial C}{\partial \eta_G}(1) \qquad (9.14)$$

as functions of the initial liquid concentration are studied ($\bar{c}$ and $\bar{C}$ are averaged values). It is found that the inlet liquid concentration does not affect the gas phase Sherwood number, but the liquid-phase Sherwood number shows peculiar behavior. For values near zero, it decreases sharply to a minimum for $\zeta_L < 0.01$ and then increases to an asymptotic value of 2.89 for $\zeta_L > 1$. For the case of larger values of the inlet liquid concentration, the liquid-phase Sherwood number decreases monotonically to its asymptotic value. The calculation shows that the asymptotic Sherwood number is almost independent of the Reynolds number Re_L.

3. The study of the effect of the reaction rate constant k shows that dimensionless liquid-phase concentration rises sharply from zero to a maximum value that depends on k. As the reaction proceeds, the liquid concentration decreases.

Solutions of two other conjugate problems are given in the same work. The heat transfer from a thick-walled tube is considered as the first problem, and the falling-film reactor is simulated in the second article. The similar solution technique is used in both studies.

Example 9.2: A Transient Response of a Heat Thermal Energy Storage Unit [3]

The study of flow and heat transfer in energy storage units is stimulated by growing use of solar systems due to its periodic nature. The model under consideration consists of parallel rectangular plates forming the channels with flowing heat transfer fluid (HTF) which exchanges heat with the phase-change material (PCM) located on the other side of the plates.

The following assumptions are used:

1. The density and specific heat of liquid and solid phases of the PCM are the same.
2. Longitudinal and transversal heat conduction (in the x and z directions) in the PCM are negligible in comparison with that in the normal y direction.
3. Longitudinal and transversal heat conduction in the channels are negligible compared to convective heat transfer.
4. The effects of natural convection are negligible.
5. The flow is fully developed and viscous dissipation is negligible.

The problem is solved using the enthalpy method [4]. The system of governing equations and initial, boundary, and conjugate conditions for the HTF, container walls, and the PCM are, respectively,

$$\frac{\partial \theta}{\partial t} + u \frac{\partial \theta}{\partial x} = \frac{\partial}{\partial y}\left(\alpha \frac{\partial \theta}{\partial y}\right), \quad \frac{\partial \theta_w}{\partial t} = \frac{\partial}{\partial y}\left(\alpha_w \frac{\partial \theta_w}{\partial y}\right),$$

$$\frac{\partial \theta_m}{\partial t} = \frac{\partial}{\partial y}\left(\alpha_m \frac{\partial \theta_m}{\partial y}\right) - \frac{1}{\text{Ste}}\frac{\partial fi}{\partial t} \tag{9.15}$$

$$t=0, \ \ \theta = \theta_w = \theta_m = 0, \ \ x=0 \ \theta = 1, \ \ y=0, \ \ \frac{\partial \theta}{\partial y}=0, \ \ y=\frac{1}{2}, \ \ \lambda \frac{\partial \theta}{\partial y} = \lambda_w \frac{\partial \theta_w}{\partial y} \tag{9.16}$$

$$y=\frac{1}{2}, \ \lambda_w \frac{\partial \theta_w}{\partial y} = \lambda \frac{\partial \theta}{\partial y}, \ y = \frac{1}{2}+\frac{\Delta_w}{\Delta}, \ \lambda_w \frac{\partial \theta_w}{\partial y} = \lambda_m \frac{\partial \theta_m}{\partial y} \quad \theta = \frac{T-T_{melt}}{T_{in}-T_{melt}} \tag{9.17}$$

$$y=\frac{1}{2}+\Delta_w, \ \lambda_m \frac{\partial \theta_m}{\partial y} = \lambda_w \frac{\partial \theta_w}{\partial y}, y = \frac{1}{2}+\Delta_w+\frac{\Delta_m}{2},$$

$$\frac{\partial \theta_w}{\partial y}=0, \quad \begin{matrix} \alpha_m = f+(1-f)\alpha_s \\ \lambda_m = f+(1-f)\lambda_s \end{matrix}$$

$$\tag{9.18}$$

$$u = \frac{1}{2}\alpha \Pr \text{Re}[1-(2y)^n][1-(2az)^m]\left(1+\frac{1}{n}\right)\left(1+\frac{1}{m}\right)(1+a),$$

$$m = 1.7+0.5a^{-1.4}$$

$$n = 2 \ [a \le (1/3)] \ or \ 2+0.3a-0.1$$

Here all variables are dimensionless and are scaled by channel thickness Δ for coordinates and other dimensions; Δ^2/α_L for t, α and λ_L for other thermal diffusivities and thermal conductivities, respectively; $\text{Re} = \bar{u}D_h/v$, $\text{Ste} = c_m(T_{in}-T_{melt})/\Lambda$ are Reynolds and Stephan numbers; Λ is latent heat; $a = \Delta/W$ is channel aspect ratio; W is the width of channel; $\Delta, \Delta_w, \Delta_m$ are the thicknesses of channel, channel walls, and PCM; and f denotes the liquid fraction. Subscripts w, m, L, in and $melt$ refer to plate, PCM, liquid, initial, and melting qualities.

The numerical solution of this set of equations is obtained using the finite difference technique detailed in Reference [4]. The value of the liquid fraction changes from 1 in a fully liquid region to 0 in a fully solid region. At each grid point, the value of this fraction is determined using the numerical solutions of the temperature equation for PCM and is updated following the condition $f^{k+1} = 0$ if $f^k < 1$ or $f^{k+1} = 1$ if $f^k > 1$. The set of algebraic equations is solved iteratively by TDMA. The details of the grid, step sizes, conservation criterion, and results of validation are given in Reference [3].

The numerical results are obtained for n-octadecane as a PCM ($\rho = 776.5 \ kg/m^3 \ c = 2165 \ J/kg \ K$, $\lambda_L = 0.148$, $\lambda_s = 0.358 \ W/m \ K$, $\Lambda = 243.5 \cdot 10^3 \ J/kg$, $T_{melt} = 300 \ K$), for water as HTF, and for walls with properties $\rho = 8933 \ kg/m^3$, $c = 385 \ J/kg \ K$, $\lambda_w = 401 \ W/mK$. The dimensions of the unit are $L = W = 0.5 \ m$ and $\Delta = 0.01$ m. The results are as follows:

1. At the inlet, the temperature is uniform across channel thickness for all time due to the high thermal conductivity of the walls. At small times ($t = 0.1$) the temperature of PCM decreases along y and reaches zero. It then increases with time reaching almost the inlet HTF temperature. With time, the thermal boundary layer develops, which leads to increasing HTF temperature.

2. Nusselt number is high at small x, decreasing along the channel and increasing with time. At small distances from the inlet, the local Nusselt numbers are close to the values corresponding to the constant heat flux boundary condition. For larger distances, the Nusselt number becomes close to the isothermal value at small time ($t = 0.1$) and in time ($t = 0.8$) reaches values corresponding to constant heat flux on the surface. However, for the full length of the channel for all time, the local Nusselt numbers differ significantly from those obtained at constant temperature or heat flux on the surface, so that using such values for prediction leads to remarkable errors.

3. The outlet HTF temperature increases with time during the heat charge period approaching the values close to the inlet temperature. During this period, regardless of the Reynolds number value, the outlet temperature first rapidly changes due to noticeable heat storage. Then the outlet temperature changes slowly due to latent heat storage. Finally, as the melting is completed, the data show the rapid change of the outlet temperature, the reason of which is sensible heat storage. On the whole, increasing the Reynolds number yields a growing heat transfer rate, which results in rapid melting of the PCM. This is confirmed by variation of liquid fraction in time. As a result, the system rapidly reaches the steady state, and the temperature of HTF reaches its inlet value.

4. The melting front penetrates more quickly at smaller x for any Reynolds number and time because close to the inlet, the Nusselt number and temperature of HTF are higher than at larger distances from the starting channel section.

The same problem of behavior of the energy storage unit is considered in Reference [5]. In this study, the hybrid scheme combining analytical and numerical techniques is used. The heat transfer in the PCM is studied by the source-and-sink method employing a heat source at the freezing front and a heat sink at the melting front. This work contains a review of the methods studying the phase-change processes (Stephan problems).

Example 9.3: Transient Solidification in an Enclosed Region [6]

This problem is concerned with the system containing a moving boundary between two phases that exchange thermal energy. The studied model consists of a mold with a liquid lump undergoing solidification and the solid around it. The process of phase change is described using the conservation equations with enthalpy as a basic variable. The system of governing equations and boundary conditions takes into account thermal buoyancy as a driving force:

$$\frac{D\mathbf{u}}{Dt} = \Pr\nabla^2\mathbf{u} - \frac{\partial p}{\partial x} + S_u, \quad \frac{Dv}{Dt} = \Pr\nabla^2\mathbf{v} - \frac{\partial p}{\partial y} + S_v + \mathrm{Ra}\,\Pr\theta,$$

$$\frac{D\rho J}{Dt} = \nabla\cdot\left(\frac{\lambda}{c_p}\nabla J\right) - S_h$$

$$\theta_m = \theta_S, \quad \nabla\theta_{S'} = \lambda_m\cdot\nabla\theta_m\cdot\mathbf{n}, \quad \nabla\theta_m\cdot\mathbf{n} = -\mathrm{Bi}\theta_m,$$

$$\theta = \frac{T - T_a}{T_i - T_a} \quad \mathrm{Ra} = \frac{g\beta\left(T_i - T_a\right)B^3}{v\alpha}$$

(9.19)

The symbol D/Dt denotes the substantive derivative that consists of the local and convective contributions. The first two momentum equations are supplemented by the usual continuity equation and are written in dimensionless variables as well as the boundary conditions. The third energy equation is written in dimension variables in terms of enthalpy J determined as integrated differential $c_p dT$. The dimensionless variables are scaled by mold width B for coordinates, α_L/B for velocities, λ_S for thermal conductivity, and B^2/α_L for time. The first two boundary conditions are the conjugate equalities at the inner surfaces of the mold, and the third is the heat transfer condition at the outer surfaces of the mold. The others are the no-slip condition and thermal conditions at the boundaries given in the authors' paper [7]. The subscripts L, S, m, i and a denote liquid, solid, mold, initial, and ambient, respectively.

The phase-change control volume is considered using the model of a porous medium with a liquid fraction f depending on the latent heat content. The source terms S_u incorporated in the momentum equations are expressed by applying the Darcy law according to which the velocity is proportional to the pressure gradient. The change of phase is regarded by the source term S_h included in the energy equation. For more details, see Reference [7].

The numerical solution is performed using the SIMPLE algorithm. The marker and cell scheme that are similar to the weighted residual method and staggered grid are employed for discretization (Sections 7.2 and 7.3). The grid refinement test for a typical problem, code validation, and other details of the numerical approach may be found in Reference [7].

The following results and conclusions are obtained:

1. One of the reasons to use the enthalpy method is that in this approach a simple examination of the numerical solution shows the location of grid points involved in the phase-change process. It is known that in the enthalpy formulation, the change-phase front takes up several grid points [5]. In this case, it occupies two points that can correspond to one control volume if a grid is fine enough. Extrapolation or interpolation of these results leads to unrealistic stepwise behavior of the moving boundary and the time history of the temperature.

2. The Nusselt number is calculated for all four walls at the solid-mold interface. The influence of the mold material, width, and aspect ratio on the solidification process is studied. The following parameters are used for the reference computation: aspect ratio $a = 2$, Stephan number Ste = 2.4, Pr = 0.011 (tin), Re = 10^4.

   The average Nusselt number at the right wall is computed for the tin as a cast metal in copper, steel, and sand molds. The copper mold showed the largest and the sand mold the smallest rates of heat transfer. Similar results are obtained for the top wall. On the curves Nu(t) obtained for metal molds, waviness can be seen, which is less for the sand mold due to low conductivity. At the same time, high thermal resistance of the sand mold results in higher vertical velocity and the highest circulation. The smallest velocity is observed in the copper mold. The interface movement is less in the sand mold than that in copper and steel molds, which yields more effective dependence Ra(t) and higher velocity.

3. The temperature-time history obtained for a point at $x = 3$, $y = 5$ shows that due to high heat transfer in a copper mold, the process of solidification is about 75% completed when the temperature at this point achieves the freezing level. In the steel mold at the same time, the solidification is less completed and in the sand mold only about 10% is solidified. In addition, a highly conducting mold heats up very gradually. Temperature profiles across the section indicate that temperature gradients in the liquid are lower than those in the solid. In the copper mold, the liquid temperatures are isothermal across the section, and the profiles in two other molds show temperature variation across the section.

4. The effect of three values 0.5, 2, and 4 of the aspect ratio is investigated in a steel mold. The interface is basically flat for low aspect ratio (width to height) and curves at the top and bottom due to high heat transfer from the liquid–solid region to the mold. This confirms the usefulness of a one-dimensional model for molds with a small aspect ratio. At the same time, the one-dimensional approach in the case of a large aspect ratio yields significant errors. The solidification proceeds faster in cavities with a small aspect ratio, because under an adiabatic top and bottom, only the side walls conduct the heat. The temperature profiles are observed to be parabolic in the solid and almost uniform in the liquid. The streamlines and isotherms in the cavity with small aspect ratio show two tall counterrotating cells in the melt region that decrease in size as the solidification progresses. The isotherms also show high-temperature gradients near the solid–liquid interface and approximately constant in time temperature gradients in the mold.

5. The data obtained for three mold widths indicate that as width increases, the heat transfer through the mold grows due to increased energy storage. The circulation in thinner cavities is higher because the heat transfer in those is less. The isotherms are curved at the top and bottom at early times and later become flat.

6. It is concluded that the problem of solidification must be considered in the conjugate formulation to obtain accurate results.

Example 9.4: Czochralski Crystal Growth Process [8–10]

The Czochralski process is used for growing semiconduction (e.g., silicon) crystals. The process is performed in an apparatus with a cylindrical crucible heated in a furnace above the melting temperature of the melt. The crystal and the crucible are rotated in opposite directions. The resulting cylindrical crystal is vertically pulled from the crucible.

Because the temperature in this process is high, the radiative heat transfer in addition to conduction and convection should be taken into account. The three-dimensional axisymmetric heat transfer model is studied. Due to axial symmetry, only the cross-section containing schematic liquid, solid, and filling gas areas is considered. In addition to the usual assumptions, the following are used [10]:

1. The diameter of the crystal pulled from the crucible is constant.
2. The studied region with liquid, solid, and gas is an enclosure.
3. The effects of rotation are negligible.
4. The crystal and liquid have the same radiative properties.

The three-dimensional steady-state momentum (with buoyancy) and energy (with radiative term)

$$\rho(\mathbf{v}\cdot\nabla)\mathbf{v} = -\nabla p + \mu\nabla^2\mathbf{v} + \rho g\beta(T-T_\infty), \quad \nabla T(\mathbf{r}) = (1/\alpha)v\cdot\nabla + (1/\lambda)q(\mathbf{r}) \tag{9.20}$$

and continuity $\nabla\cdot\mathbf{v} = 0$ equations governed the problem [9]. The radiative heat source is calculated considering the liquid and solid phases as a semitransparent medium with constant absorption coefficients k_L and k_S, while the surrounding gas is considered as a transparent medium with $k = 0$. The radiative source q^r is defined as a radiant energy absorbed by an infinitesimal surface within an infinitesimal time. In the three-dimensional case, this amount of energy is determined by the following coupled integral equations [9,10]:

$$q^r(\mathbf{p}) + \varepsilon(\mathbf{p})e_b[T(\mathbf{p})] = \varepsilon(\mathbf{p})\int_S \left\{ e_b[T(\mathbf{r})] + \frac{1-\varepsilon(\mathbf{r})}{\varepsilon(\mathbf{r})}q^r(\mathbf{r}) \right\}\tau(\mathbf{r},\mathbf{p})K(\mathbf{r},\mathbf{p})dS(\mathbf{r})$$

$$+ \varepsilon(\mathbf{p})\int_S \left\{ \int_{L_{rp}} k(\mathbf{r}')e_b[T^m(\mathbf{r}')\tau(\mathbf{r}',\mathbf{p})]dL(\mathbf{r}') \right\}K(\mathbf{r}.\mathbf{p})dS(\mathbf{r}) \tag{9.21}$$

$$q_v^r(\mathbf{p}) + 4k\left(\mathbf{p}\right)e_b[T^m(\mathbf{p})] = k(\mathbf{p})\int_S \left\{ e_b[T(\mathbf{r})] + \frac{1-\varepsilon(\mathbf{r})}{\varepsilon(\mathbf{r})}q^r(\mathbf{r}) \right\}\tau(\mathbf{r},\mathbf{p})K(\mathbf{r},\mathbf{p})dS(\mathbf{r})$$

$$+ k(\mathbf{p})\int_S \left\{ \int_{L_{rp}} k(\mathbf{r}')e_b[T^m\left(\mathbf{r}'\right)\tau(\mathbf{r}',\mathbf{p})]dL(\mathbf{r}') \right\}K_r(\mathbf{r}'.\mathbf{p})dS(\mathbf{r}) \tag{9.22}$$

Here **r** and **P** refer to current and observed points, whereas point **r'** is located on the line L_{rp} connecting points **r** and **P**; the integration is performed over the surface of the computational domain S and along the line L_{rp}; ε is the emissivity; $e_b(T)$ and $e_b(T^m)$ are the black-body emissivities of the computational domain surface and of the semitransparent medium, respectively. The kernel functions and transmissivity are

$$\mathbf{K}(\mathbf{r}, \mathbf{p}) = \frac{\cos\phi_r \cos\phi_p}{\pi |\mathbf{r} - \mathbf{p}|^2}, \quad \mathbf{K}_r(\mathbf{r}, \mathbf{p}) = \frac{\cos\phi_p}{\pi |\mathbf{r} - \mathbf{p}|^2} \tau(\mathbf{r}, \mathbf{p})$$

$$= \exp\left[-\int_{L_{rp}} k(\mathbf{r}') dL_{rp}(\mathbf{r}') \right]$$

(9.23)

The derivation of these equations and other details may be found in Reference [10].

The usual boundary conditions should be satisfied on the external surfaces of the studied enclosure. The conjugate condition on the phase-change front involves continuity of the melting temperature, a jump in heat flux, and a no-slip condition for the melt velocity:

$$T_L(\mathbf{r}) = T_S(\mathbf{r}) = T_{ph} \quad -\lambda_L \frac{\partial T_L}{\partial n_{ph}} + \lambda_S \frac{\partial T_S}{\partial n_{ph}} = -\Lambda_{ph}\rho_S(v \cdot \mathbf{n}_{ph}), \quad \mathbf{v}_L = \mathbf{v}_x \quad (9.24)$$

On the solid–gas and liquid–gas interfaces, equalities of temperatures and heat fluxes including the radiative component q_r should be satisfied:

$$T_L(\mathbf{r}) = T_G(\mathbf{r}), \quad \lambda_L \frac{\partial T_L}{\partial n} = \lambda_G \frac{\partial T_G}{\partial n} + q_{rL}, \quad T_S(r) = T_G(r), \quad \lambda_S \frac{\partial T_S}{\partial n} = \lambda_G \frac{\partial T_G}{\partial n} + q_{rS} \quad (9.25)$$

In these conditions, T_{ph}, $\mathbf{n}_{ph}$ and Λ_{ph} are the phase-change temperature, vector normal to phase-change surface, and the latent heat.

The numerical solution is performed using the commercial package FLUENT which is based on the finite-volume approach and other soft-ware based on the boundary element method (Chapter 7). The FLUENT is applied for determining velocity, pressure, and temperature fields under known distribution of the radiative heat source. The computational domain is divided into liquid, solid, and gas subdomains that are numerically analyzed separately. To calculate the radiative heat fluxes and source, the BEM code is used. The iterative procedure is employed to couple both numerical results. More details are given in Reference [10]. As an example, the velocity and temperature profiles in liquid phase and radiative heat source in a typical three-dimensional Czochralski process are given [9].

Example 9.5: Concrete at Evaluated Temperatures [11,12]

The solid skeleton of the hardened cement paste is porous hygroscopic material including various chemical compounds and pores filled with liquid and vapor water and dry air. On the condition of high temperature, heat changes the chemical compounds and fluid content resulting in changes in physical structure that affect the mechanical and other properties of the concrete. The articles in question consider the heat and moisture transfer in concrete exposed to high temperatures using comprehensive three-dimensional models. Both works take into account the basic phenomena, but additional effects of capillary pressure and adsorbed water are studied in Reference [12].

The basic assumptions are as follows [11]:

1. There is a thermal equilibrium between phases.
2. Vapor and air behave as ideal gases.
3. The temperature dependence of mass fluxes is negligible.

The conservation of mass and energy equations for dry air and moisture are as follows:

$$\frac{\partial \varepsilon_G \tilde{\rho}_A}{\partial t} = -\nabla \cdot \mathbf{J}_A \quad \frac{\partial \varepsilon_G \tilde{\rho}_V}{\partial t} + \frac{\partial \varepsilon_{FW} \rho_L}{\partial t} - \frac{\partial \varepsilon_D \rho_L}{\partial t} = -\nabla \cdot (\mathbf{J}_V + \mathbf{J}_{FW}) \tag{9.26}$$

$$\rho c \frac{\partial T}{\partial t} - \Lambda_E \frac{\partial \varepsilon_{FW} \rho_L}{\partial t} + (\Lambda_E + \Lambda_D) \frac{\partial \varepsilon_D \rho_L}{\partial t} = \nabla \cdot (\lambda \nabla T) + \Lambda_E \nabla \cdot \mathbf{J}_{FW} - (\rho c v) \cdot \nabla T \tag{9.27}$$

Here ε and $\mathbf{J}$ are volume fraction and mass flux of phase, subscripts A, D, G, L, S, V and FW denote dry air, chemically bound water released by dehydration, gas, liquid, solid, vapor, and free water (combined liquid and adsorbed); Λ_E and Λ_D are the specific heat of evaporation and dehydration, respectively. The mass fluxes of dry air, vapor, and free water are defined using Darcy's and Fick's laws and ignoring the diffusion of the adsorbed water on the surface:

$$\mathbf{J}_A = \varepsilon_G \tilde{\rho}_A \left[\mathbf{v}_G - D_{AV} \nabla (\tilde{\rho}_A / \tilde{\rho}_G) \right], \quad \mathbf{J}_V = \varepsilon_G \tilde{\rho}_A \left[\mathbf{v}_G - D_{VA} \nabla (\tilde{\rho}_F / \tilde{\rho}_G) \right],$$

$$\mathbf{J}_{FW} = \varepsilon_{FW} \rho_L \mathbf{v}_G$$

$$\mathbf{v}_G = (KK_G / \mu_G) \nabla p_G, \quad \mathbf{v}_L = (KK_L / \mu_L) \nabla p_G,$$

$$\varepsilon_{FW} = (\varepsilon_{CM} / \rho_{CM}) f[(p_V / p_{ST}), T]$$

$$\tag{9.28}$$

In these relations, $D_{AV} \approx D_{VA}$ are diffusion coefficients of dry air and water vapor; $\tilde{\rho}$ is the mass of a phase per unit volume; K, K_G, and K_L are the permeability of dry concrete and the relative permeabilities of the gas and the liquid. The pressure of air p_A and of vapor p_V are determined using the state equation of ideal gas $p = R\tilde{\rho}T$ as well as the pressure in a gas and water $p_L \approx p_G = p_A + p_V$, where R is the gas constant. The volume fraction of free water ε_{FW} is determined from the equation of sorption isotherms that relates the free water content to the cement content $\varepsilon_{CM}/\rho_{CM}$, temperature and relative humidity p_V/p_{ST}, where p_{ST} is the saturation pressure of the vapor. For temperatures above the critical point of water (374.14°C) the free water content is zero and the gas volume fraction can be defined from $\phi = \varepsilon_{FW} + \varepsilon_G$, where ϕ is the concrete porosity.

Using these basic relations, the system of differential equations is formulated which consists of 10 equations in Reference [11] and an additional 6 equations in Reference [12]. The boundary conditions that are the same in both models include the following expressions:

$$\frac{\partial T}{\partial n} = \frac{h_{GR}}{\lambda}(T_\infty - T), \quad \mathbf{J} \cdot \mathbf{n} - \gamma(\tilde{\rho}_V - \tilde{\rho}_{V,\infty}) \tag{9.29}$$

$$\frac{\partial \tilde{\rho}_V}{\partial n} + \frac{K_{VT} h_{GR}}{K_{VV}\lambda}(T_\infty - T) + \frac{\gamma}{K_{VV}}(\tilde{\rho}_{V,\infty} - \tilde{\rho}_V) \tag{9.30}$$

where h_{GR} and γ are coefficients of heat transfer and of vapor mass transfer on the boundary. The first and the second equations are the energy and gaseous mixture mass balances, and the third expression determines the vapor content gradient on the boundary.

The basic results may be formulated as follows:

1. In both studies, a steep drying front is observed. The vapor and liquid water content of the phase mixture changes from high and low on the hot side to low and high on the cold side. The water content increases ahead of the drying front due to the recondensation in the cooler zone which is named the "moisture clog zone." The maximum of peaks in a gas pressure and vapor content obtained in the modified model [12] are lower than for the initial model [11]. This may be significant in analyzing potential spalling, because the internal pore pressure is considered as the cause of it. The modified model also predicts more extensive moisture clog zones in which the liquid pressure gradient drives water away from the face exposed to fire. These results show that ignoring the adsorbed water flux can significantly affect the predicted values of free water flux, vapor content, and gas pressure.

2. The values of liquid pressure predicted by two models are considerably different. The capillary pressure is zero according to the initial model [11], and it increases rapidly with a decrease in relative humidity according to the modified model [12]. However, the overall results given by both models are similar, showing that the

capillary pressure has little or no effect on the transport in concrete under intense heating.

3. If both capillary pressure and adsorbed water are taken into account, the gas pressure and vapor content are higher than shown by the initial model or by the modified model, including only capillary pressure. In this case, the results are physically realistic in contrast to the unrealistic behavior of capillary pressure as the adsorbed water is ignored.

9.2 Drying and Food Processing

Example 9.6: Drying a Suspended Wood Board [13]

The basic assumptions are as follows:

1. There is equilibrium at each point and time.
2. The board is unsaturated and homogeneous.
3. Gravity's effects are negligible.
4. The characteristic length scales of the drying medium are much smaller than those of the external fluid.
5. The thickness of the interfacial surface is negligible.

The problem is governed by continuity, momentum, energy, and conservation of species equations and boundary and conjugate conditions:

$$\nabla \cdot \mathbf{u} = 0, \quad \mathbf{u} \cdot \nabla \mathbf{u} = -\nabla p + \frac{1}{Re} \nabla^2 \mathbf{u}, \quad \mathbf{u} \cdot \nabla T = \frac{1}{Re\,Pr} \nabla^2 T, \tag{9.31}$$

$$u \cdot \nabla \omega = \frac{1}{Re\,Sc} \nabla^2 \omega$$

$$\lambda_a \nabla T \cdot \mathbf{n} + \Lambda \rho_a D \nabla \omega \cdot \mathbf{n} = (\lambda_a + \varepsilon \Lambda \alpha_m \gamma) \nabla T \cdot \mathbf{n} + \varepsilon \Lambda \alpha_m \nabla M \cdot \mathbf{n} \tag{9.32}$$

$$T = T_w, \quad \omega = \phi(T_w, M_w), \qquad \rho_a D \nabla \omega \cdot \mathbf{n} = \alpha_m \gamma \nabla T \cdot \mathbf{n} + \alpha_m \nabla M \cdot \mathbf{n} \tag{9.33}$$

where ω is the water vapor mass fraction. All variables are dimensionless and scaled by $L, u_\infty, \rho u_\infty^2, T_\infty, L/u_\infty$ and ω_∞ for the coordinates, velocity, pressure, temperature, time, and water vapor mass fraction, and subscripts a and m denote air and moisture.

The function $\phi(T_w, M_w)$ relates absolute humidity of the air to equilibrium values of temperature and moisture content on the solid (Example 8.8), which in this study is as follows [14]:

$$\phi = 1 - \exp(AM^B), \quad A = -\frac{0.13}{T} \left(1 - \frac{T}{T_c}\right)^{-6.46}, \quad B = 110T^{-0.75}, \quad \omega = 0.62198 \frac{\phi p_w}{p - \phi p_w} \tag{9.34}$$

where $T_c = 647.14K$ is the critical temperature of water. The last relation gives the vapor mass fraction, where p is the total pressure of the humid air, and p_w is the saturation pressure. The finite-element method is used to solve the set of governing equations. Two examples of stream-lined rectangular board are considered. The initial conditions of the stream of air and solid are $T_\infty = 60°C$, $\omega_\infty = 0.116kg/kg$, Re $= 200$, $Sc = 0.6$ and $T_0 = 25°C$, $M_0 = 222° M(m.c. 40\%)$, respectively.

1. In the first example, the drying of only the upper surface of the board is studied. Both vertical surfaces are adiabatic and imper-meable. The solid temperature and moisture potential distribu-tions at 3 h time drying show that both the heating and the drying of the board are nonuniform. The leading part of the board heats much faster than the rest of it. As the air flows along the surface, the temperature gradient between the flow and solid decreases. Due to this, the heat fluxes decrease in the flow direction. For the longer drying time due to evaporation, the heat fluxes increase in the flow direction. A dry zone appears close to the leading edge, while the remainder remains wet. As air flows, it becomes more saturated, and both the humidity gradients and moisture fluxes of the surface decrease in the flow direction.

2. In the second example, the drying of the entire board under the same conditions is investigated. In this case, the results are close to those outlined for the first example. Only the temperature and moisture potential distribution are different due to the fact that the vertical surfaces are also involved in the process. Both heat and mass transfer are less intensive on the vertical surfaces because the air velocities at the leading and aft edges are significantly reduced. The drying front near the aft edge is less penetrated than that in the area close to the leading edge. The average moisture potential variations show that board drying in the second example occurs faster than that in the first case. Despite the fact that the difference in final mois-ture content is only about 1%, the difference in moisture distribution is considerable, which can be important for using the wood board.

These results are obtained under the assumption of a homogeneous wood board and are pertinent only to this case. For longer drying times, the observed difference between the final average moisture content is expected to increase.

Example 9.7: Drying Rectangular Brick [15]

A saturated water brick of aspect ratio 2 is placed in airflow with the same initial temperature, and the concentration difference between the brick surface and ambient causes the evaporation of water from the brick. The reduction of brick temperature caused by evaporation is continued until the brick reaches the wet bulb temperature at which it remains for a long period of time. Finally, the airflow heats the brick.

To study this drying process, the two-dimensional Navier-Stokes equa-tions including buoyancy terms are solved for flow domain because the boundary layer approximation is not applicable for a short brick. This solution is conjugated with solutions of energy and moisture transport equations for the brick.

The governing system includes continuity, momentum, energy, and concentration equations for flow

$$\frac{\partial u}{\partial x}+\frac{\partial v}{\partial y}=0 \quad \frac{\partial u}{\partial t}+u\frac{\partial u}{\partial x}+v\frac{\partial u}{\partial y}=-\frac{1}{\rho}\frac{\partial p}{\partial x}+\frac{\mu}{\rho}\left[\frac{\partial^2 u}{\partial x^2}+\frac{\partial^2 u}{\partial y^2}\right]$$

$$\frac{\partial v}{\partial t}+u\frac{\partial v}{\partial x}+v\frac{\partial v}{\partial y}=-\frac{1}{\rho}\frac{\partial p}{\partial y}+\frac{\mu}{\rho}\left[\frac{\partial^2 v}{\partial x^2}+\frac{\partial^2 v}{\partial y^2}\right]+g\beta(T-T_\infty)+g\beta(C-C_\infty)$$

(9.35)

$$\frac{\partial T}{\partial t}+u\frac{\partial T}{\partial x}+v\frac{\partial T}{\partial y}=\frac{1}{\rho c_p}\left[\frac{\partial^2 T}{\partial x^2}+\frac{\partial^2 T}{\partial y^2}\right]$$

$$\frac{\partial C}{\partial t}+u\frac{\partial C}{\partial x}+v\frac{\partial C}{\partial y}=D\left[\frac{\partial^2 C}{\partial x^2}+\frac{\partial^2 C}{\partial y^2}\right],$$

and energy and conservation for total (liquid and vapor) moisture equations:

$$\rho_0\frac{\partial}{\partial t}(\Sigma M_j I_j)=\frac{\partial}{\partial x}\left(\lambda\frac{\partial T}{\partial x}\right)+\frac{\partial}{\partial y}\left(\lambda\frac{\partial T}{\partial y}\right)-\frac{\partial}{\partial x}(\Sigma J_j I_j)-\frac{\partial}{\partial y}(\Sigma J_j I_j)$$

$$\frac{\partial M}{\partial t}=\frac{\partial}{\partial x}\left[(D_{M1}+D_{M2})\frac{\partial M}{\partial x}+(D_{T1}+D_{T2})\frac{\partial T}{\partial x}\right]$$

(9.36)

$$+\frac{\partial}{\partial y}\left[(D_{M1}+D_{M2})\frac{\partial M}{\partial y}+(D_{T1}+D_{T2})\frac{\partial T}{\partial y}\right]$$

Here D_T and D_M are isothermal and nonisothermal diffusion coefficients; I and J are enthalpy and mass flux; and j is 0, 1, 2 representing dry solid, liquid, and vapor, respectively. The last part of Equation (9.36) is obtained by applying Darcy's law for capillary liquid mass flow and Fick's law for diffusive mass flux. The boundary conditions on the interface consist of a no-slip velocities condition, continuity of temperature and moisture conditions, and heat and species balance (f refers to flow domain). At the inlet and far away from the solid, the airflow has the following ambient parameters:

$$u=v=0, T_w=T_f, C_f=C(T,M)_w, \quad [\lambda+(I_2-I_1)D_{T2}]\frac{\partial T}{\partial n}+(I_2-I_1)D_{M2}\frac{\partial M}{\partial n}$$

$$=\lambda_f\frac{\partial T_f}{\partial n}+(I_2-I_1)D\frac{\partial C_f}{\partial n}, \quad D_{T2}\frac{\partial T}{\partial n}+D_{M2}\frac{\partial M}{\partial n}=D\frac{\partial C_f}{\partial n},$$

(9.37)

$$u=U, \ v=0, \ T=T_a, \ C=C_a$$

At outflow, a smooth extrapolation is used and the bottom is adiabatic.

Finite element Galerkin's method (Chapter 7) is applied to solve the set of governing equations. First, Equation (9.36) is simplified using some assumptions (see details in Reference [15]). Because the characteristic time of the flow field is much smaller than that of the heat and moisture transfer, the quasi-steady state approximation is applicable (Chapter 6). Due to this, solutions of energy and concentration Equation (9.35) are obtained using steady-state velocities. To calculate fluxes at the interface from Equation (9.37), the concentration of the fluid at the interface is needed. It is determined by knowing the local relative humidity obtained from the equation of desorption isotherm [16] $\phi = 1 - \exp(-17M^{0.6})$. The fluxes on the interface are used as boundary conditions for solving Equations (9.36) for the brick. With known temperature and concentration distribution on the surface of the solid, the Equation (9.35) for air is solved for the next time step. This calculation procedure of drying is continued until the final parameters of the brick are achieved. The full drying time of 36 h takes 252 h calculation time.

The following basic results and conclusion are obtained:

1. Different examples are computed to compare with available literature data. In particular, the same problem as in Reference [13] (Example 9.6) is simulated. The comparison shows that the differences in the temperature distributions between results obtained by Navier-Stokes and by boundary layer equations are small, but the differences in moisture distributions are more considerable. These differences are seen to increase away from the leading edge. The studying of conjugation effects shows that the use of constant heat and mass transfer coefficients overpredicts the moisture removal.

2. The conjugate drying result for mixed convection indicates that in the initial period, the surface temperature decreases because the heat for vaporization is absorbed from the brick. In time, the convective heat flux from ambient goes into solid. However, the amount of this heat is still not enough for vaporization, and the brick temperature decreases, finally reaching the uniform wet bulb temperature. The distribution of the moisture content along the brick surface shows that the evaporation is higher at the leading edge than that in other regions. This occurs due to a thin concentration boundary layer developed at the leading edge. With time, the rate of evaporation decreases, but finally, when the solid attains uniform temperature, the significant moisture gradients still exist due to a smaller mass transfer diffusion coefficient.

3. The result for mixed drying is compared with that for forced drying obtained by ignoring the buoyancy terms in Equation (9.35). The comparison shows that the temperature drop in the case of a mixed regime is more than that for the forced convection. As a result, the wet bulb temperature is reached much earlier in the case of mixed convection. The same qualitative result is observed for moisture content.

4. Variations of average Nusselt and Sherwood numbers in time are studied. Two definitions of Nusselt and Sherwood numbers are considered: using instantaneous temperature and concentration differences between the ambient and the surface of the brick, and the similar differences between the ambient and wet bulb conditions. It is shown that the Nusselt and Sherwood numbers of the second type asymptotically approach the values of Nusselt and Sherwood numbers of the first type. The values of the Nusselt number increase with time until the wet bulb conditions are reached. The Sherwood number is constant for most of the drying period or decreases due to diminishing moisture potential. It is observed that the analogy between heat and mass transfer is applicable only after wet bulb conditions are reached.

Example 9.8: Drying of Capillary Porous Material [17]

The external airflow is flowing parallel to porous material that is assumed to be unsaturated, homogeneous, and nondeformable. This relatively simple model is used to simulate various situations studying the behavior of heat and mass transfer coefficients at the interface during the drying process.

The system of governing equations and boundary conditions is almost the same as in Example 9.6 with small variations. The system for a solid in both studies is based on the Luikov theory [18]. The finite element method is used to calculate the heat and mass transfer coefficients. The first (the moisture content is above the maximum sorption value) and the second periods are considered.

The main results are as follows:

1. For the first period, the initial moisture content is assumed to be 8%. The results show the thermal and mass leading edge effect, which leads to high heat and mass fluxes at the interface. Intensive evaporation and mass transfer toward the interface occur.

   For the second period, the initial moisture content is assumed to be 3%. The obtained temperature and moisture fields show a dry or sorption zone linked to recession of an evaporation front. In the dry zone, the basic moisture transport occurs by vapor. In the sorption zone, the bound water exists. The leading edge effect is also seen in this case.

2. The heat and mass transfer analogy is observed in the first drying period. This result can be expected because the specific humidity at the interface depends only on the temperature. However, depending on the situation, this analogy may or may not be valid (Example 8.8). In the second period, the specific humidity on the interface depends on the temperature and moisture content. Therefore, the boundary conditions at the interface for heat and mass transfer may be different. Thus, in this case, the analogy between heat and mass transfer may not be valid.

3. Because the temperature and the specific humidity at the interface are not uniform, the obtained heat and mass transfer coefficients differ from standard values corresponding to the case of a plate with constant temperature and moisture content. However, variations from the reference values in this study are only about 10%. This result is in conflict with huge variations from standard values obtained by many authors using one-dimensional simulations (for instance, Reference [19] and Example 8.8).

It is shown by comparing two-dimensional and one-dimensional solutions of the same problem that the one-dimensional approach cannot generally give realistic results because the one-dimensional simulation ignores the thermal and mass boundary layer leading edge effects.

4. The length scales of external flow are usually large compared with microscopic scales of a porous medium. Due to this fact, the porous medium is assumed to be homogeneous. To estimate the effect of macroscopic heterogeneities of moisture content on the mass transfer coefficient, the moisture content distribution obtained in the problem in question is used. The same problem is solved by applying this nonuniform moisture content distribution and mass transfer coefficient. The average mass flux at the interface obtained by using this mass transfer coefficient and nonuniform moisture distribution differs from the standard by about 10%.

Example 9.9: Multidimensional Freeze-Drying of Slab-Shaped Food [20]

Unlike the conventional drying process that is based on capillary motion and evaporation of water, the freeze-drying process uses sublimation of ice to dry the object. The low temperature and pressure below the triple point in freeze-drying provide a high-quality freeze-dried product. Despite the high quality of the final product, conventional drying methods are basically used in food production due to the long drying time and the high cost of the freeze-drying process. There are many investigations of drying, but most of them studied the alternative conventional drying methods.

The model consists of planar and slab-shaped products that are divided by a sublimation interface into the dried and frozen parts. The moisture or ice in the products sublimate under vacuum pressure and developed vapor m_v diffuses through pores to exit. The energy for sublimation comes from the bottom by conduction and by radiation from the upper heating plate. As a result, the uniform sublimation interface below the top forms in the planar product. In the slab-shaped product, the additional radiation energy comes through the lateral surface opened to the drying chamber. Due to this, the sublimation interfaces formed below the top and beside the lateral surface are nonuniform. Such a drying process in a slab-shaped product in contrast to that in a planar product requires multidimensional analysis.

The problem is governed by the mass and energy conservation equations which are derived and detailed in Reference [20]:

$$\varepsilon\frac{(1-S)\rho_V-(1-S^0)\rho_V^0}{\Delta t}\Delta V+\sum_{j=E,W,N,S}(\mathbf{m}_V)_j\cdot\mathbf{n}_j=-\varepsilon\rho_I\frac{S-S^0}{\Delta t}\Delta V$$

$$(\rho c_p)^0\frac{T-T^0}{\Delta t}\Delta V+\sum_{j=E,W,N,S}(-\lambda\nabla T)_j\cdot\mathbf{n}_j=\varepsilon\rho_I\Lambda_S\frac{S-S^0}{\Delta t}\Delta V$$

(9.38)

Here ΔV is the volume of the grid cell; ε and ε_I denote the porosity and the fraction of the ice volume to the total volume; $S=\varepsilon_I/\varepsilon$ is ice saturation; Λ_S is heat of sublimation of ice; $\mathbf{n}_j$ is the outward normal vector; the subscript j denotes the control surface of a cell with East, West, North, and South faces; and superscript 0 refers to initial values. The vapor flux is determined by summing the diffusion and flow through porous products:

$$\mathbf{m}_V=-(1-S)\frac{m_M}{RT}\left[D+p_V\frac{K}{\mu_V}\right]\nabla p_V$$

(9.39)

where m_M, D, and K are molecular mass, the effective diffusivity, and Darcy permeability, respectively.

The first term of the mass conservation, Equation (9.38), represents the change in the vapor contained in a particular cell, and the second term determines the vapor flow out of this cell. These changes yield the reduction of the rate of ice saturation S defined by the right term of the mass conservation equation. Similarly, the first and second terms of the energy conservation, Equation (9.38), represent the change of energy inside the cell and heat flux through the control surface. The sum of these two terms is equal to the last term of the energy equation which defines the latent heat arising due to the sublimation of ice.

The temperature boundary conditions for the top, bottom, and lateral surfaces opened to the drying chamber are

$$q_T=\sigma F_T(T_H^4-T_T^4)\qquad q_B=h\,(T_H-T_B)\qquad q_S=\sigma F_S(T_H^4-T_S^4)$$

(9.40)

where subscripts T, B, S, and H denote top, bottom, and side surfaces of product and heating plates, respectively; F is the radiation shape factor; and h is the overall heat transfer coefficient for the bottom. The pressure boundary condition for surfaces opened to the drying chamber (subscript C) is $p_V=p_{VC}$. As the initial condition at $t=0$, the uniform temperature, pressure, and rate of ice saturation are used.

The numerical procedure starts from computing the temperature distribution by solving the second part of Equation (9.38). Then, the distribution of vapor pressure is obtained from the first part of Equation (9.38) only for dried cells with $S=0$. To calculate the pressure in the frozen and sublimation cells, those are treated as Dirichlet boundaries assuming the local thermodynamic equilibrium when the vapor is saturated and its pressure is defined as $p_V = p_S(T)$. The evolution of ice saturation S in the frozen and sublimation cells is calculated by solving the mass conservation equation (the first part of Equation 9.38). Iterative calculations are repeated until all variables are obtained with desired accuracy.

The numerical results are obtained using beef as the product:

1. The average sublimation temperature of the slab-shaped product is 5% to 10% lower than that of the planar product. The reason for this is that curved sublimation interfaces in a slab-shaped product caused by literal surface opened to the drying chamber. As a result, the diffusion length is decreased and the interfacial area is increased. These two effects enabled the shorter drying time with a lower sublimation temperature, and the primary direction of drying changes from vertical from top to bottom in a planar product to radial in a slab-shaped product from lateral surface to inner core.

2. The distribution of ice saturation, temperature, and vapor pressure in a planar product shows the existence of the second sublimation interface near the bottom. The vapor from the secondary interface is transported out of a product by diffusion or is deposited in the frozen region as ice. The maximum temperature and vapor pressure increase in time. The ice saturation in the slab-shaped product in the frozen region remains relatively constant at about 0.7. The distribution of the temperature and vapor pressure in the slab-shaped product shows that the heat and mass transfer during the frozen drying in this case is a multidimensional process. Despite the fact that spatial temperature gradients in the dried area are larger than those in the frozen region, the heat transfer through this area is small due to low thermal conductivity. The spatial vapor pressure gradients in the frozen region caused by the temperature gradients according to dependence for saturated pressure $p_S(T)$ are also relatively small because of a small pore space in the frozen area.

3. The main source of energy is the convection from the bottom, while the radiation from the top and lateral surface supplies about 40% of total energy. Because a dried region is developed near the bottom, the heat flow from the bottom decreases from the initial 2.86 to the final 0.93 J/s. The initial vapor flow through the lateral surfaces is about two times larger than that through the top surface. At the end of the drying process, almost 80% of vapor flows through the lateral surface.

4. The results obtained for product of heights 5, 10, 15, and 20 mm indicate the relatively constant drying rate for the planar product. The slab-shaped product shows more nonlinear behavior. The high drying rate at the beginning changes to a lower drying rate in the latter parts of the drying process. The reason for this is that the insulating dried region surrounded the frozen area. For the

height of 5 mm, both planar and slab-shaped products exhibit almost the same drying time, but the difference increases with time because the drying time of the slab-shaped product is less sensitive to the product height than that of the planar product. The configuration of the sublimation interfaces of the slab-shaped product with different heights indicates that the primary drying direction is the radial one from the lateral surface to the inner core.

5. It is concluded that the lateral surface of the slab-shaped product is favorable for the reduction of both the drying time and the sublimation temperature by increasing the vapor diffusion and the interfacial area for sublimation.

References

1. Davis, E. J., and Venkatesh, S., 1979. The solution of conjugated multiphase heat and mass transfer problems, *Chem. Engn. Sci.* 34: 775–787.
2. Davis, E. J., 1973. Exact solution for a class of heat and mass transfer problems, *Can. J. Chem. Engn*, 51: 562–572.
3. El Qarnia, H., 2004. Theoretical study of transient response of a rectangular latent heat thermal energy storage system with conjugate forced convection, *Energy Conver. Manag.* 45: 1537–1551.
4. Voller, V. R., 1990. Fast implicit finite-difference method for the analysis of phase change problems, *Numer. J. Heat Transfer, Part B* 17: 155–169.
5. Li, H., Hsieh, C. K., and Goswami, D. Y., 1996. Conjugate heat transfer analysis of fluid flow in a phase-change energy storage unit, *Int. J. Numer. Meth. Heat Fluid Flow* 6: 77–90.
6. Viswanath, R., and Jaluria, Y., 1995. Numerical study of conjugate transient solidification in an enclosed region, *Numer. Heat Transfer, Part A* 27: 519–536.
7. Viswanath, R., and Jaluria, Y., 1993. Comparison of different solution methodologies for melting and solidification problems in enclosures, *Numer. Heat Transfer, Part A* 24: 77–105.
8. Nowak, A. J., Biaecki, R. A., Fic, A., Wecei, G., Wrobel, L. C., and Sarier, B., 2002. Coupling of conductive, convective and radiative heat transfer in Czochralski crystal growth process, *Comput. Material Sci.* 25: 570–576.
9. Nowak, A. J., Biaecki, R. A., Fic, A., and Wecei, G., 2003. Analysis of fluid and energy transport in Czochralski's process, *Comput. Fluid* 32: 85–95.
10. Wecel, G., 2006. BEM-FVM solution of the conjugate radiative and convective heat transfer problems, *Arch. Comput. Meth. Engn.* 13: 171–248.
11. Tenchev, R. T., Li, L. Y., and Purkiss, J. A., 2001. Finite element analysis of coupled heat and moisture transfer in concrete subjected to fire, *Numer. Heat Transfer, Part A* 39: 685–710.
12. Davie, C. T., Pearce, C. J., and Bicanic, N., 2006. Coupled heat and moisture transport in concrete at evaluated temperatures — Effects of capillary pressure and adsorbed water, *Numer. Heat Transfer, Part A* 49: 733–763.
13. Oliveira, L. S., and Haghighi, K., 1998. Conjugate heat and mass transfer in convective drying of porous media, *Numer. Heat Transfer, Part A* 34: 105–117.

14. Zuriz, C., Singh, R. P., Moini, S. M., and Henderson, S. M., 1979. Desorption isotherms of rough rice from 10°C to 40°C. *Trans. ASME* 22: 433–440.
15. Murugesan, K., Suresh, H. N., Seetharamu, K. N., Narayana, P. A. A., and Sundararajan, T., 2001. A theoretical model of brick drying as a conjugate problem, *Int. J. Heat Mass Transfer* 44: 4075–4086.
16. Fortes, M., and Okos, M. R., 1981. Heat and mass transfer in hygroscopic capillary extruded products, *AIChE J.* 27: 255–262.
17. Masmoudi, W., and Prat, M., 1991. Heat and mass transfer between a porous medium and parallel external flow, *Int. J. Heat Mass Transfer* 34: 1975–1989.
18. Luikov, A. V., 1966. *Heat and Mass Transfer in Capillary Porous Bodies*. Pergamon Press, Oxford.
19. Chen, P., and Pei, D., 1989. A mathematical model of drying processes, *Int. J. Heat Mass Transfer* 32: 297–310.
20. Nam, J. H., and Song, C. S., 2007. Numerical simulation of conjugate heat and mass transfer during multi-dimensional freeze drying of slab-shaped food products, *Int. J. Heat Mass Transfer* 50: 4891–4900.

10

Manufacturing Equipment Operation

10.1 Heat Exchangers and Finned Surfaces

Example 10.1: Laminar Flow in a Double-Pipe Heat Exchanger [1]

Two flows in steady state with the fully developed velocity profiles inside a double pipe are considered. The double pipe consists of an inner central tube and an outer annualar channel. Concurrent and countercurrent cases are studied. The thermal conduction in the fluids and viscose dissipation are neglected. The governing system includes energy equations for the inner and outer streams (indices 1 and 2) and for the separating wall, boundary and conjugate conditions:

$$u_1 \frac{\partial \theta_1}{\partial x} = 4\left(\frac{1}{r}\frac{\partial \theta_1}{\partial r} + \frac{\partial^2 \theta_1}{\partial r^2}\right) \quad \theta_1(0,r) = \theta_{01}, \quad \frac{\partial \theta_1}{\partial r}(x,0) = 0 \tag{10.1}$$

$$u_2 \frac{\partial \theta_2}{\partial x} = 4\frac{\lambda_2}{(mc_p)_2}\left[R_i^2 - (1+\Delta)^2\right]\left(\frac{1}{r}\frac{\partial \theta_2}{\partial r} + \frac{\partial^2 \theta_2}{\partial r^2}\right) \tag{10.2}$$

$$\theta_2(0,r) = \theta_{20}, \quad \theta_2(L,r) = \theta_{20}, \quad \frac{\partial \theta_2}{\partial r}(x,R_i) = 0 \tag{10.3}$$

$$Pe_1^2\left(\frac{1}{r}\frac{\partial \theta_s}{\partial r} + \frac{\partial^2 \theta_s}{\partial r^2}\right) + \frac{\partial^2 \theta_s}{\partial x^2} = 0 \quad \frac{\partial \theta_s}{\partial x}(0,r) = \frac{\partial \theta_s}{\partial r}(L,r) = 0 \tag{10.4}$$

$$\theta_1(x,1) = \theta_w(x,1), \quad \frac{\partial \theta_1}{\partial r}(x,1) = \lambda_w \frac{\partial \theta_w}{\partial r}(x,1) \tag{10.5}$$

$$\theta_2(x,1+\Delta) = \theta_w(x,1+\Delta), \quad \frac{\partial \theta_2}{\partial r}(x,1+\Delta) = \frac{\lambda_w}{\lambda}\frac{\partial \theta_w}{\partial r}(x,1+\Delta) \tag{10.6}$$

The nondimensional variables are scaled as follows: linear sizes by internal radius of the inner duct R_i, velocities by mean axial velocity U, and temperatures by the inlet temperature of the inner fluid $T_{0,1}$, thermal conductivity λ_2, and capacity $(mc_p)_2$ by corresponding values of the inner flow. The first and the second relations of boundary conditions in Equation (10.3) pertain to concurrent and countercurrent cases, respectively.

The only concurrent case is investigated in detail. The superposition method is used to solve the energy equations for fluids. In this case, the dependence between heat flux and temperature on the solid surface is given by Duhamel's integral (Section 1.5):

$$\theta_w - \theta_{b1} = \frac{2q_1(0)}{\mathrm{Nu}_q} + \int_0^x \frac{2}{\mathrm{Nu}_q(x-\xi)} \frac{dq_1}{d\xi} d\xi \qquad (10.7)$$

where Nu_q and θ_b are the Nusselt number for the uniform heat flux and bulk temperature. This expression for both flows along with the energy equation for a body governed the solution of the problem.

The energy equation for the wall is solved numerically by applying the finite element method. The iterative procedure starts with guessing the distributions of the bulk temperature and Nusselt number that are used as boundary conditions for the wall energy equation. As a result, new distributions of the wall temperature and Nusselt number are obtained. Then, the updated values of the Nusselt numbers and the bulk temperatures are calculated from conjugate conditions. Finally, the refined wall temperature distribution is obtained using Equation (10.7). This and corresponding values of the Nusselt number allow one to start a new run. The convergence is achieved in less than 14 iterations.

The numerical results are obtained for the following parameters: $L = 10$ and 100, $\Delta = 5$ and 2, $R_i = 3$ and 6, $\lambda_w = 1, 10, 100, 1000$ and $10,000$, $\lambda_2 = 0.1, 1$ and 5, $\mathrm{Pe}_1 = 500, 1000$ and $10,000$, $(mc_p)_2 = 0.5, 1$ and 2 to get the following results:

1. The conjugate effect of the two streams is studied by comparing results obtained with and without [2] the thermal wall conduction effect. In contrast to the latter case, two isothermal zones are created at the interface, and the wall temperatures do not coincide with inlet fluid temperatures. The length of these zones, as well as the wall-to-fluid temperature difference, increases due to axial wall conduction. With increasing wall conductivity, the wall temperature becomes more uniform, and the outlet temperature of the internal stream decreases, while the outlet temperature of the external stream increases correspondingly. For relatively small wall conductivities up to $\lambda_w = 100$, the two streams are uncoupled and a central zone with uniform and equal heat fluxes for both sides exists. This is similar to the case without wall conduction effect. For high wall conductivities, the situation completely changes, and the heat fluxes monotonically decrease from the inlet to the outlet with one crossing point. Near the inlet, the Nusselt number values coincide with those for the isothermal condition. Downstream, both the Nusselt number and heat flux reach the asymptotic values corresponding to isothermal or to uniform heat flux conditions. For low wall conductivities, the values of Nu and heat flux may be higher than those for uniform heat flux, but these approximate the isothermal values for increasing wall conductivities.

2. Distribution of the local entropy production in the wall and streams calculated as suggested in Reference [3] are sensitive to wall conductivity. For $\lambda_w = 10,000$, the entropy production monotonically decreases downstream from the maximum at the inlet. For reducing wall conductivities, the entropy production decreases progressively at the inlet, increasing in the outlet regions. Analyzing the distribution of entropy generation rate in the fluids by relating it to flow parameters indicates that the maxima in entropy production in streams correspond to high values of heat flux or wall-to-fluid temperature difference. Distribution of entropy production in the wall shows that the wall radial thermal resistance is dominant when the wall conductivity is low. For $\lambda_w = 1$, the entropy production distribution is similar to that for the wall with zero conductivity. For the high conductivities, the situation is completely revised, and the maximum is found at the middle of the wall instead of minimum existence here in the case of low conductivities.

3. Accounting for the wall conduction changes the character of dependence of the heat exchanger effectiveness on the wall conductivity. Instead of monotonically increased effectiveness with growing wall conductivity, for the intermediate λ_w, the maximum effectiveness exists at any given values of Pe_1 and capacity ratio $(mc_p)_2$. For low wall conductivities, the entropy production is concentrated in the wall. In the short device, the increasing λ_w leads to monotonically decreasing the wall contribution. In a long exchanger, the minimum of the entropy production is observed. The reduction of effectiveness due to the wall conduction effect increases for increasing wall thickness, but it is slightly affected by variation of the pipe diameter ratio. The effectiveness also reduces with increasing Peclet number. Increasing the fluids conductivity ratio has a strong and positive effect on effectiveness.

4. The proper choice of the wall material is needed for optimization. For example, in the case of countercurrent water streams separated by a copper wall, the wall conductivity yields a small reduction in effectiveness. The optimum can be achieved by using a steel wall with $\lambda_w \approx 100$. For the glass wall, the order of magnitude drops to 1, and the effectiveness decreases due to the high radial resistance. For gaseous fluids, the wall conduction effect is more pronounced, indicating, for instance, that the corresponding value of λ_w becomes higher than 10,000 for two streams of air separated by a copper wall.

The same problem for countercurrent flows is considered in Reference [4]. The system of governing equations is solved numerically using Galerkin's method (Chapter 7). The results are compared with a one-dimensional approach when the overall heat transfer coefficient is used and with two-dimensional solutions based on slug approximation for streams. It is shown that errors of such approximation approaches depend on Biot number and are applicable only for small Biot numbers, $Bi < 0.1$.

Example 10.2: Horizontal Fin Array [5]

The two adjacent internal long fins on a base are considered. The constant temperature of the base $T_{w,0}$ is higher than ambient air temperature T_∞. Heat is transferred from the fins to ambient air by convection and radiation. The problem is formulated as for a closure formed by two vertical fins and a horizontal base, and it is governed by the two-dimensional mass, momentum, and energy equations for the fluid and one-dimensional conduction equation for the fins.

The radiation heat transfer part is included in the energy equation as a heat source that is calculated as follows. The radiation exchange occurs between the left (subscript 1) and right (2) fins' surfaces, the base (3), and the walls of the room through the open top (4), front side (5), and rear side (6) of closure. The open top and sides are considered as imaginary surfaces. The black body irradiations of the fins and base are $E_{b1} = E_{b2} = \sigma T_w^4$ and $E_{b3} = \sigma T_{w,0}^4$, while $E_{b4} = E_{b5} = E_{b6} = J_4 = J_5 = J_6 = \sigma T_\infty^4$, where J is the radiosity. The radiation heat fluxes from fin and base are present as

$$q_{R1} = \frac{1}{1}[S_{13}(J_1 \quad J_3) + (S_{14} + 2S_{15})(J_1 \quad E_{b4})], \quad J_1 = \frac{a_{22}b_1 + S_{11}b_2}{a_{11}a_{22} \quad 2S_{13}S_{34}} \tag{10.8}$$

$$q_{R3} = \frac{3}{1}[2S_{31}(J_3 \quad J_1) + (S_{14} + 2S_{15})(J_3 \quad E_{b4})], \quad J_3 = \frac{2S_{31}b_1 + a_{11}b_2}{a_{11}a_{22} \quad 2S_{13}S_{31}}$$

$$S_{ij} = \frac{1 - \varepsilon_i}{\varepsilon_i} F_{ij}, a_{11} = 1 + S_{13} + S_{14}, a_{22} = 1 + 2S_{31} + S_{34}, \tag{10.9}$$

$$b_1 = E_{b1} + S_{14}E_{b4}, b_2 = E_{b3} + S_{34}E_{b4}$$

where ε is emissivity; S is spacing between adjacent fins; $i = 1, 3$; and $j = 1-6$. The shape factors F_{ij} are calculated by formulae given in Reference [6].

The final governing system written in nondimensional variables consists of the energy equation, the momentum equation in vorticity-stream function ($\zeta - \psi$) form, and the two equations for fins:

$$\frac{\partial \theta}{\partial t} + u\frac{\partial \theta}{\partial x} + v\frac{\partial \theta}{\partial y} = \frac{1}{Pr}\left(\frac{1}{Gr^{1/2}}\frac{\partial^2\theta}{\partial x^2} + \frac{\partial^2\theta}{\partial y^2}\right) + \frac{1}{PrGr^{1/4}}\left(2q_{R1} + \frac{q_{R3}}{A_R}\right) \tag{10.10}$$

$$\frac{\partial \zeta}{\partial t} + u\frac{\partial \zeta}{\partial x} + v\frac{\partial \zeta}{\partial y} = \frac{\partial \theta}{\partial y} + \frac{1}{Gr^{1/2}}\frac{\partial^2\zeta}{\partial x^2} + \frac{\partial^2\zeta}{\partial y^2} \qquad \zeta = \frac{1}{Gr^{1/2}}\frac{\partial^2\psi}{\partial x^2} + \frac{\partial^2\psi}{\partial y^2} \tag{10.11}$$

$$\frac{PrGr^{1/2}}{\alpha}\frac{\partial\theta_w}{\partial t} = \frac{\partial^2\theta_w}{\partial x^2} + \Lambda\frac{\partial\theta}{\partial y}\bigg|_{y=0}, \quad \frac{PrGr^{1/2}}{\alpha}\frac{\partial\theta_w}{\partial t} = \frac{\partial^2\theta_w}{\partial x^2} - \Lambda\frac{\partial\theta}{\partial y}\bigg|_{y=Gr^{1/4}} \tag{10.12}$$

The radiation heat fluxes q_{R1} and q_{R3} in the energy equation are given by Equations (10.8) and (10.9). The following scales are used to form the dimensionless variables in Equations (10.10), (10.11), and (10.12): $S, S/\mathrm{Gr}^{1/4}, S^2/\nu\mathrm{Gr}^{1//2}, \nu\mathrm{Gr}^{1//2}/S, \nu\mathrm{Gr}^{1//4}/S, T_{w0} - T_\infty, \nu\mathrm{Gr}^{1//4}, \nu\mathrm{Gr}^{3/4}/S^2 \alpha, \lambda(T_{w,0} - T_\infty)$ $\mathrm{Gr}^{1/4}/S$, and σT_∞^4 for variables $x, y, t, u, v, T - T_\infty, \psi, \zeta, \alpha_w, q_R$ and E_b, respectively, and the Grashof number is $\mathrm{Gr} = g\beta(T_{w,0} - T_\infty)S^3/\nu$. The conduction-convection parameter, the radiation parameter, the aspect ratio, and the temperature ratio are system parameters defining the solution:

$$\Lambda = \frac{\lambda PS}{\lambda_w A_w}, \quad N_R = \frac{S\sigma T_\infty^4}{\lambda\mathrm{Gr}^{1/4}(T_{w,0} - T_\infty)}, \quad A_R = \frac{H}{S}, \quad A_w = \frac{\Delta W}{2}, \quad P = \Delta + W \quad (10.13)$$

where P, A_w, H and W are half perimeter, half section area, height, and width of the fin. The boundary conditions include $T = T_{w,0}$ and the no-slip condition for the fins and the base and $\partial T/\partial x = 0$ for the fin tips in addition. For the open top, the boundary conditions suggested in Reference [7] are applied $v = \partial u/\partial x = \partial\psi/\partial x = \zeta = 0$ and $\partial T/\partial x = 0$. The other types of boundary conditions for the open top are discussed in the Appendix in Reference [5].

The alternating direction method is used to solve numerically the system of Equations (10.10), (10.11), and (10.12). The temperature distributions are obtained in the fluid from the first and in the fins from the fourth and fifth equations, respectively. The vorticity is calculated using Equation (10.11), and then the stream function is computed from the third equation, and finally, velocity components are obtained knowing the stream function. The procedure is continued step by step until steady-state fields for all variables are obtained. The heat fluxes from the inner fins, ends of fins, and from the base are calculated as follows:

$$q_1 = \frac{1}{A_R}\int_0^{A_R}\left[\left.\frac{\partial T}{\partial y}\right|_{y=0} + N_R\frac{\varepsilon_1(E_{b1} - J_1)}{1 - \varepsilon_1}\right]dx,$$

$$q_5 = \frac{1}{A_R}\int_0^{A_R}\left\{-\left.\frac{\partial T}{\partial y}\right|_{y=0} + \varepsilon_1 N_R\left[(\gamma T_w + 1)^4 - 1\right]\right\}dx \quad (10.14)$$

$$\gamma = \frac{T_{w,0} - T_\infty}{T_\infty} \quad q_3 = \frac{1}{\mathrm{Gr}^{1/4}}\int_0^{\mathrm{Gr}^{1/4}}\left[-\frac{1}{\mathrm{Gr}^{1/4}}\left.\frac{\partial T}{\partial y}\right|_{x=0} + N_R\frac{\varepsilon_3(E_{B3} - J_3)}{1 - \varepsilon_3}\right]dy$$

The numerical results are obtained for various fin arrays under different thermal characteristics and are compared with available experimental data:

1. Calculations of average Nusselt numbers for a four-fin array for small and high emissivities agree with experimental data presented in References [8] and [9], respectively. Analysis of these

results shows that the contributions of the fins, base, and end fins to total heat transfer are 36%, 13.5%, and 50.5%, respectively, which agrees with the observation in Reference [9]. The effect of fin spacing on heat fluxes is studied for arrays with different numbers of fins over a fixed base. As the number of fins increases from four to sixteen, and the value of spacing S decreases from 20 to 2.8 mm, the heat fluxes from fin and from base decrease from 149 to 44 W/m² and from 379 to148 W/m². Despite increased numbers of fins, the heat transfer rate and effectiveness remain almost the same, but the average heat transfer coefficient decreases remarkably from 5.29 to 1.48 W/m² K. The effect of the base temperature is studied for an array experimentally investigated in Reference [10]. The numerical and experimental data are in agreement and indicate that the total heat transfer rate increases as the base temperature grows for any studied values of spacing and heights. The effectiveness increases as well for all heights, but it is found that for small values of S, effectiveness decreases as the base temperature grows. The results obtained for different fin thicknesses indicate that in the case of low heights and high thermal conductivities, the heat flux from the fin practically does not depend on thickness. For instance, as the thickness increases from 1.5 to 6 mm, the heat flux changes only from 146.6 to 150.6 W/m². The role of thermal conductivity and emissivity of the fin is also studied. It is observed that decreasing thermal conductivity leads to reduction of the fin heat flux, and increasing emissivity yields growing heat flux due to increase of the radiation component.

2. The temperature profiles obtained for two different spacings show that the temperature far away from the fins (at the distance $S/2$) is lower for higher spacing than that for smaller spacing. At the same time, the velocity profiles indicate that there is a greater recirculation at larger S, resulting in higher velocities near the wall and lower velocities at the distance $S/2$. The isotherms and streamlines for the same two enclosures indicate that the air temperature is high in the middle of the enclosure with smaller spacing, while in the other enclosure with larger spacing, the heating is confined to the air near the fins and the base. It can be seen from the streamline distribution that the streamlines travel upwards along the fins, where the temperature is high compared to that of the air in the enclosure.

3. The following equations for heat fluxes, average Nusselt number, and effectiveness are obtained by the nonlinear regression on the numerical results:

$$\frac{q_i S}{\lambda(T_{w,0} - T_\infty)} = K \mathrm{Ra}^{K_1} \left(\frac{S}{H}\right)^{K_2} \left(\frac{1+\varepsilon_i}{1+N_R}\right)^{K_3} \tag{10.15}$$

$$\mathrm{Nu}_m \text{ and } E = K \mathrm{Ra}^{K_1} \left(\frac{S}{H}\right)^{K_2} \left(\frac{1+\varepsilon_i}{1+N_R}\right)^{K_3} N^{K_4} \tag{10.16}$$

The constants are as follows: $K = 0.047, 0.023, 0.52, 0.102, 0.022$; $K_1 = 0.42$, $0.55, 0.23, 0.36, 0.33$: $K_2 = 0.4, 1.32, 0.29, 0.4, -0.41$; $K_3 = 0.022, 0.087, 0.42, 0.1$, -0.16; $K_4 = -0.04, 0.9$ for $q_1, q_3, q_5 (\varepsilon_5 = \varepsilon_1)$, Nu_m and E, respectively, and N is the number of fins. These expressions predict the numerical results with deviation of $\pm 6.1\%$ for a range of $10 \le S \le 25$ mm, $60 \le T_{w,0} \le 104°C, 30 \le H \le 70$ mm, $2300 \le Ra \le 60,000, 0.05 \le \varepsilon_1 \le 0.85$, and $0.3 \le N_R \le 1$.

Example 10.3: Flat Finned Surface in a Transverse Flow [11]

An incompressible fluid flows along a finned surface transversely to the fins. Because the flow is normal to the fins, the eddy forms in each interfin space (Figure 10.1). On side 1 of the fin, the temperature gradient increases, while on side 2 it decreases in the direction of the eddy flow. On base 3, the temperature gradient may be assumed to be constant to a first approximation. It is assumed that at intersections, the surfaces are rounded, owing to which stagnant zones or secondary eddy flows do not form in the corners. It is also assumed that the boundary layer formed on the finned surface makes the main contribution to the thermal resistance to heat transfer between the surface and the potential eddy flow. This boundary layer develops from the end face to the base on the front (in the direction of the flow) fin surface 1 and increases from the base to the end on the back fin surface 2 (Figure 10.1). These presentations of the dynamic boundary layer on the wall of the cavern under the conditions of eddy flow inside it are developed by Batchelor [12].

Assuming that the conditions in all cells on the surface are identical, the model of the problem of heat transfer in fins is formulated as bilateral flow over the body, as schematically shown in Figure 10.1. Here the body surface represents the surface of the fin and of two adjacent cells.

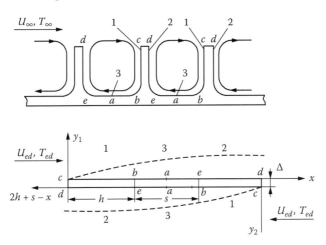

FIGURE 10.1
Diagram of a transverse flow over a finned surface.

The upper body surface represents the right cell surface, the ends of the model body correspond to the ends of the two adjacent fins, and the lower body surface represents the surface of the left cell. The numbers in the model correspond to those in the scheme of the finned surface. Thus, the numbers 1, 2, and 3 correspond to sections with increasing, decreasing, and constant temperature gradients, respectively, and the model represents the case of countercurrent flows with complicated velocity and temperature distributions.

Because the boundary layer in the real system is formed on a curvilinear surface, the longitudinal curvature and the gradient of the external flow should be taken into account. In this model, the effect of the longitudinal curvature is neglected, and the velocity gradient can be taken into account using Görtler's variable (Chapter 3).

For a thermally thin fin, the governing equation and boundary condition are

$$\lambda_w \Delta \frac{d^2 T_w}{dx^2} = q_1(x) + q_2(x), \qquad \frac{dT}{dx}\bigg|_{x=0} = 0, \qquad T\big|_{x=H} = T_0 \tag{10.17}$$

where $q_1(x)$ and $q_2(x)$ are heat fluxes from surfaces of the fin, H is the height of the fin, and T_w and T_0 are the fin and the base temperatures. To calculate the heat fluxes, Equation (3.38) for gradientless flow or Equation (3.40) in the case of an arbitrary pressure gradient may be used. For the case of gradientless flow, both heat fluxes may be presented as

$$q_2(x) = h_*(x)(2H + s - x)\left[T_E - T_{ed} + \int_{2H+s}^{H+s} f\left(\frac{2H + s - \xi}{2H + s - x} \right) \frac{\partial T_w}{\partial \xi} d\xi \right.$$

$$\tag{10.18}$$

$$\left. + \int_s^x f\left(\frac{2H + s - \xi}{2H + s - x} \right) \frac{\partial T_w}{\partial \xi} d\xi \right], \qquad q_1(x) = h_*(x)\left[T_E - T_{ed} + \int_0^x f\left(\frac{\xi}{x} \right) \frac{\partial T_w}{\partial \xi} d\xi \right]$$

Here T_E and T_{ed} are temperatures of the end of the fin and inside the eddy flow, which plays the role of the temperature of the external flow for the boundary layer on the fin; s is the distance between the fins; and the influence function is given by Equation (3.39):

$$f\left(\frac{\xi}{x} \right) = \left[1 - \left(\frac{\xi}{x} \right)^{C_1} \right]^{-C_2} \tag{10.19}$$

The expression for heat flux $q_2(x)$ takes into account that the boundary layer on the back surface of the fin develops starting from the end point c (Figure 10.1) of the front surface. Therefore, the heat transfer on the back surface depends on the temperature distribution on the front surface (section bc) and the length s of the interfin section. This fact is taken into account by the first integral in the equation for $q_2(x)$, while the second integral determines the effect of the section ex on the back surface.

Substituting the expressions for the heat fluxes into Equation (10.17) yields the integro-differential equation determining the temperature distribution over the height of the fin. In dimensionless variables, this equation and the boundary conditions have the following form:

$$\frac{d^2\theta_w}{d\eta^2} = N^2\left\{ \varphi(\eta)\left[\theta_E + \int_0^\eta f\left(\frac{\zeta}{\eta}\right)\frac{d\theta_w}{d\zeta}d\zeta \right] + \varphi(\eta_0 - \eta)\left[\theta_E + \int_0^1 f\left(\frac{\zeta}{\eta_0 - \eta}\right) \right.\right.$$

$$\left.\left. \frac{d\theta_w}{d\zeta}d\zeta - \int_\eta^1 f\left(\frac{\zeta}{\eta_0 - \eta}\right)\frac{d\theta_w}{d\zeta}d\zeta \right] \right\} \qquad \frac{d\theta}{d\eta}\bigg|_{\eta=0} = 0, \quad \theta\big|_{\eta=1} = 1$$

$$\tag{10.20}$$

$$\theta_w = \frac{T_w - T_D}{T_0 - T_D}, \qquad N^2 = \frac{\overline{h}_* H^2}{\lambda_w \Delta} \qquad \varphi(\eta) = \frac{h_*(\eta)}{\overline{h}_*}$$

$$\overline{h}_* = \frac{1}{H}\int_0^S [h_*(\zeta) + h_*(2H + H_B - \zeta)]d\zeta$$

Here $\eta = x/H$, $\eta_0 = 2 + H_B/H$, $\overline{h}_*$ is the average heat transfer coefficient of an isothermal fin, and the parameter N^2, in fact, is the Biot number (note that H^2/Δ is the linear characteristic determining the conjugation effect (Section 6.1).

Equation (10.20) is solved using the method of reduction of the conjugate problem to the conduction problem and the successive differential-integral numerical approach outlined in Section 6.3. According to this approach, applying to heat fluxes the equivalent differential forms (3.32) with the first two derivatives, transforms Equation (10.20) to the system of an ordinary differential equation of the second order and equation for error

$$\frac{d^2\theta_w}{d\eta^2} = N^2\left\{ \varphi(\eta)\left[\theta_w + g_1\eta\frac{d\theta_w}{d\eta} + g_2\eta^2\frac{d^2\theta_w}{d\eta^2} \right] + \varphi(\eta_0 - \eta) \right.$$

$$\tag{10.21}$$

$$\left. \times \left[\theta_w - g_1(\eta_0 - \eta)\frac{d\theta_w}{d\eta} + g_2(\eta_0 - \eta)^2\frac{d^2\theta_w}{d\eta^2} \right] \right\} + \varepsilon(\eta) \quad \varepsilon(\eta) = \theta_w^{int} - \theta_w^{diff}$$

The first approximation is obtained assuming $\varepsilon(\eta) = 0$ and solving the ordinary differential equation. Using the first approximation results for $\theta_w(\eta)$, the next approximation for $\theta_w(\eta)$ is obtained by integrating the right-hand part of the intego-differential Equation (10.20) and calculating the error $\varepsilon(\eta)$. Then, solving the refined by including $\varepsilon(\eta)$ differential Equation (10.21), the next approximation is obtained. The procedure is continued until a desired accuracy is achieved.

Numerical results are obtained for laminar flow using the following values of constants: $C_1 = 3/4$, $C_2 = 1/3$, $g_1 = 0.63$, $g_2 = -0.14$ (Chapter 3):

1. The solutions of the conjugate problem are compared with the results of approximate calculations using local and average heat transfer coefficients of the isothermal surface. The comparison shows that in the range $0 \leq N^2 \leq 2$, the results obtained using the conjugate model and both simplified methods are in agreement. However, for $N^2 > 2$, the difference between the results for both the local and integral characteristics of the fin becomes more significant. The values of the fin efficiency and the total heat flux removed by the fin obtained using simplified methods are too low, and the error increases as the conjugate parameter N^2 increases, reaching for large N^2 the values of 60% to 70%. The local heat transfer coefficients and heat fluxes obtained using the conjugate model are substantially nonuniform over the height of the fin and reach maximum values near the end face on the front side of the fin. For $N^2 > 2$, the local characteristics obtained using the conjugate model differ significantly from those computed by simplified methods. The greatest differences are observed on the front part near the base of the fin.

2. On the back side of the fin, for $N^2 \geq 1.9$, the heat flux inversion is observed when the heat flux is directed toward the fin despite the temperature head remaining positive (Section 5.3). As the value of N^2 increases, the absolute magnitude of the inverse heat flux and the length of the heated section end (Figure 10.1) increase. The heat transfer coefficients in this region become negative values. This effect is explained by inertia of fluid: when the fluid temperature near the wall exceeds the wall temperature and the heat flux becomes directed toward the fin, the external fluid temperature due to inertia is still lower than the wall temperature, and the temperature head remains positive (see Section 5.3). Because the inversion effect cannot be obtained with the use of the simplified methods, neglecting the conjugation of the problem in this case not only yields quantitative errors, but also leads to qualitatively incorrect results. The reason for this is that on the back side of the fin, the temperature head decreases in the flow direction. In such a case, the conjugate effect should always be taken into account (Chapter 5).

3. It is found that for large values of conjugate parameter $N^2 > 2$, the heat flux removed by fin is the maximum for fins with $1 < s/H < 1.5$. This result is a consequence of the nonisothermicity of the finned surface and cannot be obtained using the simplified methods.

Example 10.4: Vertical Fin with a Rounded Tip Embedded in a Porous Medium [13]

The fin is composed of a rectangular part of thickness Δ and a half cylindrically shaped part of the same diameter which is attached to its lower section. The cylindrically shaped part is used to avoid the prescription of the boundary condition at the tip of the fin, because it is not clear how this condition should be formulated (see Example 10.2). The vertical fin is attached to the base and is surrounded by porous medium. Heat transfer occurs by natural convection.

The governing system consists of boundary layer equations for porous medium and two-dimensional heat conduction equations for both parts of the fin. These equations are presented in variables connected with the form of Darcy equation suggested in Reference [14]:

$$u = \frac{\partial \psi}{\partial y} = \frac{K\rho_\infty \beta g}{\mu}(T - T_\infty)S(x) \tag{10.22}$$

$$S(x) = \sin\left(\frac{2x}{\Delta}\right) \text{ if } 0 \le x \le \frac{\pi\Delta}{4}, \quad S(x) = 1 \text{ if } \frac{\pi\Delta}{4} \le x \le \frac{\pi\Delta}{4} + L$$

$$F(x,\eta) = \psi(x,y)\left(2\int_0^x S(\xi)d\xi\right)^{-1/2}, G(x,\eta) = \theta(x,y) \ \eta = yS(x)\left(2\int_0^x S(\xi)d\xi\right)^{-1/2} \tag{10.23}$$

The boundary layer equations for the porous medium in these variables take the form

$$F' - G = 0, \quad G'' + FG' = a\tan\left(\frac{x}{a}\right)\left(F'\frac{\partial G}{\partial x} - G'\frac{\partial F}{\partial x}\right) \text{ if } 0 \le x \le \frac{a\pi}{4}$$

$$G'' + FG' = \left[a\left(1 + \frac{\pi}{2}\right) + 2x\right]\left(F'\frac{\partial G}{\partial x} - G'\frac{\partial F}{\partial x}\right) \text{ if } \frac{a\pi}{4} \le x \le \frac{a\pi}{4} + 1 \tag{10.24}$$

with boundary conditions at the solid $\eta = 0, F' = 0, G = \theta_s; \eta \to \infty, F' \to 0$, $G \to 0$. The equations for cylindrical and rectangular parts of the fin are, respectively, as follows:

$$\left(y + \frac{a}{2}\right)^2 \frac{\partial^2 \theta_s}{\partial y^2} + \left(y + \frac{a}{2}\right)^2 \frac{\partial \theta_s}{\partial y} + \frac{a^2}{4}\frac{\partial^2 \theta_s}{\partial x^2} = 0, \quad \frac{\partial^2 \theta_s}{\partial x^2} + \frac{\partial^2 \theta_s}{\partial y^2} = 0 \tag{10.25}$$

Both equations are subjected to the conjugate conditions at the porous medium–solid interface and at the line between the cylindrical and rectangular parts and in addition to the symmetry condition of the fin. These conditions for first and second equations are similar:

$$0 \le x \le \frac{a\pi}{4}, \ y = 0, \ \frac{\partial \theta_S}{\partial y} = \frac{\mathrm{Ra}^{1/2}}{\Lambda} \sqrt{\frac{2}{a}} \cos\left(\frac{x}{a}\right) G', \ 0 \le x \le \frac{a\pi}{4}, \ y = -\frac{a}{2},$$

$$\theta_S^{circ} = \theta_S^{rec}, \ x = 0, \ -\frac{a}{2} \le y \le 0, \ \frac{\partial \theta_S}{\partial y} = 0, \ x = \frac{a\pi}{4}, \ -\frac{a}{2} \le y \le 0,$$

$$\theta_S^{circ} = \theta_S^{rec}, \ \left(\frac{\partial \theta_S}{\partial y}\right)_{rec} = \frac{a}{2(y+a/2)} \left(\frac{\partial \theta_S}{\partial y}\right)_{circ}, \quad 0 \le x \le \frac{a\pi}{4}, \ y = 0,$$

$$\frac{\partial \theta_S}{\partial y} = \frac{\mathrm{Ra}^{1/2}}{\Lambda} \frac{G'}{\sqrt{a(1-\pi/2)+2x}}, \quad \frac{a\pi}{4} \le x \le \frac{a\pi}{4}+1, \quad y = -\frac{a}{2}, \ \frac{\partial \theta_S}{\partial y} = 0,$$

$$x = \frac{a\pi}{4}+1, \ -\frac{a}{2} \le y \le 0, \ \theta_S = 0, \quad x = \frac{a\pi}{4}, \ -\frac{a}{2} \le y \le 0,$$

$$\theta_S^{circ} = \theta_S^{rec}, \ \left(\frac{\partial \theta_S}{\partial y}\right)_{rec} = \frac{a}{2(y+a/2)} \left(\frac{\partial \theta_S}{\partial y}\right)_{circ}$$

(10.26)

Here variables are dimensionless and are formed using the following scales: $L, L/\mathrm{Ra}^{1/2}, \alpha/\mathrm{Ra}^{1/2}$ for x, y, ψ, respectively; $\theta = (T - T_\infty)/(T_0 - T_\infty), a = \Delta/L$ is the aspect ratio of the rectangular part $\Lambda = \lambda/\lambda_w$; $\mathrm{Ra} = gK\beta\rho_\infty(T_0 - T_\infty)L/\alpha\mu$, T_0 is the base temperature; and K is the permeability of the porous medium.

The system of governing equations with appropriate boundary conditions is solved numerically using the finite difference iterative method. Details may be found in Reference [15]. First, Equation (10.25) for the cylindrical part of the fin with boundary condition (10.26) is solved in order to obtain an estimate for $(\theta_S)_{circ}$. Then this estimate is used to solve Equation (10.25) for the rectangular part of the fin with boundary conditions (10.26). Finally, knowing θ_S for the whole fin, Equation (10.24) is solved for porous medium.

The calculations are performed for different aspect ratio $a = 0.025$, 0.05, 0.1, 0.2 and 0.5 and for various values of conjugation parameter $N = 2\mathrm{Ra}^{1/2}/a\Lambda$, which in the case of natural convection determines the relation between conduction and convection heat resistances similar to the Biot number which determines this relation in the case of forced convection. The calculations show the following:

1. The results for the isothermal fin agree with other known data. The temperature distributions for two angles $0°$ and $60°$ between the y-axis of the fin and downward vertical indicate a relatively small effect of the fin inclination. The comparison between temperature distributions obtained for a fin with a rounded tip and the rectangular plate fin shows that the temperatures of the rounded fin increase by decreasing the aspect ratio and tend to the curve $a \to 0$, which is close to the temperature profile for the rectangular plate at a small aspect ratio. These results are similar for both studied values of conjugate parameter $N = 0.5$ and $N = 2$.

2. The heat transfer coefficient and heat flux distributions are obtained for $N = 0.5$, 2 and 12 and are compared to those obtained for a rectangular fin with an insulated tip. Both distributions of heat transfer coefficient and of heat flux along the fin with rounded tip for a different aspect ratio approach the data for the rectangular fin. The rapidly decreasing heat fluxes and heat transfer coefficients near the tip of the fin are observed for both fins with rounded and isolated tips. These characteristics decrease sharply at small values of the aspect ratio and become smoother as the aspect ratio increases.

3. For all values of conjugate parameters, the total heat transfer rate is always smaller for the plate fin than for the fin with a rounded tip. This result is expected because in the latter case the fin surface is larger. For relatively small values of conjugation parameter $N = 0.5$ and $N = 2$, the total heat transfer rate increases as the aspect ratio grows. However, for $N = 12$, the results show a slightly decreasing total heat transfer rate with growing aspect ratio. This can be associated with small material conductivities corresponding to large values of N when the effect of decreasing heat conductivity cannot be compensated for by the growing surface.

4. Results obtained in this work show that all thermal characteristics of the rectangular fin with rounded tip approach the corresponding values for the rectangular fin as the aspect ratio goes to zero.

Example 10.5: Mixed Convection along a Cylindrical Fin in a Porous Medium [16]

A vertical cylindrical fin of radius R and length L is considered. The fin is embedded in a porous medium at ambient temperature and is attached to the wall that has temperature higher than ambient T_∞. An external flow with velocity U is directed vertically upward; hence, the buoyancy force is acting in the same direction. It is assumed that the radius of the fin is small compared to the length, so the one-dimensional heat conduction equation is employed. This equation and the boundary layer equations for porous medium involving Darcy's law and the buoyancy force in the Boussinesq approximation complete the governing system (K is permeability) with the following boundary conditions for the fin and porous medium (T_0 is the base temperature):

$$\lambda_w R \frac{d^2 T_S}{dx^2} = h(x)(T_S - T_\infty), \quad \frac{\partial}{\partial r}\left(\frac{1}{r}\frac{\partial \psi}{\partial r}\right) = \frac{g\rho_\infty \beta K}{\mu}\frac{\partial T}{\partial r},$$

$$\alpha \frac{\partial}{\partial r}\left(r\frac{\partial T}{\partial r}\right) = \frac{\partial \psi}{\partial r}\frac{\partial T}{\partial x} + \frac{\partial T}{\partial r}\frac{\partial \psi}{\partial x}$$

(10.27)

$$x = L, \ T_S = T_0; \ x = 0, \ dT/dx = 0 \quad r = R, \ u = 0,$$

$$T = T(x, R); \ x \to \infty, \ u = U, \ T = T_\infty$$

The heat transfer coefficient $h(x)$ in the first equation should be determined because the conjugate problem is considered. The governing equations, boundary, and conjugate condition are transformed to dimensionless variables to obtain

$$\frac{\partial \theta_s}{\partial \xi} = N\bar{h}(\xi)\theta_s, \quad \frac{\partial^2 f}{\partial \eta^2} = \Omega \frac{\partial \theta}{\partial \eta}, \quad (1+\gamma\sqrt{\xi}\eta)\frac{\partial^2 \theta}{\partial \eta^2} + \left(\frac{f}{2} + \gamma\sqrt{\xi}\right)\frac{\partial \theta}{\partial \eta}$$

$$= \xi\left(\frac{\partial f}{\partial \eta}\frac{\partial \theta}{\partial \xi} - \frac{\partial \theta}{\partial \eta}\frac{\partial f}{\partial \xi}\right)$$

(10.28)

$$\xi = 1, \ \theta_s = 1, \ \xi = 0, \ \frac{\partial \theta_s}{\partial \xi} = 0; \ \xi\frac{\partial f}{\partial \xi}(0,\xi) + \frac{f(0.\xi)}{2} = 0,$$

$$\theta = \theta(0,\xi), \ f'(\infty,\xi) = 1, \ \theta(\infty,\xi) = 0,$$

$$\theta_s(\xi) = \theta(0,\xi), \ \bar{h}(\xi) = \sqrt{\Pr}[-\theta'(0,\xi)] / \sqrt{\xi}\theta_s :$$

$$f(\xi,\eta) = \frac{\psi}{R\sqrt{\alpha U x}}, \ \eta = \frac{R-r}{2R}\sqrt{\frac{U}{x\alpha}}, \ \xi = \frac{x}{L}$$

$$\theta = \frac{T-T_\infty}{T_0-T_\infty}, \ \bar{h}(\xi) = \frac{hL}{\lambda\sqrt{\Re}}, \ \Omega = \frac{g\beta K(T_0-T_\infty)}{Uv} = \frac{\Gr}{\Re},$$

(10.29)

$$N = \frac{2L\lambda\sqrt{\Re}}{\lambda_w R}, \ \gamma = \frac{2L}{R}\sqrt{\frac{\alpha}{UL}}$$

Here N and γ are the conjugate and the surface curvature parameters, respectively.

The local nonsimilarity method is used to solve the governing equations. According to this method, the equations for porous medium are approximated by a set of ordinary differential equations. Eliminating the derivatives with respect to ξ from these equations and from boundary conditions, one gets a system of ordinary differential equations and boundary conditions for the first approximation:

$$f'' = \Omega\theta', \quad (1+\gamma\sqrt{\xi}\eta)\theta'' + (f/2 + \gamma\sqrt{\xi})\theta' = 0$$

(10.30)

$$f(0,\xi) = 0, \ \theta(0,\xi) = \theta_s(\xi), \ f(\infty,\xi) = 1, \ \theta(\infty,\xi) = 0$$

A more accurate solution is obtained by taking into account the eliminated derivatives. This is achieved by introducing new functions $G = \partial f/\partial \xi$ and $\phi = \partial \theta/\partial \xi$. Differentiating the original equation for porous medium with respect to ξ gives the additional equations for unknown functions G and ϕ. These along with original equations form the system of equation and additional boundary conditions for the second approximation:

$$f'' = \Omega\theta', \quad (1+\gamma\sqrt{\xi}\eta)\theta'' + (f/2+\gamma\sqrt{\xi})\theta' = \xi(f'\phi - \theta'G), \quad G'' = \Omega\phi'$$

$$(1+\gamma\sqrt{\xi}\eta)\phi'' + (f/2+\gamma\sqrt{\xi}+\xi G)\phi' = (f'+\xi G')\phi - (\gamma\eta/2\sqrt{\xi})\theta''$$

$$-\left[(3/2)G+\gamma/2\sqrt{\xi}\right]\theta' \tag{10.31}$$

$$G(0,\xi) = 0, \quad \phi(0,\xi) = \partial\theta_s/\partial\xi, \quad G'(\infty,\xi) = \phi(\infty,\xi) = 0$$

Similarly, systems for subsequent approximations may be generated. In particular, in Reference [16] is given the system for the third approximation. More details of the local nonsimilarity method may be found in the author's article [17].

The set of equations for the third approximation for porous medium and the heat conduction equation for the fin are converted in an integral form and numerically solved using an iterative method described in Reference [18]. The calculations are performed for Pr = 5.5 (e.g., water at 300 K) and the following results are obtained:

1. The fin temperature increases monotonically from the tip to the base of the fin for all values of parameters studied: $\gamma = 2$ and $5, \Omega = 0, 1$ and $2, N = 0.1, 0.5$ and 1.5. The fin temperature is higher for larger values of conjugate parameter. This is expected because high convection and low fin conductivities correspond to larger values of the conjugate parameter. The increasing Ω also leads to temperature rising due to a greater buoyancy effect. The fin temperature increases as the curvature parameter grows.

2. The results for the heat transfer coefficient are obtained for the same set of parameters. For $\Omega = 0$ and for all values of other parameters, the heat transfer coefficient monotonically decreases from a high value at the tip to some finite value at the base. In the case of $\Omega \neq 0$ and high values of N when the fin temperature is high, the heat transfer coefficient at first decreases along the fin and after approaching a minimum increases. This behavior occurs because high fin temperature leads to enhanced buoyancy force which compensates the falling trend of the heat transfer coefficient. As the conjugate parameter increases, the location of the minimum shifts toward the tip, and the rising part of the curve increases. The higher heat transfer coefficient corresponds also to a higher curvature parameter γ.

3. At given γ, the local heat flux is higher for larger Ω due to a larger buoyancy force. For the large values of N, most of the heat is transformed to the fluid in the area close to the fin base. This effect results due to low conductivities corresponding to high values of the conjugate parameter. Local heat flux is also higher for larger curvature parameters γ. This, however, does not mean that the total heat flux is larger for the fin with higher γ, because this parameter not only depends on fin radius.

4. The total heat flux and the efficiency are computed for $\gamma = 0.5, 1$ and 2 and $\Omega = 0, 0.5$ and 1 for total heat flux and $\Omega = 0$ and 2 for

efficiency. It is found that total heat flux is higher for smaller values of curvature parameter, which is expected because smaller γ represents a larger fin radius and, hence, a larger surface. The efficiency is higher also for smaller values of γ and for smaller Ω.

10.2   Cooling Systems

10.2.1   Electronic Packages

Continuing miniaturization of electronic systems leads to increasing the amount of heat flux per unit volume, which must be removed to provide stable and reliable operation conditions. The solution to this problem requires highly accurate methods for predicting heat transfer characteristics and designing the appropriate cooling systems.

Example 10.6: Horizontal Channel with Protruding Heat Sources [19]

The model consists of a horizontal channel with four volumetric sources mounted on the bottom wall. Because the objective of this work is to steady the heat transfer characteristics and the effect of thermal parameters, the geometric parameters are fixed. The heat transfer occurs by mixed convection and radiation. The air is considered as a cooling agent and is assumed to be a nonparticipating medium. The flow is assumed to be incompressible, laminar, and hydrodynamically and thermally developing. The Boussinesq approximation is applied for buoyancy force.

Two-dimensional Navier-Stokes and energy equations (Equation 10.32) for fluid and two-dimensional heat conduction (Equation 10.33) are used for flow, walls, and heat sources:

$$\frac{\partial u}{\partial x}+\frac{\partial v}{\partial y}=0, \quad \frac{\partial u}{\partial t}+u\frac{\partial u}{\partial x}+v\frac{\partial u}{\partial y}=-\frac{\partial p}{\partial x}+\frac{1}{Re}\left(\frac{\partial^2 u}{\partial x^2}+\frac{\partial^2 u}{\partial y^2}\right)$$

$$\frac{\partial v}{\partial t}+u\frac{\partial v}{\partial x}+v\frac{\partial v}{\partial y}=-\frac{\partial p}{\partial y}+\frac{1}{Re}\left(\frac{\partial^2 v}{\partial x^2}+\frac{\partial^2 v}{\partial y^2}\right)+\frac{Gr}{Re}\theta, \tag{10.32}$$

$$\frac{Pe}{Lu_w}\frac{\partial \theta}{\partial t}=\frac{\lambda_w}{\lambda}\left(\frac{\partial^2 \theta}{\partial x^2}+\frac{\partial^2 \theta}{\partial y^2}\right)$$

$$\frac{\partial \theta}{\partial t}+u\frac{\partial \theta}{\partial x}+v\frac{\partial \theta}{\partial y}=\frac{1}{Re}\left(\frac{\partial^2 \theta}{\partial x^2}+\frac{\partial^2 \theta}{\partial y^2}\right) \quad \frac{Pe}{Lu_h}\frac{\partial \theta}{\partial t}=\frac{\lambda_h}{\lambda}\left(\frac{\partial^2 \theta}{\partial x^2}+\frac{\partial^2 \theta}{\partial y^2}\right)+\frac{R^2}{L\Delta} \tag{10.33}$$

Here $Re = U_\infty R/v$, $Pe = U_\infty R/\alpha$, and subscript h denotes the heat source. The third and fifth energy equations correspond to channel walls and to heat sources. The following scales are used to create basic dimensionless variables: $R, U_\infty, \rho U_\infty^2$ for x and y, for u and v and for p, respectively. R and Δ, L are channel height, height, and width of heat sources; $\mathrm{Lu}_w = \rho c/(\rho c)_w, \mathrm{Lu}_h = \rho c/(\rho c)_h$ are Luikov numbers for channel wall and for heat sources (Section 6.6.1). The dimensional temperature and the modified Grashof number are defined in terms of the volumetric heat generation q_v as $\theta = (T - T_\infty)/\Delta T_{ref}$ and $\mathrm{Gr} = g\beta q_v L\Delta R^3/v^2$ where $\Delta T_{ref} = q_v L\Delta/\lambda$. Due to these definitions, the source term in energy equation contains only geometric parameters.

The uniform inlet conditions for fluid and no-slip boundary conditions at the walls are used. The outer wall surfaces are assumed to be adiabatic. At the outlet of the channel an extended domain is used to avoid the influence of large recirculation occurring at the last heat source. It is assumed that at the outlet and on the walls of extended domain the following boundary conditions are appropriate: $\partial^2 u/\partial x^2 = \partial^2 v/\partial x^2 = \partial^2 \theta/\partial x^2 = 0$ and $\partial u/\partial y = v = \partial \theta/\partial y = 0$.

Radiative heat transfer is calculated using the radiosity/irradiation approach. The surfaces are considered as opaque, diffuse, and gray, and the inlet and outlet of the channel are treated as black surfaces at ambient temperature. The dimensionless radiative heat flux q_{Ri} from the discrete surface and radiosity $J_i/\sigma T_\infty^4$ are determined as

$$q_{Ri} = \frac{\varepsilon_i}{1-\varepsilon_i}\left[\left(\frac{T_i}{T_\infty}\right)^4 - \frac{J_i}{\sigma T_\infty^4}\right] \qquad \frac{J_i}{\sigma T_\infty^4} = \varepsilon_i\left(\frac{T_i}{T_\infty}\right)^4 + (1-\varepsilon_i)\sum_{j=1}^{n}\frac{J_i}{\sigma T_\infty^4}F_{ij} \qquad (10.34)$$

where ε and F_{ij} are emissivity and shape factor, respectively. Temperatures at the channel walls–fluid interface and at heat source–fluid interface are determined from energy balance equations:

$$-\frac{\lambda_w}{\lambda}\left(\frac{\partial\theta}{\partial y}\right)_w = -\frac{\partial\theta}{\partial y} + N_{RC}q_R \qquad N_{RC} = \frac{\sigma T_\infty^4}{\lambda}\frac{R}{\Delta T_{ref}}$$

$$\frac{\lambda_h}{\lambda}\left(\frac{\partial\theta}{\partial x}\right)_w \Delta y - \frac{\lambda_h}{\lambda}\left(\frac{\partial\theta}{\partial y}\right)_w \Delta x + \left(\frac{\partial\theta}{\partial y}\right)_n \Delta x + \frac{\lambda_h}{\lambda}\left(\frac{\partial\theta}{\partial x}\right)_e \qquad (10.35)$$

$$\times \Delta x + \frac{R^2}{L\Delta}\Delta x\Delta y - N_{RC}q_R\Delta x = \mathrm{Pe}\frac{\partial\theta}{\partial t}\Delta x\Delta y$$

where Δx and Δy are dimensionless width and height (scaled by R) of an element taking into account the balance, and subscript e refers to the extended domain.

The SIMPLE (Semi-Implicit Method for Pressure-Linked Equations) algorithm (see Section 7.3.1) and the point-by-point Gauss-Siedel iteration method are used to solve the governing equations for velocity component, pressure, temperature, and radiosity. The following basic results and conclusions are obtained:

1. The results obtained show that an increasing Reynolds number leads to shifting the center of circulation to the right side wall of the cavities. Beyond the last heat source, the intensity of circulation increases as Re grows. The temperature of the first chip is lower than others due to contact with incoming fresh air. The maximum temperature of the last chip is lower than that of the preceding one due to high recirculation at the last chip. The first and the last chips are exposed to the atmosphere. As a result, the radiation heat transfer from these chips is higher than from the interior chips. The radiation effect and the maximum temperature become smaller as the Reynolds number grows.

2. To study the buoyancy effect, different Grashof numbers are obtained by changing the values of the volumetric heat generation q_v. The results show that with increasing Grashof number, the dimensionless temperature decreases linearly, from which follows that the effect of buoyancy is negligible for the range of parameters studied. Despite the dimensionless temperature decreases, the actual dimension temperature increases as the Grashof number increases, as is expected.

3. The dimensionless temperature decreases as the emissivity of heat sources and of the walls increase at all Reynolds numbers analyzed. The comparison shows that the effect of the wall emissivity is more significant than that of the heat sources. The temperature decreases also when the emissivity of the substrate grows. As the emissivity of substrate changes from 0.02 to 0.85, the maximum temperature decreases in 11°C, while the same change of emissivity of heat sources gives only 4°C drops in maximum temperature. When the emissivity of the top wall increases, the convective heat transfer from the wall also increases due to radiation interaction increasing.

4. The dimensionless temperature decreases as the thermal conductivities of the heat sources and substrate increase. At $\lambda_h/\lambda = 500$, the heat sources become isothermal. When the thermal conductivity ratio of heat sources and fluid λ_h/λ changes from 50 to 500, the calculation shows a 20% drop in the maximum temperature. Similar behavior of the dimensionless temperature is observed when the wall–fluid conductivity ratio is varied.

5. As the emissivity of substrate and top wall increases, the contribution by radiation decreases and the convective contribution increases. Increasing of the heat source's emissivity leads to increasing radiation contribution. In particular, at Re = 250, the radiation contribution increases from 10.5% at emmisivity of heat sources $\varepsilon_h = 0.02$ to 19% at $\varepsilon_h = 0.85$. As the Reynolds number increases, the radiation fraction of heat transfer becomes less.

6. The correlation expressions for maximum temperature are obtained in the form $(A/A_{ref})^m$ for $A = $ Re, Gr, λ_h/λ, λ_w/λ, ε_h and ε with $A_{ref} = 500, 8.65 \cdot 10^5, 100, 50, 0.55, 0.55$, and $m = 0.2, 1, 0.6, 0.6, 0.65\ 0.56$.

Example 10.7: A Microchannel Heat Sink as an Element of a Heat Exchanger [20]

The microchannel heat sink studied in this work is considered as an element of the microchannel heat exchanger. The heat transfer characteristics of such an exchanger should significantly increase due to great reduction of the thickness of the thermal boundary layer and overall notable capacity based on large surface–volume ratio. The studied model is a rectangular silicon channel with hydraulic diameter $D_h \approx 100\mu m$, which is a basic element of an experimentally investigated pattern in Reference [21].

The following assumptions are made:

1. The range of Knudsen number $Kn = l/D_h$ (l is a free path) lies in a continuum flow regime, and hence, Navier-Stokes equations are appropriate.
2. The flow is incompressible, laminar, and steady state. The thermophysical properties are temperature dependent.
3. The largest temperature gradients and thermal stresses are expected to occur at the inlet of the channel. Therefore, the development of the flow and temperature at the inlet should be carefully resolved.
4. Thermal radiation is negligible because the typical operation temperature is below 100°C.

The three-dimensional Navier-Stokes and energy equations for fluid with temperature-dependent properties and energy equation for walls are employed:

$$\nabla(\rho\mathbf{V}) = 0, \quad \mathbf{V}\cdot\nabla(\rho\mathbf{V}) = -\nabla p + \nabla\cdot(\mu\nabla\mathbf{V}) \tag{10.36}$$

$$\mathbf{V}\cdot\nabla(\rho c_p T) = \nabla\cdot(\lambda\nabla T), \quad \nabla(\lambda_w \nabla T_S) = 0 \tag{10.37}$$

A uniform heat flux is imposed at one of the channel walls, but the others are isolated. The entering flow velocity and temperature are given, and gradients of velocity and temperature at the exit are taken to be zero. The no-slip condition at the walls and conjugate conditions at the fluid–solid interface should be satisfied.

Patankar's technique of discretization and SIMPLER (SIMPLE Revised) algorithm (Chapter 7) are used to solve the system of governing equations. The predicted and experimental data from Reference [21] are in agreement. The following results and conclusions are deduced:

1. The local temperature distribution shows that the walls are isothermal, but the temperature field in fluid is essentially nonuniform. Initially, the high temperature gradient zone forms at the inlet of the channel and then increasing fluid core temperature is observed. Three basic conclusions can be stated:
   a. The maximum heat fluxes occur at the inlet of the channel.
   b. The heat flux imposed at the wall is spread out by conduction within walls and finally is transferred to the fluid.

    c. Due to the effect of conjugation, thermal development occupies the entire channel. The temperature distributions at the inner and outer wall surfaces show a very complex pattern resulting from convective heat transfer and three-dimensional conductions.

2. The distribution of the local heat fluxes on the inner walls, which are the fluid–walls interface, confirms the observation deduced by the temperature distribution analysis in (1). In particular, the local heat fluxes are the greatest at the inlet of the channel where the temperature gradients are high. The local heat flux inside the channel is distributed high nonuniformly so that magnitude variation in the heat fluxes ranges several orders. The reason for this is the difference in spacing between channel walls. The channel cross-section is a stretched rectangle such that the distance between two walls in one direction is about three times smaller than that between two other walls. As a result, the boundary layer between walls with small spacing is much thinner, and consequently, the convective heat transfer is much larger. The complicated heat flow structure is observed in the corners where the heat flux is a product of interaction between the boundary layers developed along both adjacent walls. The resulting configuration of heat flow in corners shows the negative heat fluxes directed from the fluid to the walls. In such a case, the heat transfer coefficients are also negative (Section 5.3) when traditional methods cannot be used and only conjugate formulation of the problem can give realistic results (Chapter 5).

3. The average characteristics in general conform to local distribution quantities. The average wall temperature increases significantly in the entering portion due to high local temperatures in this area. In contrast to this, the fluid bulk temperature grows gradually along the channel approaching the wall temperature at the exit. Large temperature gradients in the inlet channel portion may result in significant thermal stresses, which is important to take into consideration during design. The average heat fluxes and average heat transfer coefficient gradually decrease along the channel. The average heat flux of all walls of the channel is smaller than the initially imposed heat flux everywhere except the inlet portion, where the average heat flux is greater than the imposed one. This occurs because the area where the heat flux is imposed is much smaller than that of the surfaces of the walls inside the channel.

10.2.2 Turbine Blades and Rockets

Example 10.8: Film-Cooled Turbine Blade [22]

Film cooling is used to protect turbine blades and vanes of the first and second rows from direct contact with streams of hot gas. Injected cold air covers the surface of the blade or vane producing a layer of cold air between the protected object and hot gas.

This example presents a method for computing conjugate heat transfer in a film-cooled turbine blade or vane. The temperature field of a model that corresponds to a real vane of a film-cooled engine [23] obtained by this method is given and discussed.

The NASA explicit finite volume code Glenn-HT is used to solve the Navier-Stokes equation for fluid domain, and the boundary element method is employed in a solid domain for solving the Laplace equation. Such a combination approach as applied to a conjugate problem provides two advantages arising from the fact that the boundary element method (BEM) requires only body surfaces discretization. The first is evident consisting of computation time and storage saving, because there is no need in BEM to involve in calculation the whole body volume. This feature of the BEM method is especially important for the conjugate problem that has a solution usually carried out by iterations. In this process, the temperature distribution on the interface obtained from the Laplace equation is used as a boundary condition for the Navier-Stokes equation, while the heat flux distribution on the interface resulting from the Navier-Stokes solution gives the boundary condition for the Laplace equation. Thus, only the surface temperature distribution given in BEM under the Laplace equation solution is needed for the iteration procedure. Another advantage to using BEM in a conjugate problem follows from the same feature. Because in BEM the temperature and heat fluxes on the interface are directly obtained, those can be used to satisfy the conjugate conditions and so there is no need to differentiate the solid temperature field usually employed to get heat fluxes on the interface (Chapter 7).

A three-dimensional density-based integral form of the Navier-Stokes equation for fluid and heat conduction for a solid constitutes the system of governing equations:

$$\int_{\Omega} \frac{\partial \mathbf{W}}{\partial t} d\Omega + \int_{\Gamma} (\mathbf{F} - \mathbf{G})\hat{n}d\Gamma = \int_{\Omega} S d\Omega, \quad \nabla \cdot \left[\lambda_s(T_s) \nabla T_s \right] = 0, \quad f(T) = \frac{1}{\lambda_0} \int_{T_0}^{T} \lambda_s(T)dT \quad (10.38)$$

where Ω is the volume, Γ is the surface bounded by the volume Ω, and $\hat{n}$ is outward normal. The vector $\mathbf{W}$ contains variables $\rho, \rho u, \rho v, \rho w, \rho e, \rho k, \rho \omega$, where e, k, ω are the specific total energy, the kinetic energy turbulent fluctuations, and the specific dissipation rate; vectors $\mathbf{F}$ and $\mathbf{G}$ represent the convective and diffusion fluxes; and $\mathbf{S}$ is a vector containing noninertial terms, turbulent production, and dissipation quantities. The kinetic energy turbulence and the specific dissipation rate determine the two equations $k - \omega$ turbulence model [24]. The turbulence viscosity μ_{tb} and turbulence thermal conductivity λ_{tb} are defined in this model as

$$\mu_{\Sigma} = \mu + \mu_{tb} = \mu + \rho k / \omega \qquad \lambda_{\Sigma} = \lambda + \lambda_{tb} = c_p \left(\frac{\mu}{\text{Pr}} + \frac{\mu_{tb}}{\text{Pr}_{tb}} \right) \qquad (10.39)$$

where $Pr_{tb} \approx 0.9$ is turbulent Prandtl number. The last, Equation (10.38), is the Kirchhoff transform that converts the conductivity equation to the Laplace equation in the case of temperature-dependent thermal conductivity. In Equation (10.38), λ_0 and T_0 are the reference values.

The Glenn-HT code solves the Navier-Stokes equation by the time marching scheme until the steady-state conditions are achieved. The fourth-order explicit Runge-Kutta algorithm is used. The heat conduction equation is solved using the BEM procedure based on converting the Laplace equation into a boundary integral equation:

$$C(\xi)T(\xi) + \int_B T(x)\hat{q}(x,\xi)dB(x) = \int_B q(x)\hat{T}(x,\xi)dB(x)$$

$$C(\xi) = \int_B \left[-\lambda_s \frac{\partial\hat{T}(x,\xi)}{\partial n} \right] dB(x), \quad \lambda_s \nabla^2 \hat{T}(x,\xi) = -\delta(x,\xi), \tag{10.40}$$

$$\hat{T}(x,\xi) = \frac{1}{4\pi\lambda_s r(x,\xi)}$$

Here, $B(x)$ is the surface bounding the domain of interest, ξ is the source point, x is field point, $q(x)$ is heat flux, $\hat{T}(x,\xi)$ is a fundamental or Green solution which is a response of an adjoint governing differential operator at any field point x due to a perturbation of Dirac delta function acting at the source point ξ. In the case considered, this fundamental solution is defined by the second, Equation (10.40), and may be presented in the form of the third, Equation (10.40), in which $r(x,\xi)$ is the Euclidean distance from the source point ξ. The other details of solution of the Laplace equation are given in Reference [22].

The numerical calculations are performed under conditions that match the planning experiment at the NASA Glenn Research Center. The geometry of the model vane is based on the scaled factor 2.943 to match the exit Mach number $M = 0.876$ and Reynolds number $Re = 2.9 \cdot 10^6$ based on real chord 0.206 m. The vane has two plena that feed twelve rows of film-cooling holes. A special program GridPro is adopted to model this complex geometry. The stagnation temperature of the inlet free stream flow is given as 3109 R, the average blade temperature is estimated to be 2164.9 R, and the nondimensional temperature 0.5 corresponds to 1554.5 R.

The temperature distribution along the vane obtained from the conjugate problem solution is compared with that calculated by the standard two-temperature method. Two solutions obtained for isothermal temperatures 2174.9 R and 2485.6 R are used to calculate the temperature distribution along the vanes by the traditional approach for the same condition as in the conjugate problem. The comparison shows a significant difference between those two results. The temperature span across the surface obtained by the traditional approach is 1720 to 2420 R, and the temperature span predicted by the conjugate solution is remarkably different at 1620 to 2620 R. Thus, the conjugate prediction for the minimum temperature is 100 R lower, and for the maximum temperature it is 200 R higher. These two temperature distributions differ not only in separate values or at discrete points, but are entirely different. In particular, according to the conjugate results, the trailing part of the blade is much hotter, and the forward blade section as it follows from the conjugate solution is considerably cooler. It is evident that such temperature changes are important for stress analysis and design of the highly loaded turbine blades.

Example 10.9: Turbine Blade with Radial Cooling Channels [25]

The other type of cooling system for turbine blades is performed using radial channels with flowing cold air. Such channels are made through the whole blade in the radial (vertical) direction perpendicular to the blade cross-section. The cold air flowing through these channels cools the blade or vane.

The set of governing equations describing a model of such a cooling system in this study is similar to that employed in the previous example. The two-dimensional Navier-Stokes equation is used for the fluid domain as well as the Laplace equation is applied to compute the heat conduction in a solid. The Navier-Stokes equation is written in curvilinear coordinates using the same as in previous problem density-based approach, but in the initial differential form without integration. The finite-volume method is applied for the numerical solution of both the Navier-Stokes and Laplace equations. The turbulence is taken into account by standard Baldwin-Lomax model [24]. To calculate the heat fluxes from the blade to cooling air in the radial channels, the heat transfer coefficient is recommended.

To conjugate solutions of different domains, the iteration procedure is employed. As usual, the Dirichlet problem is considered for fluid, and the Neumann problem formulation is used for the heat conduction equation. It is noted that such an approach gives stable solutions, but also, vice versa, when the heat fluxes are considered as a boundary condition for the Navier-Stokes equation, and the surface temperatures are employed to solve the Laplace equation, unstable results may be obtained.

Three blade configurations with different size and location of cooling channels at two exit Mach numbers $M = 0.59$ and $M = .0.95$ were investigated. Parameter distributions around different blades are calculated showing the following:

1. Cooling duct configurations have little effect on pressure distribution around the blade, but the temperature distribution strongly depends on the size and location of the cooling ducts.
2. The blades with small channels and relatively uniform positions show better cooling effects and smaller mass flow rate. This effect is more pronounced on the pressure blade side.
3. On both blade sides, the temperature decreases approaching the minimum close to the stagnation point and then rapidly increases to its maximum.

4. Mach number gradually increases on the suction side of the blade, but on the pressure side, Mach number increases almost until the exit and then goes down. For higher Mach number at the exit (M = 0.95), this behavior is more pronounced and leads to the supersonic values at the maximum.

5. Accordingly, the gauge pressure on the suction side starting from the stagnation point is almost constant except in a small area close to the exit where it drops to zero. At the same time, on the other side in conformity with Mach number behavior, the pressure decreases, approaches the negative values of gauge pressure at the minimum and then grows to zero at the exit.

Example 10.10: Charring Material in a Solid Propellant Rocket [26]

Charring materials exposed to high temperature is a process of decomposition and loss of surface material by ablating to absorb the heat. Such processes are used, for instance, for internal thermal protection of rocket combustion chambers or for thermal shield of reentry vehicles.

This example studies the charring material process using the three-dimensional model composed of three zones: the virgin zone, the decomposition zone, and the char zone. These zones are disposed one over another so that on the top, the char zone appears along which the working fluid flows. In the first zone, the material changes are negligible; in the second zone, the material undergoes chemical and physical changes and energy is absorbed by decomposition; and in the third zone, composed mainly of char, the heat is transferred by conduction and convection. The changes in the material proceed in two ways: by the free material surface recession and by its decomposition, when the surface does not move, but the material properties are changed. During this process due to heating, the material releases pyrolysis gas, which passes through the solid into the fluid that flows along the upper char zone. To simplify the mathematical description of the problem, two basic assumptions are used: (1) the pyrolysis gas velocity is approximately orthogonal to the receding surface, and (2) the surface regression is locally uniform and occurs along the normal direction to the surface.

The governing equations in the fluid region are the conservation of mass, momentum, and energy equations:

$$\frac{D\rho}{Dt} + \rho \nabla \cdot \mathbf{v} = 0, \qquad \rho \frac{D\mathbf{v}}{Dt} = \rho g - \nabla p + \mu \nabla^2 \mathbf{v} + \frac{\mu}{3} \nabla(\nabla \cdot \mathbf{v}) \tag{10.41}$$

$$\rho \frac{DT}{Dt} = \rho \frac{Dp}{Dt} + \nabla \cdot (\lambda \nabla T) + \mu \Phi$$

The energy equation for decomposition charring material and its derivation on the base of the Arrhenius decomposition law are given in Reference [26]:

$$\rho c_p \left(\frac{\partial T}{\partial t} \right)_\xi = \nabla \cdot (\lambda \nabla T) + \rho c_p \mathbf{v} \cdot \nabla T + (J_g - \hat{J}) \left(\frac{\partial \rho}{\partial t} \right)_x + \rho_g \mathbf{v}_g \cdot \nabla J_g$$

$$\hat{J} = \frac{\rho_v J_v - \rho_c J_c}{\rho_v - \rho_c}, \qquad J_c(T) = J_c(T_r) + \int_{T_r}^{T} c_{pC}(t)dt,$$

(10.42)

$$J_v(T) = J_v(T_r) + \int_{T_r}^{T} c_{pv}(t)dt$$

$$\lambda = x\lambda_v + (1-x)\lambda_c, \qquad c_p = xc_{pv} + (1-x)c_{pC}, \qquad x = \frac{\rho_v}{\rho_v - \rho_c} \left(1 - \frac{\rho_c}{\rho} \right)$$

Here $\mathbf{v}$ is the thermal protection recession velocity vector; Φ is the dissipation function; J is solid enthalpy; x is a fraction of virgin value; the subscripts g,c,v denote pyrolysis gas, charred, and virgin values; and the subscripts x and ξ refer to derivatives at constant x in a fixed frame and at constant ξ in a moving frame.

In Equation (10.42), the term on the left-hand side is the sensible energy accumulation; the first term on the right-hand side is the conduction term, the second term is the energy convected by the motion of the reference frame, the third term is the difference between the energy convected away by pyrolysis gas and the chemical energy accumulation, and the fourth term is the energy convected by pyrolysis gases passing through the solid. The turbulence is taken into account using the $k-\omega$ turbulence model [24] (Example 10.8).

The governing equations are solved numerically by applying three-dimensional code Phoenics. This software uses a finite volume element approach and, as well as other similar programs, like SIMPLE or SIMPLER, can be used for solving the energy equation of both the fluid and the solid by accounting for the corresponding boundary and velocity conditions (Chapter 7). The other details of numerical performance are given in Reference [26].

The method developed in this example is used to predict the results of several heat transfer problems for which analytical, experimental, or numerical solutions are known:

1. Results of simulation of heat transfer in a blast tube with thermal protection are compared with data obtained by the CMA program [27]. The initial protection temperature is 300 K. The laminar flow of combustion gas in the tube is at temperature 3600 K. The temperature and density profiles in the fluid and solid obtained in both studies are in agreement.
2. The process in Material Test Motor (MTM) for testing new ablative materials with protection is simulated for the charring material ES59A with a low-density thermal protection ESA-ESTEC

developed for space rockets [28]. The dimensions of the model cross-section were extrapolated from the test section in MTM and the suitable curvilinear mesh is used. The velocity and pressure as well as turbulent viscosity and conductivity show low numerical errors which indicates that these are computed correctly. The results indicate that assumptions of negligible propellant reduction and constant pressure in the chamber during burning are possible. The temperature profiles obtained in this case are similar to those for the blast tube with a typical sudden derivative variation arising due to passage through the different computation domains. The density changes suddenly between charred and virgin zones at approximately one half of the solid thickness. The comparison of Material Affected Depths (MAD) obtained experimentally and predicted by three different approaches shows a reasonable agreement, indicating that after 25.3 seconds burning time, the MAD is about 4 mm. The mass flow rate and heat fluxes predicted by the conjugate heat transfer approach are slightly underestimated, but the MAD predictions are basically overestimated.

3. The heat exchange analysis of the igniter of the solid propellant rocket during turbulent combustion is performed. Some approximations are employed to simplify the problem. The pressure at the rocket exit is assumed to be constant, the external thermal protection of the igniter is considered as nondecomposing, and the igniter switching time is assumed to be negligible. Two thermal protections are considered: one from aramidic fiber and another by using silica phenolic with reduction erosion rate. The adiabatic combustion gas temperature 3424 K is uniformly kept everywhere except for at the solid parts which are at 300 K. The velocities in the chamber are slow after a short transient period, and therefore, the buoyancy convection is taken into account, while the radiation is assumed to be negligible. The MAD and the thermal fields in gas and solid are calculated. The results indicate that the steel interface temperature is between 350 and 400 K under the initial temperature of 300 K. The temperature profiles in the internal thermal protection of the igniter are calculated as well as are the profiles of the solid density in the thermal protection.

10.2.3 Cooling by Rewetting Surfaces

The next two examples considered the cooling of hot surfaces by the rewetting process. Such cooling methods are used, for example, in quenching hot metal surfaces, for thermal control of electronic systems, and for thermal regulation of space stations. One of the most important applications is the use of the rewetting surface in nuclear reactors in the case of a loss-of-coolant accident (LOCA) when the coolant system fails to operate. In such an emergency situation, the controller rewetting process is used to bring the hot nuclear fuel to cooled conditions.

Example 10.11: Rewetting the Hot Surface by Falling Film [29]

A hot rod or flat plate cooled by falling liquid film is considered. The
model is based on two assumptions: the solid is infinite and the wet
front velocity is constant. These assumptions are typically used in many
studies of rewetting surfaces, transforming this unsteady problem into a
quasi-steady state problem with a constant solid temperature at the moving front. The additional assumptions, most of which are also typical,
are listed in this example, too. The governing equations for the solid and
the fluid in dimensionless variables are used in the following form:

$$\frac{1}{r^m}\frac{\partial}{\partial r}\left(r^m\frac{\partial\theta}{\partial r}\right)+\frac{\partial^2\theta}{\partial r^2}+\hat{u}(r)\frac{\partial\theta}{\partial z}=0,$$

$$\hat{u}_1=\frac{ur_2}{\alpha_1},\ 0\le r\le r_1,\ \hat{u}_2=\frac{(u-v)r_2}{\alpha_2},\ r_1<1$$

(10.43)

where $m=0$ for a flat plate and $m=1$ for a rod, the subscripts 1 and 2
denote the solid and the fluid, u and v are velocities of front and liquid
film, the variable r is scaled by r_2, $\theta=(T-T_{2in})/(T_{1in}-T_{2in})$, and subscript
in denotes the initial temperature.

Equation (10.43) is subjected to the following conditions: symme-
try $\partial\theta/\partial r=0$ at $r=0$, for the wet (A) part $\theta_A=0$ at $r=1$, for dry (B) part
$\partial\theta/\partial r=0$ at $r=r_1$, for $z\to\pm\infty$ $\theta_B=1$ for (+) and $\theta_A=0$ for (−), at the moving
front $(z=0)$ $\theta_A=\theta_B$, $\partial\theta_A/\partial z=\partial\theta_B/\partial z$, and $\theta_A=\theta_0$ or $\theta_B=\theta_0$ at the interface $r=r_1$ and the conjugate conditions $\theta_A^-=\theta_A^+$, $(\lambda_1/\lambda_2)(\partial\theta_A/\partial r)=\partial\theta_B/\partial r$,
where θ_0 is the rewetting temperature.

The problem is solved using separation of variables. Although it can be
seen that Equation (10.43) is not separable, there is a method suggested
by Yeh [30] that is applicable to the problem at hand. The solution in
detail is presented in the reviewed article [29] in the summary. The solution is sought as a product of two functions $R(r)$ and $Z(z)$, each of which
depends on one of the independent variables. The temperatures in wet
(A) and dry (B) regions are defined as

$$\theta_A=\sum_{n=1}^{N}R_{An}(r)\left[\sum_{j=1}^{N}a_{nj}A_j\exp(s_jz)\right]\qquad\theta_B=1+\sum_{n=1}^{N}B_nR_{Bn}(r)\exp(-\gamma_n z)$$

$$R_{A1n}(r)=\frac{b_{12}b_{23}-b_{13}b_{22}}{b_{13}b_{21}}X(\mu_n r)\qquad R_{A2n}(r)=X(\mu_n r)-\frac{b_{12}}{b_{13}}Y(\mu_n r)$$

(10.44)

$$\begin{array}{lll}X=\cos(\mu_n r) & X=J_0(\mu_n r) & R_{Bn}=\cos(\gamma_n r) \\ \text{or} & & \text{or} \\ Y=\sin(\mu_n r) & Y=Y_0(\mu_n r) & \sin(\gamma_n r_1)=0 \end{array} \quad \begin{array}{l} R_{Bn}J_0(\gamma_n r) \\ J_1(\gamma_n r_1)=0 \end{array}$$

$$b_{21}b_{32}b_{13}+b_{31}b_{12}b_{23}-b_{13}b_{22}b_{31}-b_{33}b_{12}b_{21}=0$$

The expressions for s,a,b,A,B and C are given in Reference [29].
Eigenvalues μ_n are determined as positive roots of the last equation, γ_n

are given by trigonometric functions for the flat plate or by Bessel functions for the rod, and v_n are roots of equation $v_n^2 A_n + \sum_{i=1}^{N} C_{Ani} \gamma_n A_i - \mu_n^2 A_n = 0$. The calculation procedure is iterative, and it starts from guessing the rewetting velocity u which is corrected in the next step using the following equations:

$$\sum_{n=1}^{N} R_{An}(r_1) \sum_{i=1}^{N} a_{ni} A_i = \theta_0 \quad or \quad 1 + \sum_{n=1}^{N} B_n R_{Bn}(r_1) = \theta_0 \qquad (10.45)$$

The numerical results are obtained for the flat plate:

1. The comparison between rewetting velocities obtained by prediction and experimental data for different pairs (water–stainless steel, nitrogen–copper, freon113–stainless steel) and various rewetting and initial wall temperatures shows reasonable agreement. For low flow rate, the agreement is better than for higher ones. In some cases, the predicted values of rewetting velocity are underestimated, and in others, these turn out to be overestimated. There are also several situations when theoretical results bounded corresponding experimental data.

2. The temperature distributions in the solid and in the fluid are calculated for stainless steel at an initial temperature of 700°C cooled by water at an initial temperature 20°C and a flow rate 0.33×10^{-3} kg/s. The distribution of the temperature in a fluid indicates that there is a sharp temperature increase toward the triple interline. It follows from the temperature distribution in a solid that in the traversal direction, the temperature is uniform and the temperature decreases toward the wetted surface only in the area close to the quench front. In the axial direction, the sharp temperature gradients also are observed only in small areas above and under the quench front. Comparison of temperatures on wetted and dry regions shows that in small areas close to the front temperature distribution coincides, but the temperature on the wetted side is lower than that on the unwetted one. The difference between these temperatures grows when the flow rate increases.

3. The rewetting velocity increases slightly with flow rate. This result is obtained for both stainless steel–water and nitrogen–copper pairs. For the stainless steel–water pair, at the given flow rate, the heat transfer coefficient increases sharply to a peak just above the quench front and then decreases slightly, becoming constant almost before reaching the front. As the flow rate increases, the heat transfer coefficient grows from 10^4 W/m² K at low flow rates to 10^5 W/m² K at higher flow rates. Similar behavior is observed for heat flux variation, but upstream close to the front, the heat flux becomes lower as the flow rate grows.

4. The accuracy of the one-dimensional model is estimated by comparing results obtained by such a model and conjugate approach

for the same rewetting temperature and heat transfer coefficient. For the low flow rate, the rewetting velocity predicted by a one-dimensional approach is 12% to 17% less. This error increases to 18% to 24% for high flow rates but becomes less as the nondimensional wall temperature grows. Thus, the one-dimensional model for the same rewetting temperature and heat transfer coefficient underpredict the rewetting velocity with reasonable accuracy at least for low flow rates. However, there is no way to properly choose values of the rewetting temperature and heat transfer coefficient.

5. The dependence between the rewetting temperature and heat transfer coefficient obtained in the conjugate problem resembles the experimental values found by the Yu et al. [31] formula $(T_0 - T_q)\sqrt{h} = cont.$, where T_q is the temperature of the water at the quench front. According to this dependence, these two parameters cannot be considered as independent quantities.

Example 10.12: Transient Cooling of a Semi-Infinite Hot Plate [32]

The standard model usually used in studies of rewetting surfaces is based on two main assumptions: the solid is infinite and the velocity of the wet front is constant in time. Under these assumptions, the unsteady heat conduction equation is transformed into the quasi-steady state form by using the coordinate system moving with the film front. As a result, the temperature of the solid at the moving front is also constant (Example 10.11).

In this study, a more realistic model of a semi-infinite solid is developed. There are some challenges in developing such a model compared with the situation when the solid is infinite. The length of the wet portion becomes time dependent. Thus, the heat transfer is a transient process, and the temperature of the solid at the moving film front is an unknown function of time. This temperature can be determined by conjugating the solutions describing the temperature fields in wet and dry solid portions. However, to obtain these solutions, the boundary condition at the film front is required that is the same unknown temperature at the film front.

Another distinction of the problem at hand is that in this case, in contrast to most conjugate convective problems, two solutions of the same type of equation should be conjugated. The reason for this is that there are wet and dry solid domains with different properties.

To transform the variable length of the wet portion to a constant one, a dimensionless variable is used instead of the longitudinal coordinate. To determine the unknown temperature at the moving front, the superposition method in the form of a series is applied. In this case, the solution of a differential equation with unknown boundary conditions reduces to an infinite system of differential equations with known boundary conditions.

A one-dimensional model of a semi-infinite plate covered with moving liquid film is considered. The lower plate surface is assumed to be adiabatic, and the heat transfer coefficient between the plate and the film is constant and known. It is also assumed that the film is supplied at constant velocity U and heat losses to surrounding flows are negligible. The one-dimensional conduction equation in the moving frame with the origin at the film front and boundary condition for wet and dry portion according to the assumptions are

$$\lambda_w \frac{\partial^2 T}{\partial x^2} - \frac{h}{\Delta}(T - T_f) = \rho_w c_w \frac{\partial T}{\partial t} \tag{10.46}$$

$$\text{wet } T(x, t = 0) = T_i, \quad T(x = 0, t > 0) = T_w(t), \quad \frac{\partial T}{\partial x}(x = -U_w t, t > 0) = 0$$

$$\text{dry } T(x, t = 0) = T_i, \quad T(x = 0, t > 0) = T_w(t), \quad T(x \to \infty, t > 0) = T_i$$

Here, T, T_f, T_i are plate, film, and initial plate temperatures; $U_w t$ is the length of the wet part of the plate at time t, where U_w is the front velocity which due to evaporation is less than U (see below). To convert the time-dependent length to a constant length and reduce the number of parameters, the dimensionless variables are introduced for the wet portion:

$$\eta = \frac{x}{Ut}, \quad z = \frac{h}{c\rho\Delta}, \quad \theta = \frac{T - T_f}{T_i - T_f}, \quad \theta_w = \frac{T_w - T_f}{T_i - T_f} \tag{10.47}$$

Applying these variables to the governing system of the equation and boundary conditions yields the system depending on only one parameter $\mathrm{Bi}/\mathrm{Pe}^2$. As shown below, this ratio determines the rate of the transient cooling process. The greater this ratio, the shorter is the dimensionless time required to cool the plate to a given dimensionless temperature. Because cooling by a moving thin film proceeds in the boiling transitional state, it is suggested in Reference [32] to name the ratio $\mathrm{Bi}/\mathrm{Pe}^2$ the Leidenfrost number, Ls, similar to the Leidenfrost point on the boiling curve.

The governing equation for the wet part in these dimensionless variables becomes

$$\mathrm{Ls}\frac{\partial^2 \theta}{\partial \eta^2} + z\eta\frac{\partial \theta}{\partial \eta} - z^2\frac{\partial \theta}{\partial z} - z^2\theta = 0, \quad \mathrm{Ls} = \frac{\mathrm{Bi}}{\mathrm{Pe}^2} = \frac{\lambda_w h}{\rho_w^2 c_w^2 U_w^2 \Delta}$$

$$\tag{10.48}$$

$$\theta(\eta, z = 0) = 1, \quad \theta(\eta = 0, z > 0) = \theta_w(z), \quad \frac{\partial \theta}{\partial \eta}(\eta = -1, z > 0) = 0$$

For the dry region, it is assumed that $h = 0$, and the governing equation is used in another form:

$$\frac{\partial^2 \vartheta}{\partial \xi^2} - \frac{\partial \vartheta}{\partial z} = 0, \quad \xi = x\sqrt{\frac{h}{\lambda_w \Delta}}, \quad \vartheta = \frac{T - T_i}{T_f - T_i} \quad \vartheta_w = \frac{T_w - T_i}{T_f - T_i} \tag{10.49}$$

$$\vartheta(\xi, z = 0) = 0, \quad \vartheta(\xi = 0, z > 0) = \vartheta_w(z) \quad \vartheta(\xi \to \infty, z > 0) = 0$$

The temperature fields in the wet and dry plate portions should be conjugated at the moving film front. The conjugate conditions for that can be derived using the energy balance at the moving front. The heat $q^+(t)$ conducted from the dry region is slightly absorbed by evaporation and sputtering $q_w(t)$ at the moving film front, while the majority of the heat $q^-(t)$ is transferred to the wet region. The heat absorbed by evaporation and sputtering is usually neglected. Peng et al. [33] suggested that such an assumption is acceptable only if the plate temperature is lower than the rewetting temperature. Following this suggestion, it seems reasonable to assume that this amount of heat is proportional to the difference between the plate temperature at the moving film front and the rewetting temperature. Then, the conjugate condition takes the following form:

$$x = \eta = 0 \quad q^+(t) = q^-(t) + q_w(t,), \quad q_w(t) = h_w(T_w - T_0) \tag{10.50}$$

where h_w is the heat transfer coefficient of evaporation and sputtering.

It follows from Equations (10.48) and (10.49), respectively, that $\theta = f_1(z, \eta, Ls, \theta_w)$ and $\vartheta = f_2(z, \xi, \vartheta_w)$. Using the energy balance (Equation 10.50) and expressions for heat fluxes, one gets two results for the negligible and significant heat absorbed at the front:

$$q^- = -\lambda_w \left(\frac{\partial T}{\partial x} \right)_{x=0} = -\lambda_w \frac{T_i - T_f}{U_w t} \left(\frac{\partial f_1}{\partial \eta} \right)_{\eta=0},$$

$$q^+ = -\lambda_w \left(\frac{\partial T}{\partial x} \right)_{x=0} = -\lambda_w \left(T_f - T_i \right) \sqrt{\frac{h}{\lambda \Delta}} \left(\frac{\partial f_2}{\partial \xi} \right)_{\xi=0}$$

$$\tag{10.51}$$

$$f_2'(\xi, \vartheta_w) = \frac{\sqrt{Ls}}{z} f_1'(z, \theta_w, Ls)$$

$$f_2'(\xi, \vartheta_w) = \frac{\sqrt{Ls}}{z} f_1'(z, \theta_w, Ls) + \frac{h_w}{h} \sqrt{Bi} (\theta_w - \theta_0)$$

Because $\vartheta_w = 1 - \theta_w$, the first relation indicates that if the heat absorbed at the front is negligible, the function $\theta_w(z)$ depends only on the Leidenfrost number. From the second relation, it follows that in the case of significant heat absorbed at the front, the temperature distribution in the wet plate portion is determined by Leidenfrost number and two additional parameters $(h_w/h)\sqrt{Bi}$ and rewetting temperature θ_0.

To determine the unknown temperature at the moving film front, the solution of Equations (10.48) and (10.49) is presented in the form of a series similar to the series used in Chapter 3 for the study of heat transfer from arbitrary nonisothermal surfaces:

$$\theta = \sum_{n=1}^{\infty} G_n(\eta, z) \frac{\partial^n \theta_w}{\partial z^n} \qquad \vartheta = \sum_{n=1}^{\infty} H_n(\eta, z) \frac{\partial^n \vartheta_w}{\partial z^n} \tag{10.52}$$

Substituting these series into Equations (10.48) and (10.49) yields two infinite systems of equations with constant initial and boundary conditions:

$$\text{Ls}\frac{\partial^2 G}{\partial \eta^2} + \eta z \frac{\partial G_n}{\partial \eta} - z^2 \frac{\partial G_n}{\partial z} - z^2 G_n - z^2 G_{n-1} = 0, \quad \frac{\partial^2 H_n}{\partial \xi^2} - \frac{\partial H_n}{\partial z} - H_{n-1} = 0 \quad (10.53)$$

$$G_{-1} = 0, \quad z = 0 \, G_0 = 1, G_n = 0; \quad \eta = 0 \, G_0 = 1, G_n = 0; \quad \eta = -1 \frac{\partial G_n}{\partial \eta} = 0$$

$$H_{-1} = 0; \quad z = 0 \, H_n = 0; \quad \xi = 0 \, H_0 = 1, \ H_n = 0; \quad \xi \to \infty H_n = 0$$

The solutions of Equation (10.53) for H_n are given by the error functions [34]. The first two are

$$H_0(\xi, z) = 1 - erf \frac{\xi}{2\sqrt{z}}, \quad H_1(\xi, z) = \frac{\xi}{2} erfc \frac{\xi}{2\sqrt{z}} - \xi \sqrt{\frac{z}{\pi}} \exp\left(-\frac{\xi^2}{4z}\right) \quad (10.54)$$

For $n > 1$, solutions are also given in terms of error functions [34].

Equation (10.53) for G_n with variable coefficients are solved by approximate method of moments (Chapter 7). The suitable form of solutions satisfying the initial and boundary conditions contains two unknown functions $A_n(z)$ and $B_n(z)$ that are determined according to the method of moments using integral relations containing weighting functions 1 and η:

$$G_0(\eta, z) = 1 + \left(\frac{\eta}{2} + \eta\right) A_0(z) + \left(\frac{\eta^3}{3} + \eta\right) B_0(z)$$

$$G_n(\eta, z) = \left(\frac{\eta}{2} + \eta\right) A_n(z) + \left(\frac{\eta^3}{3} + \eta\right) B_n(z) \quad (10.55)$$

$$\int_{-1}^{0} F_0(\eta, z) d\eta = 0, \quad \int_{-1}^{0} F_0(\eta, z) \eta d\eta = 0 \ \text{ for } n = 0, \quad \int_{-1}^{0} F_0(\eta, z) \eta^n d\eta = 0 \ \text{ for } n > 0$$

where $F_n(\eta, z)$ is a function resulting from substitution expression for $G_n(\eta, z)$ into the left-hand side of Equation (10.53). The ordinary differential equations defining A_0 and B_0 are

$$\frac{dA_0}{dz} = A_0(z)\left(-\frac{28\text{Ls}}{z^2} + \frac{14}{3z} - 1\right) + B_0(z)\frac{16}{3}\left(\frac{\text{Ls}}{z^2} - \frac{1}{z}\right) + 28 \quad A_0(0) = 0 \quad (10.56)$$

$$\frac{dB_0}{dz} = A_0(z)10\left(-\frac{2\text{Ls}}{z^2} + \frac{1}{3z}\right) + B_0(z)\left(\frac{20\text{Ls}}{z^2} + \frac{11}{3z} + 1\right) + 20 \quad B_0(0) = 0 \quad (10.57)$$

To conjugate solutions for wet and dry regions, the equations in Equation (10.52) are used to find heat fluxes. Considering only two first terms, and applying conjugate condition (Equation 10.50), one gets an ordinary differential equation determining the plate temperature at the moving front:

$$q^-(z) = -\frac{\lambda_w(T_i - T_f)}{U_w t} \sum_{n=1}^{\infty} \left(\frac{\partial G_n}{\partial \eta}\right)_{\eta=0} \frac{\partial^n \theta_w}{\partial z^n},$$

$$q^+(z) = \lambda_w(T_f - T_i)\sqrt{\frac{h}{\lambda\Delta}} \sum_{n=1}^{\infty} \left(\frac{\partial G_n}{\partial \xi}\right)_{\xi=0} \frac{\partial^n \vartheta_w}{\partial z^n}$$

$$(10.58)$$

$$\left[z + (g_1 - g_0)\sqrt{z\pi Ls}\right]\frac{d\theta_w}{dz} + \theta_w\left[1 + g_0\sqrt{\frac{\pi Ls}{z}} + \frac{h_w}{h}\sqrt{z\pi Bi}\right] - \left[1 + \frac{h_w}{h}\sqrt{z\pi Bi}\theta_0\right] = 0$$

$$\theta_w(0) = 1$$

where $g_0(z) = A_0(z) - B_0(z)$ and $g_1(z) = [A_1(z) - B_1(z)]/z$. If the first term in this equation is relatively small, the first approximation solution for θ_w can be obtained. Then, setting $\theta_w = \theta_0$ in the resulting expression for θ_w, one gets the rewetting temperature:

$$\theta_w = \frac{1 + (h_w/h)\sqrt{z\pi Bi}\theta_0}{1 + g_0\sqrt{\pi Ls/z} + (h_w/h)\sqrt{z\pi Bi}}, \qquad \theta_0 = \frac{1}{1 + g_0\sqrt{\pi Ls/z_0}} \qquad (10.59)$$

Here, z_0 is the onset time that defines the dimensionless time required to cool the plate at the film front to the rewetting temperature. Comparing the equations in Equation (10.59), one concludes that these relations coincide if $h_w = 0$. From this result, it follows that the onset time depends on the rewetting temperature and on the Leidenfrost number but does not depend, at least in the first approximation, on heat absorbed by evaporation and sputtering.

Due to the evaporation and sputtering, the actual film velocity U_w is less than the given supplying velocity U. In a time Δt, the film advances actually by a length that is in $\Delta x = (U - U_w)\Delta t$ shorter than it would be without evaporation and sputtering. Therefore, the average amount of heat absorbed at the film front per second is $\Delta_f \rho_f h(\Delta x/\Delta t)$. On the other hand, this amount of heat is defined by Equation (10.50). Comparing these equations and noting that $m_f = \Delta_f \rho_f U$, one gets an expression for velocity U_w:

$$\frac{U - U_w}{U} = \frac{h_w \Delta}{m_f h}(T_i - T_f)\left(\frac{1}{z}\int_0^z \theta_w d\tilde{z} - \theta_0\right) \qquad (10.60)$$

The numerical results indicate that the dimensionless temperature sharply decreases at the beginning of the cooling process. As time passes, the rate of cooling decreases and finally becomes zero at the minimum temperature. For the case of negligible heat absorbed at the front, the character of function $\theta_w(z)$ depends only on the Leidenfrost number. As the Leidenfrost number increases, the plate cools faster, the achieved minimum temperature is smaller, and the time z_{min} required reaching it decreases. It is found that the dependence between a minimum temperature and corresponding time to reach it is close to linear. At the same time, this time may be approximately related to the Leidenfrost number by a quadratic function:

$$(\theta_w)_{min} = 0.036 z_{min} - 0.03, \quad z_{min} = 5 - 6\log(\mathrm{Ls}) + 3\log^2(\mathrm{Ls}) \qquad (10.61)$$

Because both equations in Equation (10.59) coincide, if $h_w = 0$, the function $\theta_0(z_0)$ in this case is the same as the function $\theta_w(z_{min})$, and hence, the features obtained for the latter can be used to establish a relation between rewetting temperature and onset time. However, in this study, the rewetting temperature is considered as a given value. Therefore, relations $\theta_w(z_{min})$ can be used for $\theta_0(z_0)$ only if $\theta_0 \geq \theta_w$.

1. The model used here describes only the transient part of the wetting process when the plate temperature at the moving front is higher than the wetting temperature. Although the model does not describe the second part of the process with practically constant rewetting temperature, the model consists of another dry plate portion with initial high temperature $T \leq T_i$. Therefore, in this case, the plate temperature at the moving front starts to increase after it reaches the minimum value, and in conformity with the boundary condition, ultimately in the limit as $t \to \infty$ becomes equal to the initial plate temperature.

2. Although the onset time does not depend on heat absorbed by evaporation and sputtering, it follows from Equation (10.59) that this amount of heat significantly affects the dependency $\theta_w(z, \mathrm{Ls}, \gamma)$, where $\gamma = (h_w/h)\sqrt{\mathrm{Bi}}$. The calculations show that despite all the curves $\theta_w(z)$ for different values of γ having the same initial and end points, the dependency $\theta_w(z)$ varies remarkably with γ. For small values of γ, the curves $\theta_w(z)$ are close to linear, but as the absorbed heat grows, the dependency $\theta_w(z)$ changes so that the temperature sharply decreases at the beginning of the process and then continues to decrease with time with a gradually reducing cooling rate.

3. The second approximation is found in Reference [32] by solving the linear Equation (10.58) for θ_w. The numerical results are obtained only for Leidenfrost number $\mathrm{Ls} = 1$. The difference between both solutions are small for small and large values of z and reaches the maximum of 15% to 20% for the middle values of z.

References

1. Pagliarini, G., and Barozzi, G. S., 1991. Thermal coupling in laminar flow double-pipe exchangers, *ASME J. Heat Transfer* 113: 526–534.
2. Pagliarini, G., and Barozzi, G. S., 1984. Thermal coupling in laminar double stream heat exchangers, *Proc. 2nd Natl. Conf. Heat Transfer*, Bologna, Italy, 103–113.
3. Bejan, A., 1982. Second law analysis in heat transfer and thermal design. In *Advances in Heat Transfer*, edited by T. F. Irvine and J. P. Hartnett, 15: 1–58.
4. Song, W., and Li, B. Q., 2002. Finite element solution of conjugate heat transfer problems with and without the use of gap elements, *Int. J. Numer. Methods Heat Fluid Flow* 12: 81–99.
5. Rao, V. D., Naidu, S, V., Rao, B. G, and Sharma, K. V., 2006. Heat transfer from a horizontal fin array by natural convection and radiation — A conjugate analysis, *Int. J. Heat Mass Transfer* 49: 3379–3391.
6. Incropera, F. P., and Dewitt, D. P., 1996. *Fundamental Heat and Mass Transfer*, 4th ed. Wiley & Sons, New York.
7. Roache, P. J., 1985. *Computational Fluid Dynamics*. Hermosa, Albuquerque, NM.
8. Jones, C. D., and Smith, L. F., 1970. Optimum arrangement of rectangular fins on horizontal surfaces for free convection heat transfer, *ASME J. Heat Transfer* 92: 6–10.
9. Rao, R., and Venkateshan, S. P., 1996. Experimental study on free convection and radiation in horizontal fin arrays, *Int. J. Heat Mass Transfer* 39: 779–789.
10. Starner, K. E., and McManus Jr., H. N., 1963. An experimental investigation of free convection heat transfer from rectangular fin arrays, *ASME J. Heat Transfer* 85: 273–278.
11. Grechannyi, O. A., Dorfman, A. S., and Gorobets, V. G., 1986. Coupled heat transfer and effectiveness of flat finned surface in a transverse flow, *High Temperature* 24: 678–683.
12. Chang, K., 1970. *Separation of Flow*. Pergamon Press, New York.
13. Vaszi, A. Z., Elliott, L., Ingham, D. B., and Pop, I., 2004. Conjugate free convection from a vertical plate fin with a rounded tip embedded in a porous medium, *Int. J. Heat Mass Transfer* 47: 2785–2794.
14. Merkin, J. H., 1979. Free convection boundary layer on axi-symmetric and two-dimensional bodies of arbitrary shape in a saturated porous medium, *Int. J. Heat Mass Transfer* 22: 1461–1462.
15. Vaszi, A. Z., Elliott, L., Ingham, D. B., and Pop, I., 2002. Conjugate free convection above a heated finite horizontal flat plate embedded in a porous medium, *Int. J. Heat Mass Transfer* 45: 2777–2795.
16. Liu, J. Y., Minkowycz, W. J., and Cheng, P., 1986. Conjugated mixed convection-conduction heat transfer along a cylindrical fin in a porous medium, *Int. J. Heat Mass Transfer* 29: 769–775.
17. Minkowycz, W. J., and Sparrow, E. M., 1978. Numerical solution scheme for local nonsimilarity boundary layer analysis, *Numer. Heat Transfer* 1: 69–85.
18. Liu, J. Y., Minkowycz, W. J., and Cheng, P., 1986. Conjugate mixed convection heat transfer analysis of a plat fin embedded in porous medium, *Numer. Heat Transfer* 9: 453–468.

19. Premachandran, B., and Balaji, C., 2006. Conjugate mixed convection with surface radiation from a horizontal channel with protruding heat sources, *Int. J. Heat Mass Transfer* 49: 3568–3582.
20. Fedorov, A. G., and Viskanta, R., 2000. Three-dimensional conjugate heat transfer in the microchannel heat sink for electronic packaging, *Int. J. Heat Mass Transfer* 43: 399–415.
21. Kawano, K., Minakami, K., Iwasaki, H., and Ishizuka, M., 1998. Development of microchannels heat exchanging. In *Application of Heat Transfer in Equipment, Systems and Education*, edited by R. A. Nelson, Jr., L. W. Swanson, M. V. A. Bianchi, and C. Camci, HTD-Vol. 361-3/PID- Vol. 3, pp. 173–180. ASME, New York.
22. Kassab, A., Divo, E., Heidmann, J., Steinthorsson, E., and Rodriguez, F., 2003. BEM/FVM conjugate heat transfer analysis of a three-dimensional film cooling turbine blade, *Int. J. Numer. Methods Heat Fluid Flow* 13: 582–610.
23. Heidmann, J., Rigby, D., and Ameri, A., 2002. A three-dimensional coupled external/internal simulation of a film-cooled turbine vane, *ASME J. Turbomachinery* 122: 348–359.
24. Wilcox, D. C., 1993, 1998. *Turbulence Modeling CFD*. DCW Industries, La Canada, CA.
25. Croce, G., 2001. A conjugate heat transfer procedure for gas turbine blades, *Ann. NY Acad. Sci.* 934: 273–280.
26. Baiocco, P., and Bellomi, P., 1996. A coupled thermo-ablative and fluid dynamic analysis for numerical application to solid propellant rockets. AIAA 96-1811, 31 st. *AIAA Thermophysics Conference*, June 17–20, New Orleans, LA.
27. Aerotherm Corporation, Ed., 1992. User's manual non proprietary aerotherm charring material response and ablation program. CMA925, Mountain View, CA.
28. BPD, Ed., 1994. EBM motor, low density liner characterization. RE-EBM-7101. Ed. 3.
29. Olek, S., Zvirin, Y., and Elias, E., 1988. Rewetting of hot surfaces by falling liquid films as a conjugate heat transfer problem, *Int. J. Multiphase Flow* 14: 13–33.
30. Yeh, H. C., 1980. An analytical solution to fuel-and-cladding model of the rewetting of a nuclear fuel rod, *Nucl. Engn. Des.* 61: 101–112.
31. Yu, S. K. W., Farmer, P. R., and Coney, M. W. E., 1977. Methods and correlations for the prediction of quenching rats on hot surfaces, *Int. J. Multiphase Flow* 3: 415–443.
32. Dorfman A. S., 2004. Transient heat transfer between a semi-infinite hot plate and a flowing cooling liquid film, *ASME J. Heat Transfer* 126: 149–154.
33. Peng, X. F., Peterson, G. P., and Wang, B. X., 1992. The effect of plate temperature on the onset of wetting, *Int. J. Heat Mass Transfer* 35: 1605–1613.
34. Carslaw, H. S., and Jaeger, J. C., 1986. *Conduction of Heat in Solids*, 2nd ed. Clarendon Press, Oxford.

Conclusion

Should Any Convective Heat Transfer Problem Be Considered as a Conjugate?

The term "level or effect of conjugation" is used here to estimate how different the conjugate and traditional solutions of the same heat transfer problem are. The presented theory and analysis of the examples given here show that the level of conjugation of any convective heat transfer problem depends on many factors, the most basic of which are as follows:

1. Variation of the temperature head in flow direction or in time. The decreasing temperature head affects the heat transfer characteristics much more strongly than an increasing temperature head (Section 5.1).

2. Relation between thermal resistances of the body and fluid at isothermal conditions characterized by the Biot number. The level of conjugation is greater for the case of comparable resistances and usually small when one of the thermal resistances is negligible (Section 6.1).

3. Distribution of the pressure gradient determined in this study by Görtler's variable. The level of conjugation is usually higher in the flows with unfavorable gradients (Section 5.1.3).

4. A scheme of heat transfer. For example, the effect of conjugation is usually greater for the heat transfer between countercurrent fluids than that for concurrent fluids under the same conditions.

5. Parameters and conditions determining the coefficients and exponents in the differential and integral basic expressions for nonisothermal surfaces (Chapters 3 and 4): (a) flow regime — the effect of conjugation is greater in laminar flows than in turbulent flows; (b) state of heat transfer — the level of conjugation of unsteady problems is higher than that for steady-state problems; (c) Prandtl number — the higher is the Prandtl number, the smaller the level of conjugation of a problem; (d) Reynolds number — the effect of conjugation decreases with growing the Reynolds number; (e) type of coolant — for some non-Newtonian fluids, the effect of conjugation is greater than for Newtonian liquids; (f) shape of the surface — for

TABLE C.1

Effect of Different Factors on the Level of Conjugation

Decreasing temperature head		Increasing temperature head
Comparable thermal resistance		Incomparable thermal resistance
Countercurrent flows		Concurrent flows
Unfavorable pressure gradients	Gradientless flows	Favorable pressure gradients
Laminar flows	Transition flows	Turbulent flows
Small Reynolds numbers	Mean Reynolds numbers	High Reynolds numbers
Unsteady heat transfer		Steady heat transfer
Non-Newtonian fluids with $n > 1$	Newtonian fluids	Non-Newtonian fluids with $n < 1$
Small surface curvature		Large surface curvature
Porous surface with injection	Nonporous surface	Porous surface with suction
Continuously moving surface		Streamlined surface

example, increasing surface curvature leads to increasing the conjugation level of the problem; (g) type of boundary layer — for instance, the conjugation effect of the boundary layer on a continuously moving plate is greater than that of the boundary layer on a streamlined plate.

Different factors affecting the conjugation level of the heat transfer problems are listed in Table C.1 where they are arranged so that next to the right issue, each represents a subject with a lower level of conjugation. For instance, because turbulent flow is located to the right of the laminar flow, this means that the conjugation effect in the problems with turbulent flows is less than that in corresponding problems with laminar flows.

It is obvious that in reality, the answer to the question formulated in the heading largely depends on the aim of a particular heat transfer problem and the desired accuracy of results. Therefore, such qualitative considerations can be used for preliminary approximate estimations, because exact information of the conjugate effect can be obtained only by solving the particular conjugate problem.

Nevertheless, the investigation presented in this book made it possible to formulate two general conclusions to answer the heading question:

1. Convective heat transfer problems containing a temperature head decreasing in flow direction or in time should be, as a rule, considered as a conjugate, because in this case, the effect of conjugation is usually significant (Section 5.1).
2. For the turbulent flow of the fluids with high (for example, higher than 100) Prandtl numbers, the convective heat transfer problems may be solved using a traditional approach with a boundary

condition of the third kind, because for such fluids, the effect of conjugation is negligible (Figures 4.11 and 4.12).

For other cases, the error arising by using a traditional approach may be approximately estimated by computing the second term of the series for heat flux, applying the known traditional solution. Examples of such estimations are given in Section 6.2.

Author Index

Subject Index

boiling, in channels, 263
carbon dioxide absorption in, 323
in concrete, 331–333
continuously moving surface in, 165,
    290, 295
critical temperature of, 332, 334
evaporation of, 334
as HTF, 325
pipe in flowing, 258–259
Prandtl number for, 108f, 165, 295, 357
in rewetting process, 370–371
steel plate in flowing, 188
volume fraction of free, 332
Water vapor mass fraction, 333–334

Weak formulation, 238
Wedges, 4–5, 47–50; *See also* Fins
Weighted residual approach, 235–236, 327
Weighting functions, 218, 235–240, 256, 374
Wet bulb temperature, 334, 336–337
Wet front velocity, 369

X

X-ray heating, 272–274

Y

Yeh method, 369

Printed and bound by CPI Group (UK) Ltd, Croydon, CR0 4YY

21/10/2024

01777089-0017